# GibbsCAM TUTORIAL

## Version 2005

**JEFF HATLEY**

ISBN: 1-58503-270-0

Schroff Development Corporation

www.schroff.com
www.schroff-europe.com

# GibbsCAM Tutorial

## Version 2005

## COPYRIGHT



## ABOUT THE AUTHOR

Jeff Hatley is a CAD/CAM instructor for Livonia Public Schools and Schoolcraft Community College in Metro Detroit.

## SPECIAL THANKS:

**Gibbs and Associates** – For their support of manufacturing technology education.
**www.gibbscam.com**

**Great Lakes CAM** – GibbsCAM sales, training, and technical support.
**www.glcam.com**

**Advanced Technologies Consultants** – Educational sales of Z-Corp 3D printers, EMCO CNC machines, and SolidWorks CAD software.
**www.advancedtechnologies.net**

**SDC Publications** – Educational textbooks for future architects, designers, and engineers.
**www.schroff.com**

# TABLE OF CONTENTS

**Chapter 1 – Getting Started in GibbsCAM**

| | | |
|---|---|---|
| 1 | Exercise 1 | The GibbsCAM interface |

**Chapter 2 – Creating Geometry**

| | | |
|---|---|---|
| 15 | Exercise 1 | Creating Simple Shapes and Deleting Geometry |
| 23 | Exercise 2 | Creating Points for a Hole Pattern |
| 28 | Exercise 3 | Connectors |
| 42 | Exercise 4 | Geometry Expert |
| 61 | Exercise 5 | Editing Geometry with Geometry Expert |
| 66 | Exercise 6 | CAD |
| 81 | Exercise 7 | Workgroups |
| 98 | Exercise 8 | Text |
| 105 | Exercise 9 | Dimensions |

**Chapter 3 – Importing CAD Geometry**

| | | |
|---|---|---|
| 111 | Exercise 1 | Opening AutoCAD .dwg Files |
| 115 | Exercise 2 | Extracting Geometry from a 3D CAD Model |

**Chapter 4 – Mill**

| | | |
|---|---|---|
| 125 | Exercise 1 | Drilling |
| 144 | Exercise 2 | Hole Wizard |
| 154 | Exercise 3 | Hole Manager |
| 164 | Exercise 4 | Roughing and Contours |
| 185 | Exercise 5 | Machining a 3D CAD Solid Model |
| 188 | Exercise 6 | Editing Machining Operations |

**Chapter 5 – Lathe**

| | | |
|---|---|---|
| 191 | Exercise 1 | Lathe |

| | | |
|---|---|---|
| 211 | Appendix A | Post Processing |
| 212 | Appendix B | Tool List Summaries |
| 213 | Appendix C | Operation Summaries |

**Notes:**
Part files for the exercises can be downloaded from www.schroff1.com.

After downloading a file, click the right mouse button on the file and select *Properties*. Uncheck the *Read Only* box.

# NOTES:

Chapter 1

# Getting Started in GibbsCAM

## Introduction to GibbsCAM

GibbsCAM is a CAD/CAM software program that can be used to simulate CNC machining and generate CNC code. Part geometry can be created within GibbsCAM, opened directly from another CAD program, or imported from another CAD program. Some of these programs include AutoCAD, Autodesk Inventor, Solid Edge, and SolidWorks.

## The GibbsCAM Workflow:

1. Create a new part file.

2. Create part geometry or import geometry from a CAD file.

3. Create tools.

4. Create machining operations.

5. Simulate CNC machining.

6. Post process to generate CNC code.

## Using This Textbook

The chapters in this book are set up according to the GibbsCAM workflow. Therefore, it is suggested that you complete the chapters in sequence. You can learn GibbsCAM without having any CNC programming experience. However, it helps to have some basic knowledge of CNC programming before you move on to CAD/CAM software. This will allow you to better understand and work with the final output generated by GibbsCAM – CNC code.

It is also recommended to complete the Creating Geometry exercises even if you plan on importing CAD data from another program. Completing the Creating Geometry exercises will help you create, edit, and manage geometry in GibbsCAM whether it was created in GibbsCAM or not.

## Exercise 1 – The GibbsCAM Interface

### Creating a New File

Launch GibbsCAM by double-clicking on the GibbsCAM icon on the desktop.

From the *Main Menu*, Select *File – New*.

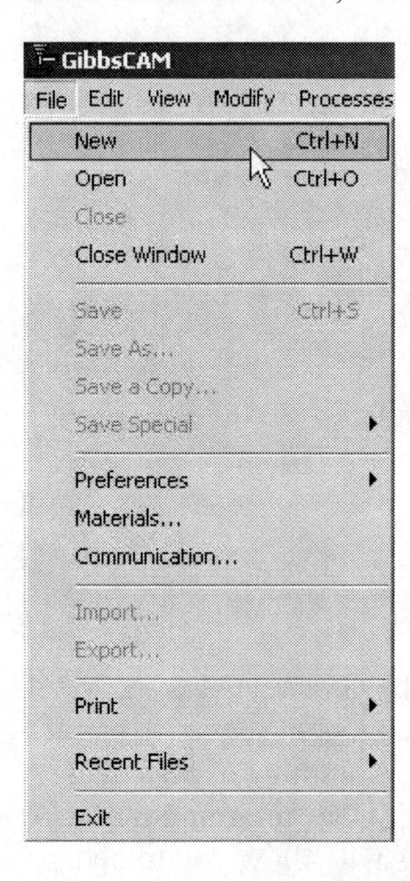


The *Save New Part File Dialog Box* will appear. In the Save in box, navigate to the storage location for your files. Enter *Part1* in the *File Name* box and click on Save.

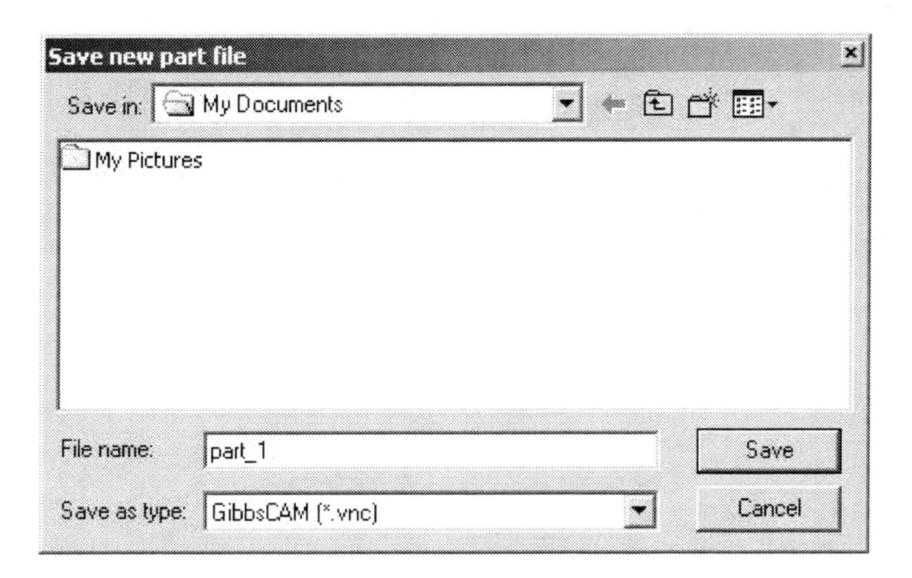


Locate the Top Level Palette in the upper right corner of the GibbsCAM interface. This palette is the main palette in GibbsCAM.

Click on the Documents icon in the Top Level Palette.

Set the machine type to 3-Axis Vertical Mill and the units to inch as shown in the next graphic. Enter the values as shown in the next graphic to set the stock size. This will set the PRZ to the center of the stock on the top face.

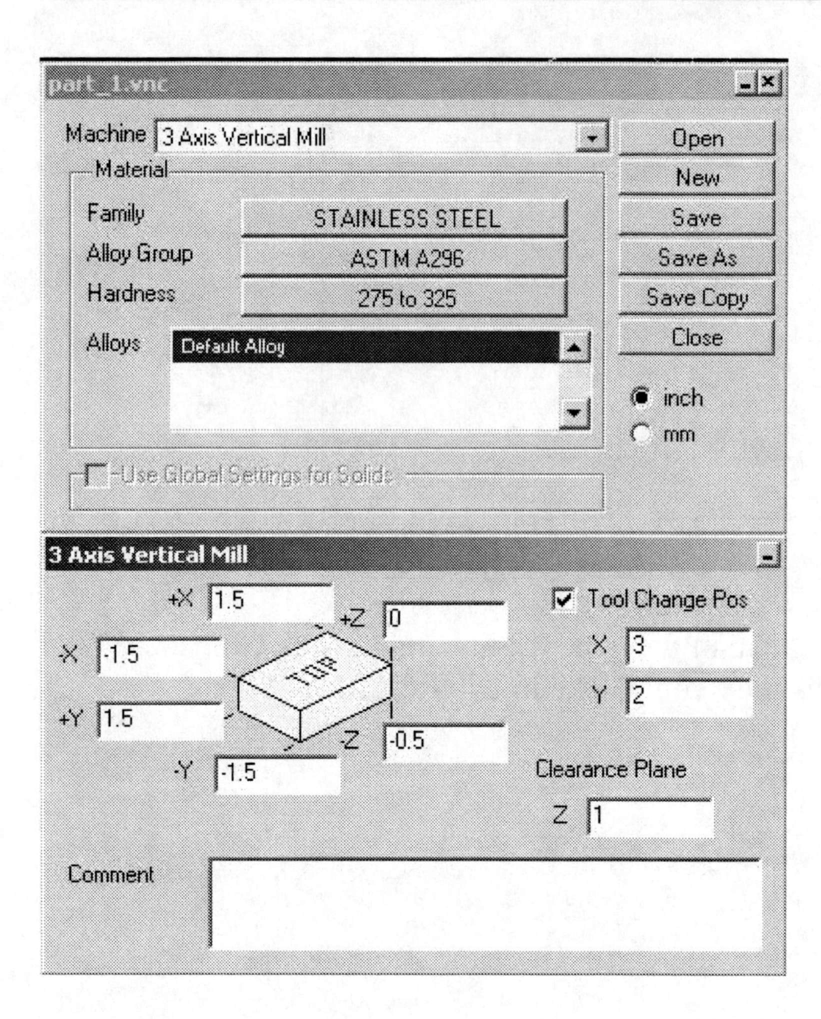


The *Tl Change Pos* box is used to specify where the tool will be sent when a tool change occurs. The *Clearance Plane* box is the Z position the tool will rapid to and from during a tool change.

Close the dialog box. Notice the location of the origin.

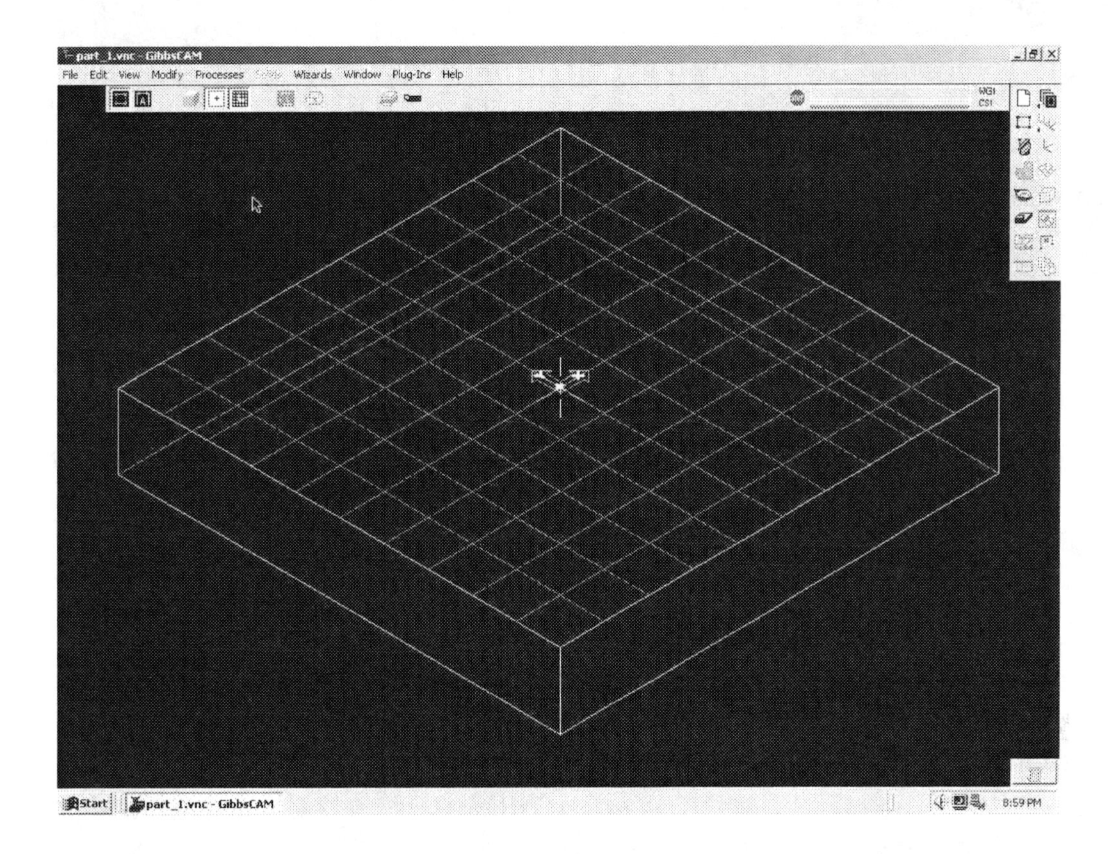

## Getting Help in GibbsCAM

From the *Main Menu*, select *Help-Tooltips*.

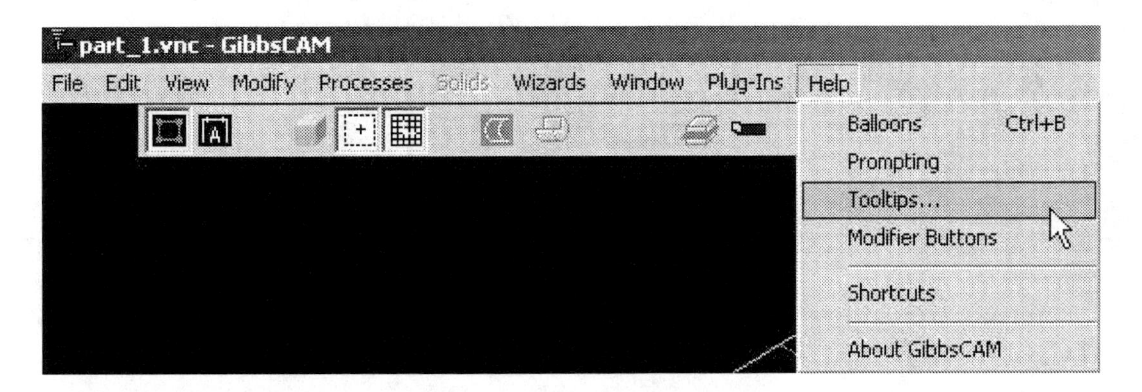


Place a check mark in the *Show Tooltips* box to active tooltips.

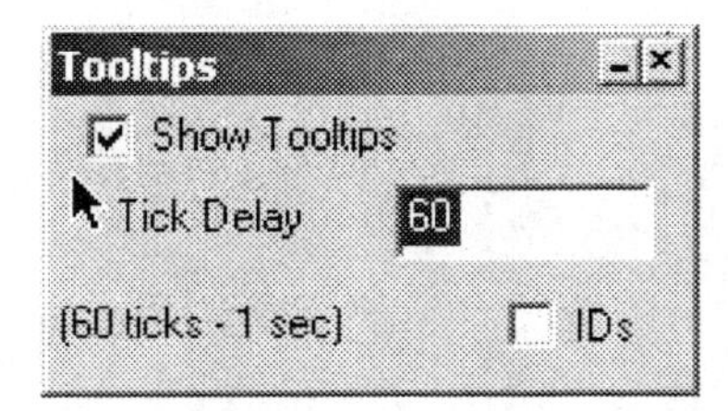

Close the Tooltips dialog box by clicking on the X.

Hold your pointer over one of the icons in the Top Level Palette and a tooltip will appear telling you the name of the icon.

To turn this feature off, select the item again and remove the check mark next to Show Tooltips.

From the *Main Menu*, select *Help-Balloons*.

As you pass your pointer over the icons in the Top Level Palette you will see a balloon with a description for each icon.

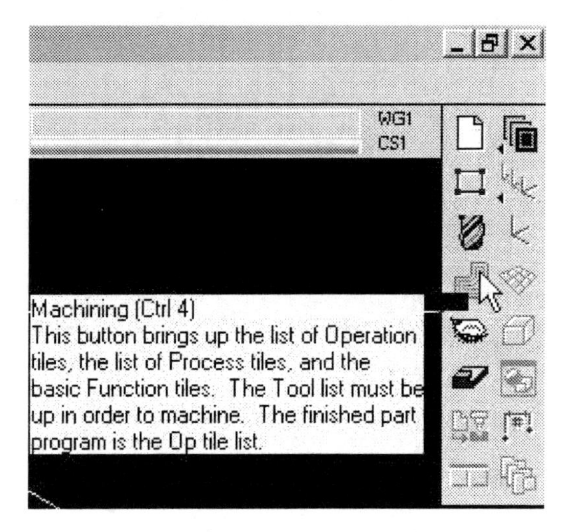


When this feature is active you will see a check mark next to *Balloons* under *Help* in the main menu. To turn this feature off, select the item again to remove the check mark.

**Viewing Objects in GibbsCAM**

Select *View-Unzoom* from the *Main Menu* to see a full view of your part.

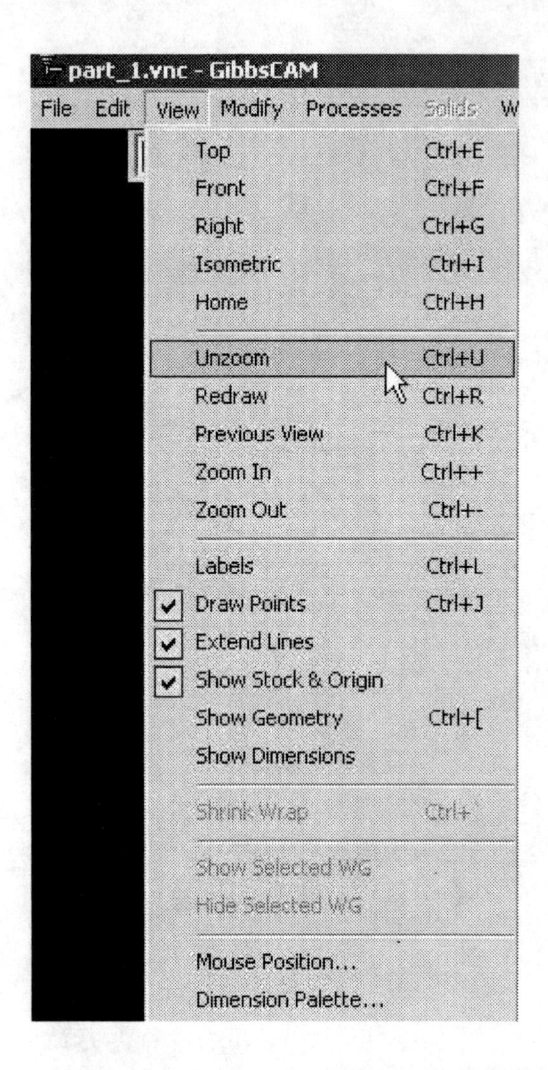


Select the *View* icon from the Top Level Palette.

The View Control Palette will appear.

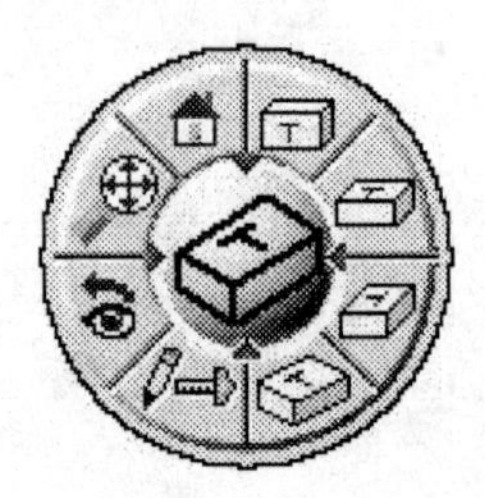

This palette contains icons for viewing standard views (Top, Front, R.Side) and also an icon in the center of the palette that acts like a rotating track ball. Click on these icons on your own to create views of the part.

Change the view to an isometric view by clicking the *isometric* icon in the View Control Palette.

Click and drag a rectangle around the origin to zoom in on that area. You can also use the scroll wheel on a 3-button mouse to zoom in and out.

Click and manipulate the free-rotate icon to rotate your part with your mouse.

Select the *Previous View* icon (PV) on the View Control Palette to return to the previous view. You can also choose *View-Unzoom* from the Main Menu.

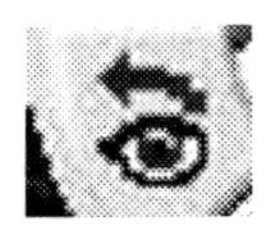

Hold down the control key and the left mouse button to pan your view

**Geometry**

Geometry tools in GibbsCAM are activated through the Geometry icon in the Top Level Palette. Click on the Geometry icon.

The Geometry Palette will appear. You will use this palette to create geometry in GibbsCAM. You will learn more about creating Geometry in GibbsCAM in Chapter 2.

The Geometry Palette

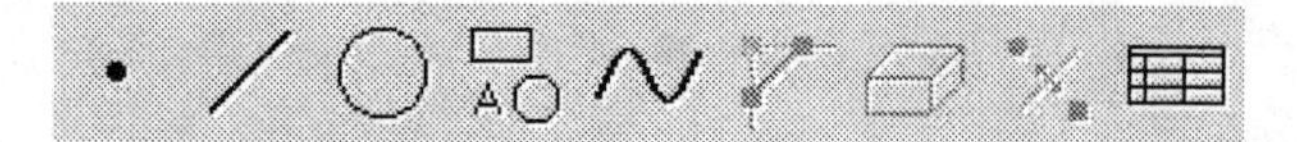

**Tools**

The tool list in GibbsCAM is activated through the Tools icon in the Top Level Palette. Click on the Tools icon.

The Tool List will appear.

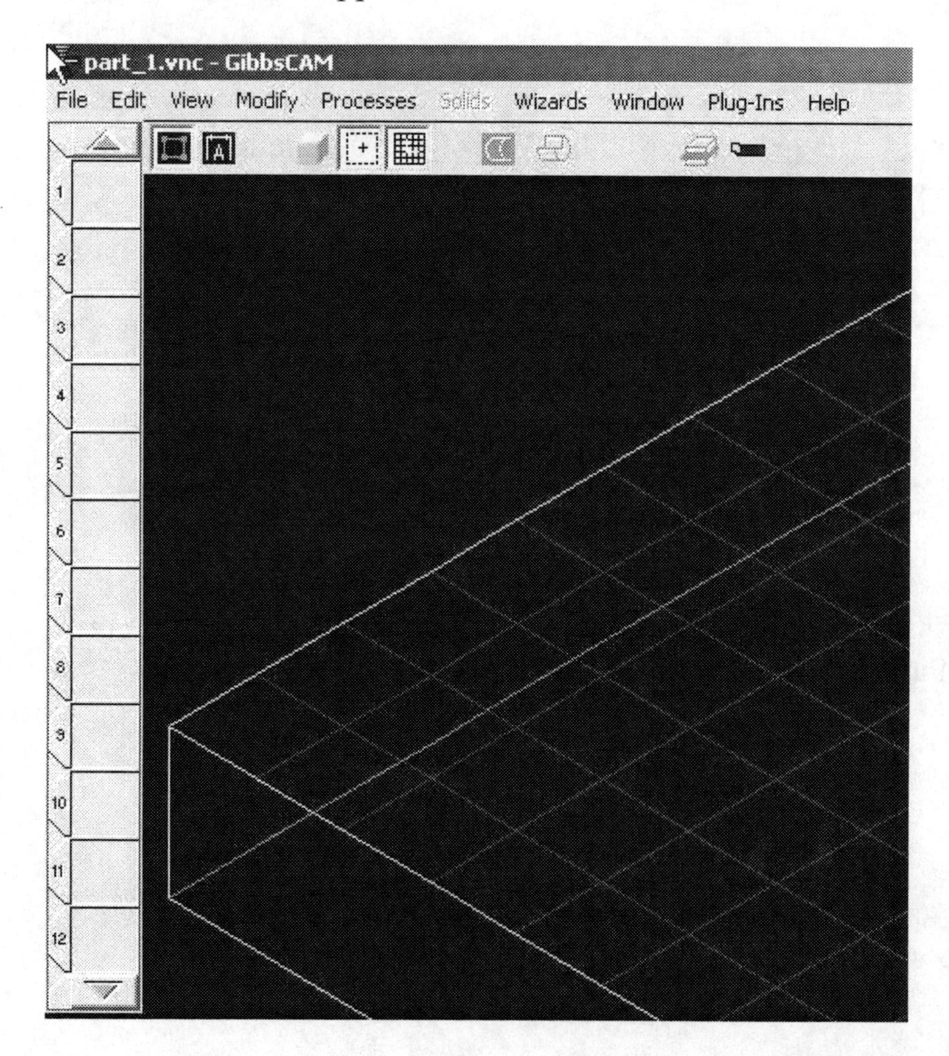

You will use the tiles in this list to create tools for machining. You will learn how to create tools in Chapter 4.

**CAM**

The Machining Palette and Operations/Process Lists in GibbsCAM are activated through the CAM icon in the Top Level Palette. Click on the CAM icon.

The Machining Palette will appear. This palette has tools for creating drilling, contour, pocket, and thread mill operations. You will learn about CAM in chapter 4.

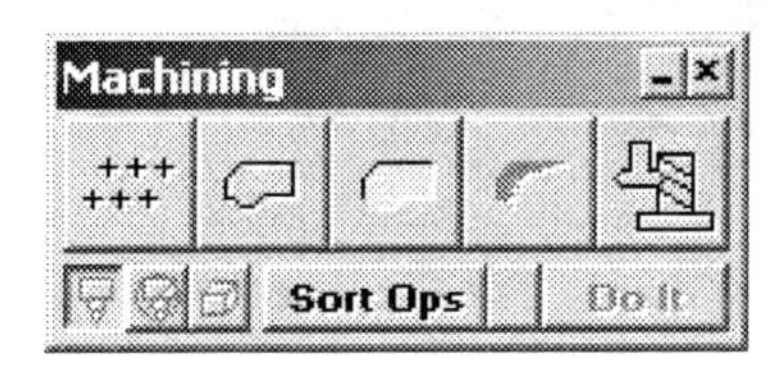


The Process List and Operations List will also appear. The Process List is used to generate machining operations. The Operations List contains the final output, which will be simulated and then processed into CNC code.

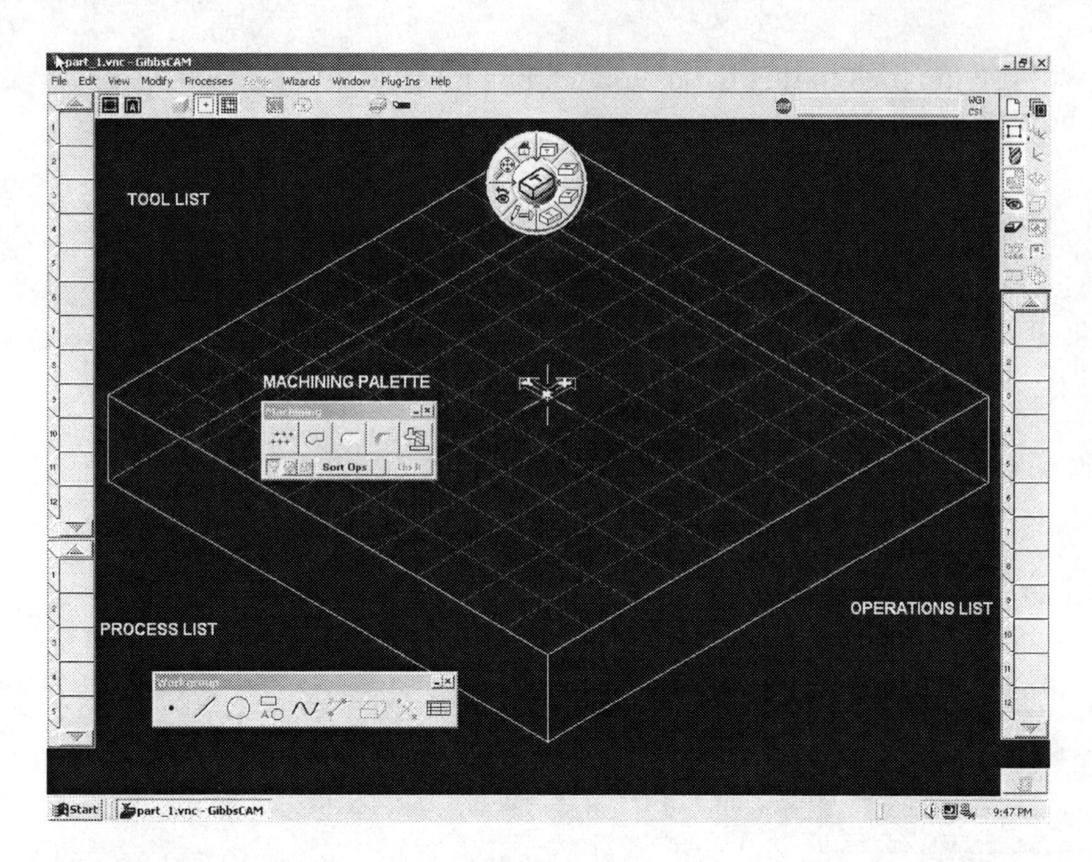


## Simulation

The Cut Part Rendering Palette in GibbsCAM is activated through the Cut Part Rendering icon in the Top Level Palette. Click on the Cut Part Rendering icon.

The Cut Part Rendering Palette will appear. This palette is used to simulate the machining of a part in GibbsCAM. You will learn about simulation in chapter 4.

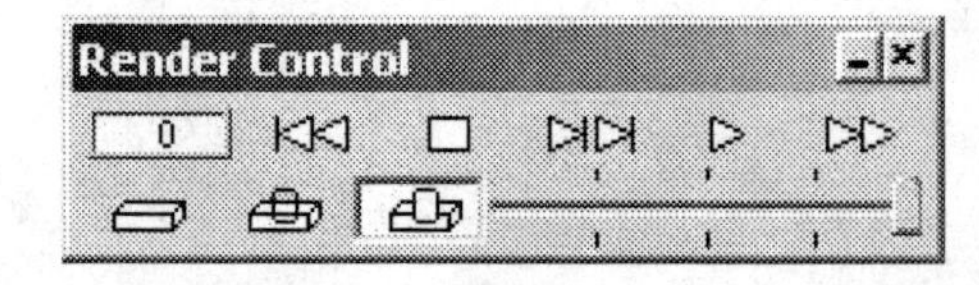


Close the Render Control dialog box.

## Post Processing

Post processing is used to generate CNC code from the GibbsCAM graphical interface. Post Processing in GibbsCAM is activated through the Post icon in the Top Level Palette. This icon will become available after geometry, tools, and machining operations are created.

## Saving Files

Select *File-Save* from the *Main Menu*.

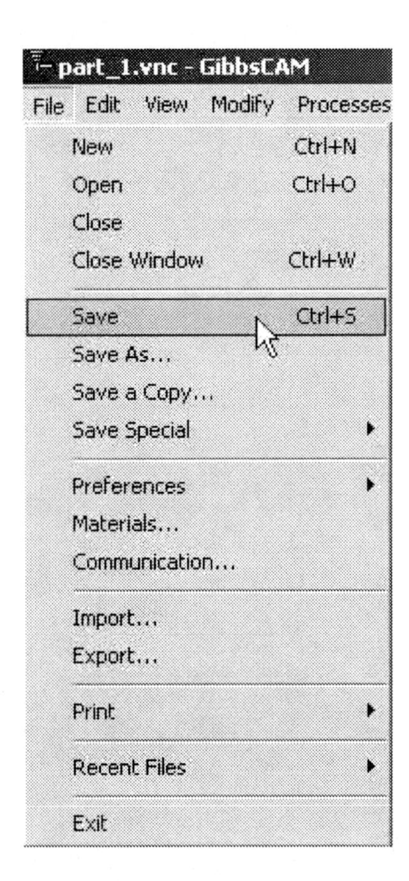


## Closing Files

Select *File-Close* from the *Main Menu*.

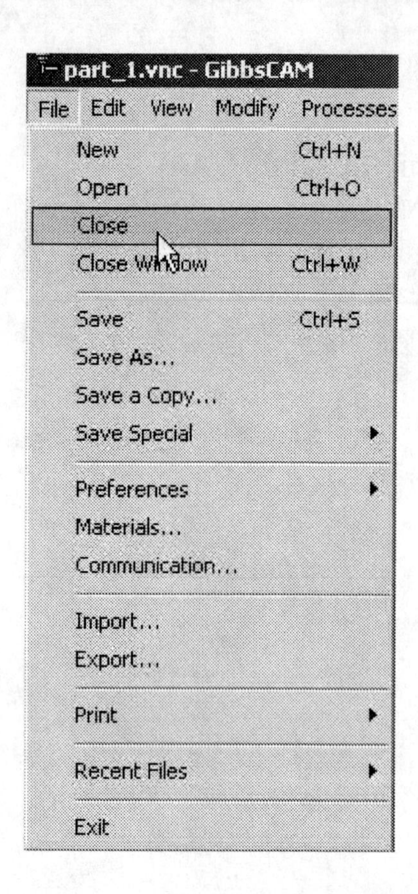


This completes exercise 1.

Chapter 2

# Creating Geometry

## Introduction

Geometry can be created in GibbsCAM by using the CAD tools or the Geometry Expert. The exercises in this chapter utilize both methods. The Geometry Expert functions much like a spreadsheet. The CAD tools work like many 2D CAD drawing programs. It is also possible to use a combination of the CAD tools and the Geometry Expert to create geometry.

## Exercise 1 – Creating Simple Shapes and Deleting Geometry

In this exercise you will create a hexagon and a rectangle using the GibbsCAM CAD tools.

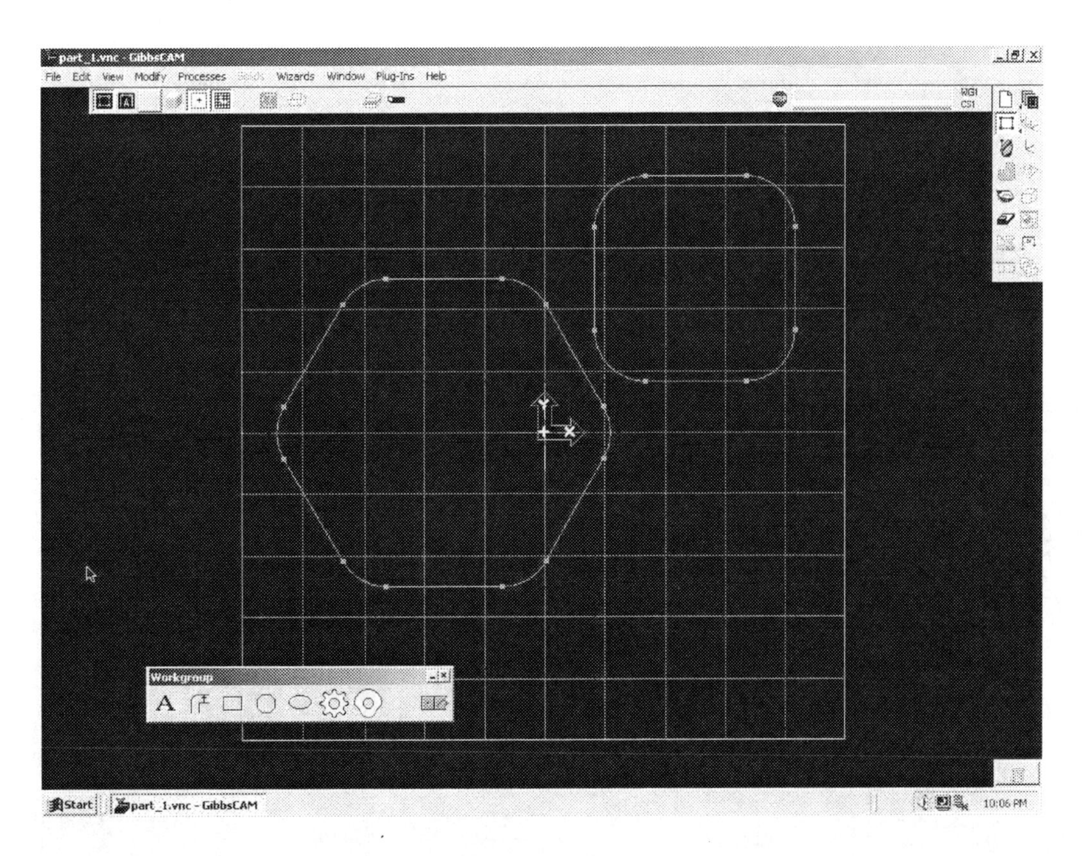

Open PART1. Choose *File-Open* from the *Main Menu.*

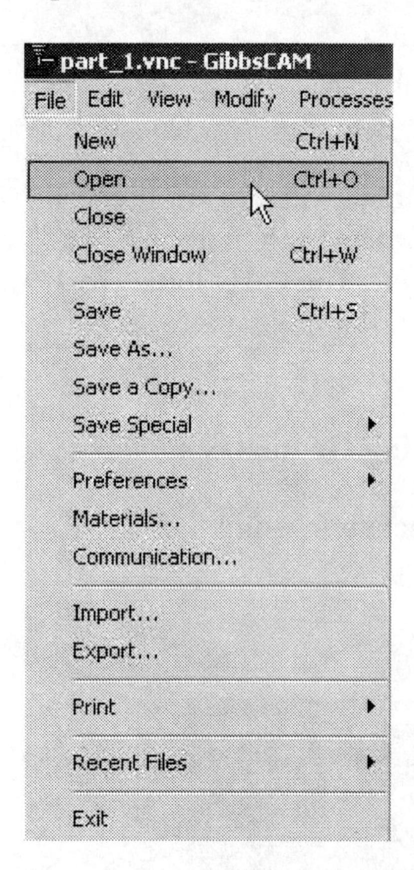


Select Part1.vnc and click on *Open*.

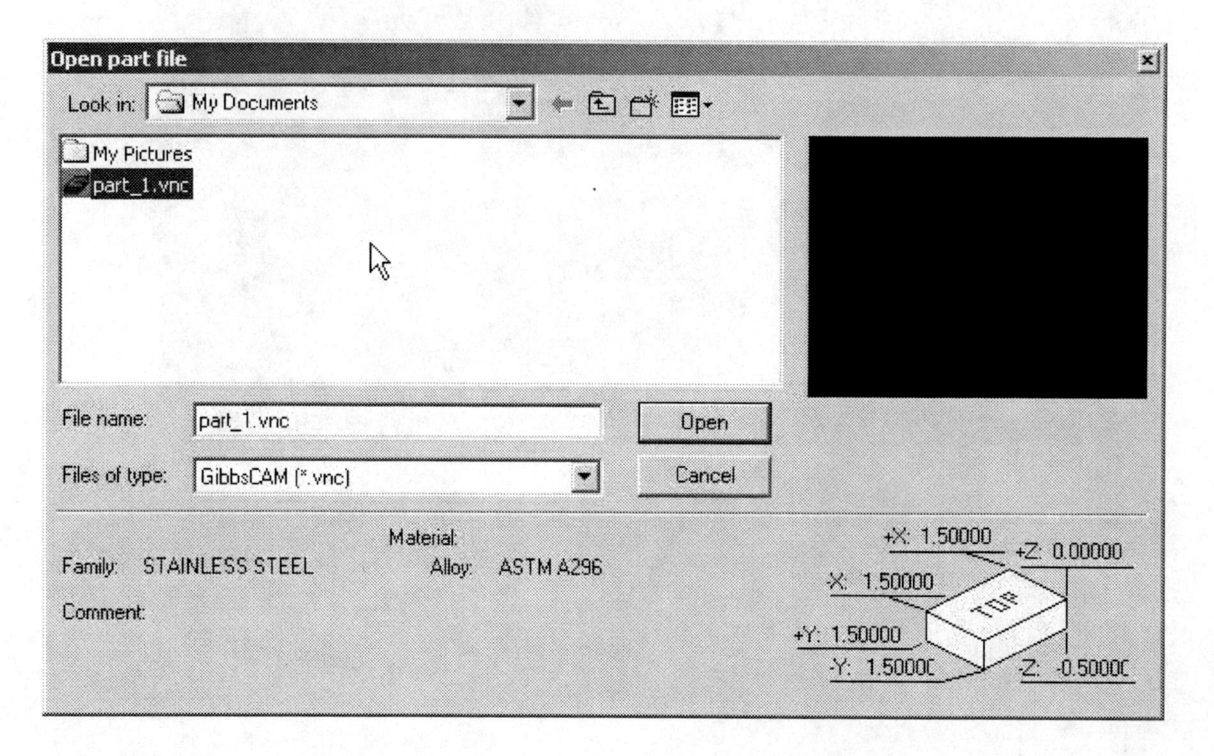

Select the Geometry icon from the Top Level Palette.

The Geometry Palette will appear.

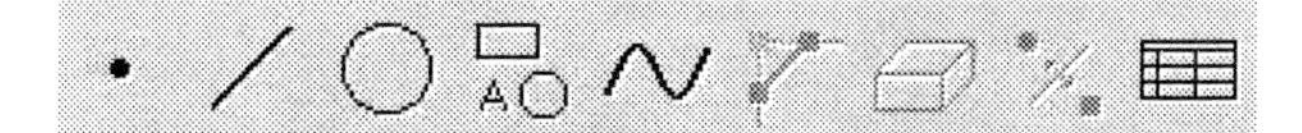

Select the Shape icon from the Geometry Palette.

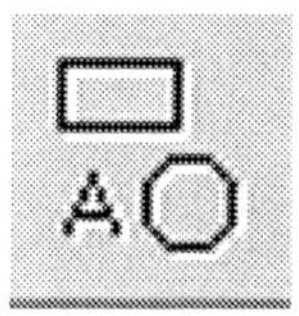

The Shapes Palette will appear.

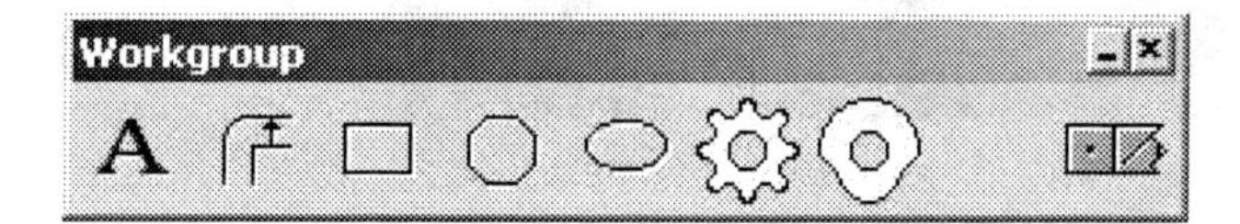


Select the Polygon icon from the Shapes Palette.

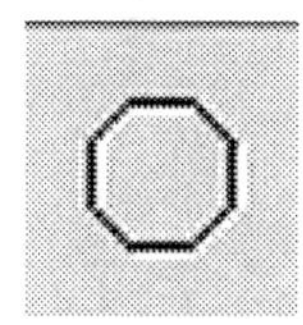

Enter the values as shown and click on Do It.

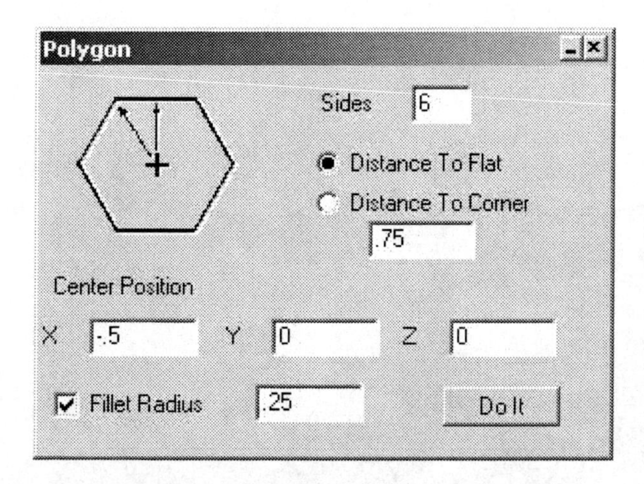

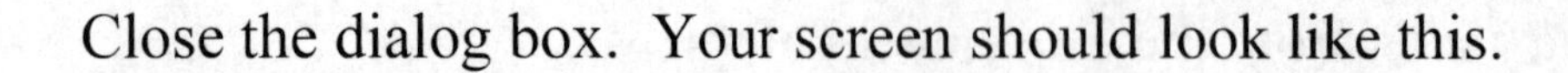

Close the dialog box. Your screen should look like this.

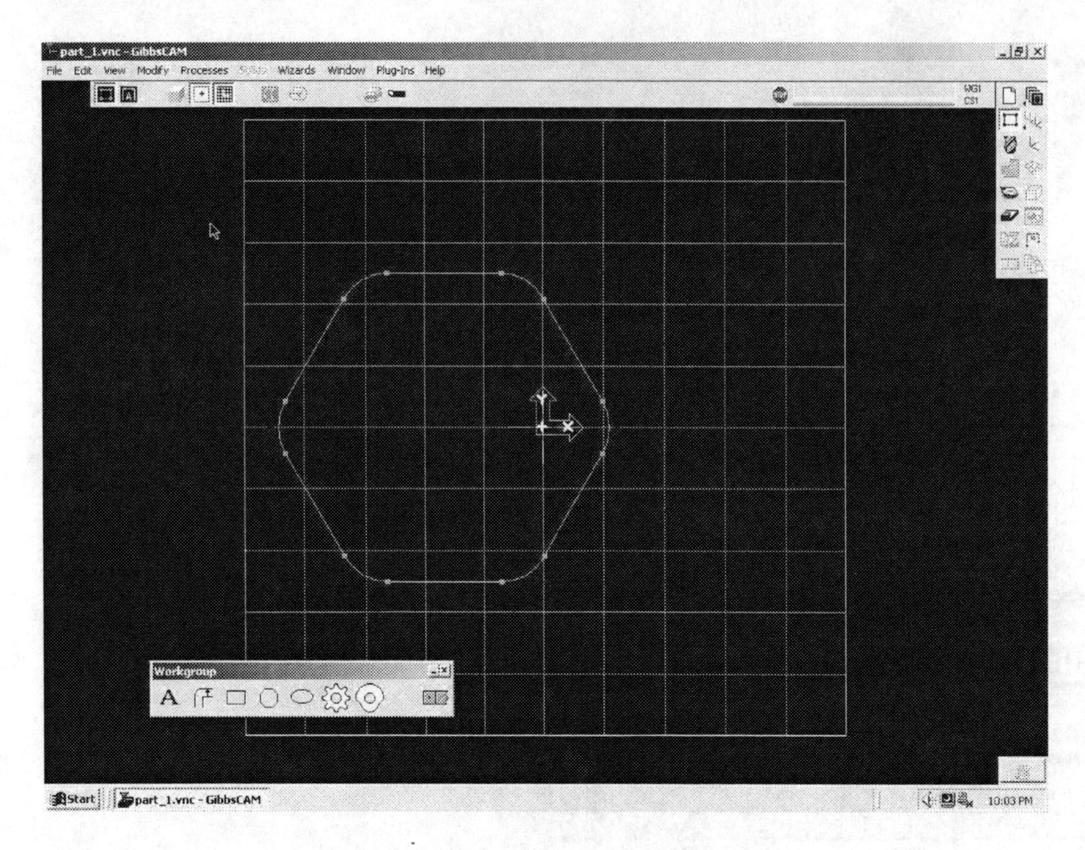


Select the Box icon from the Shapes Palette.

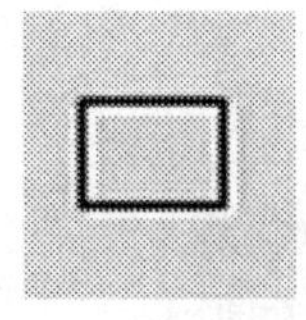

The Rectangle dialog box will appear. Enter the values as shown and click on Do It.

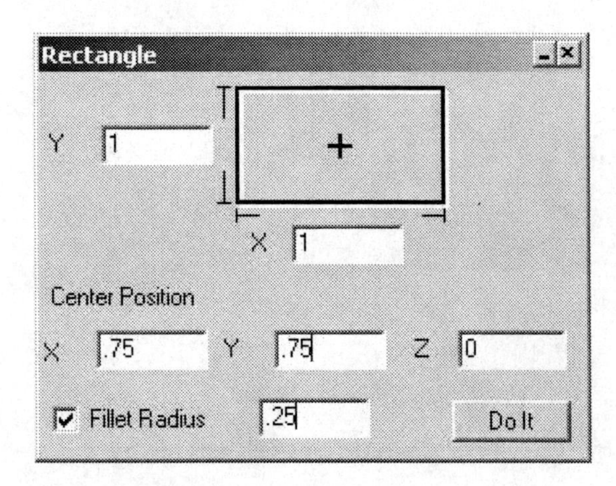

Close the dialog box. Your screen should look like this.

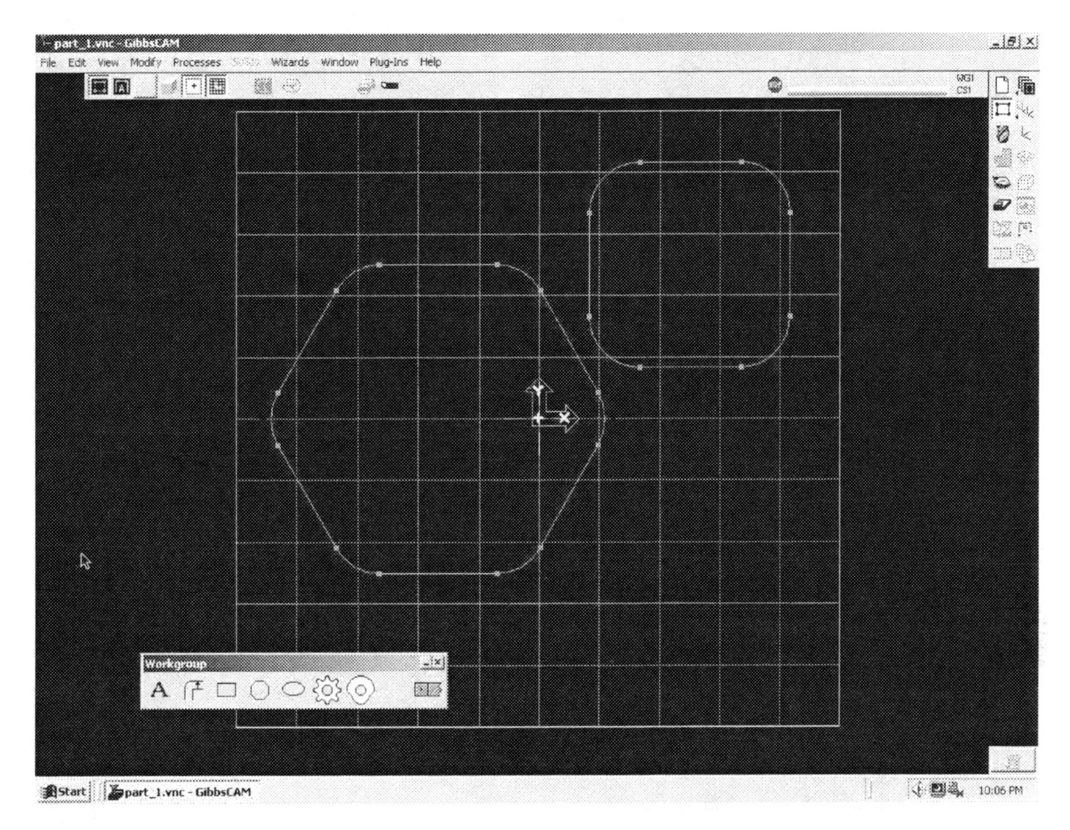


Save the file. Choose *File-Save* from the *Main Menu*.

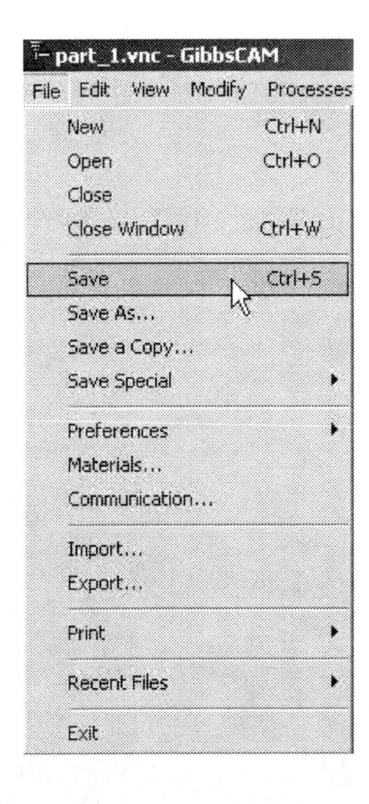

## Deleting Geometry

You will now practice deleting geometry.

Select the box icon from the Shapes Palette.

The Rectangle Dialog Box will appear. Enter the values as shown in the next graphic and click on Do It.

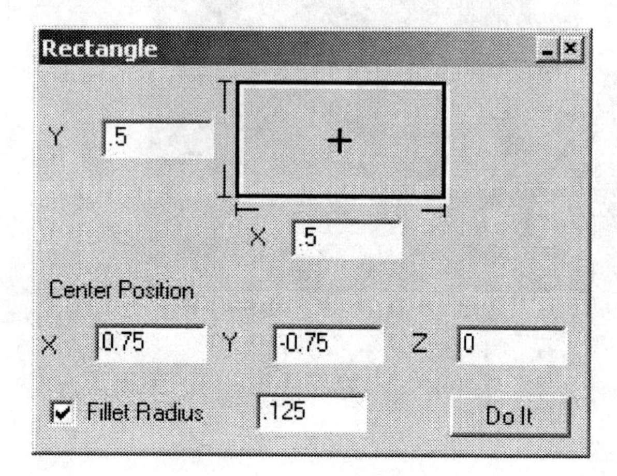

Your screen should look like this.

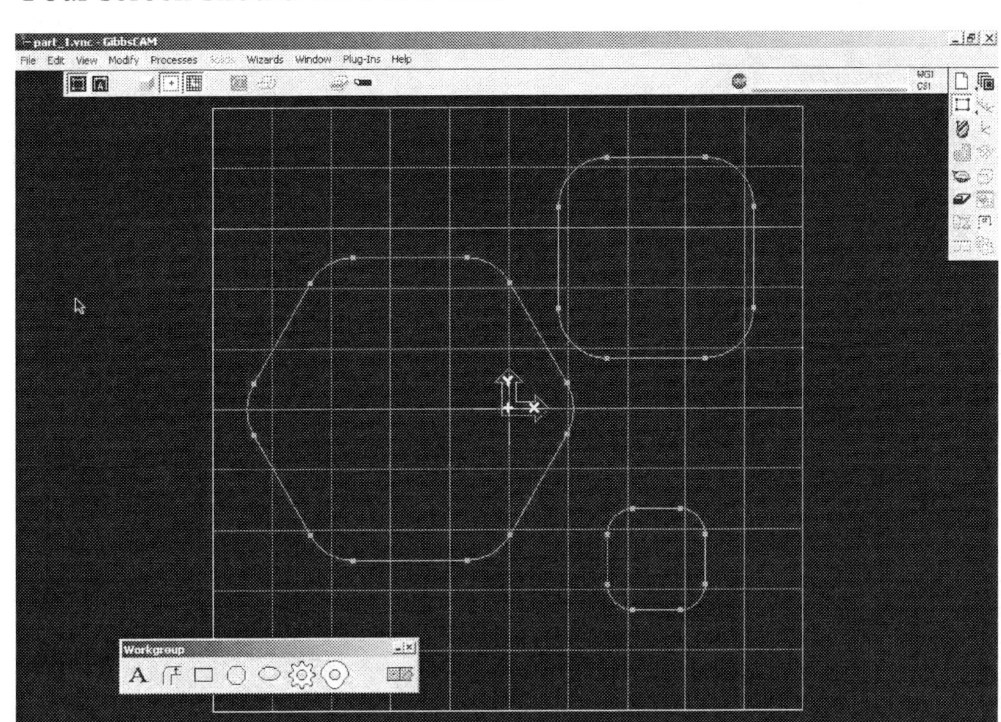

Hold down the shift key and drag a box around the profile in the lower right corner as shown.

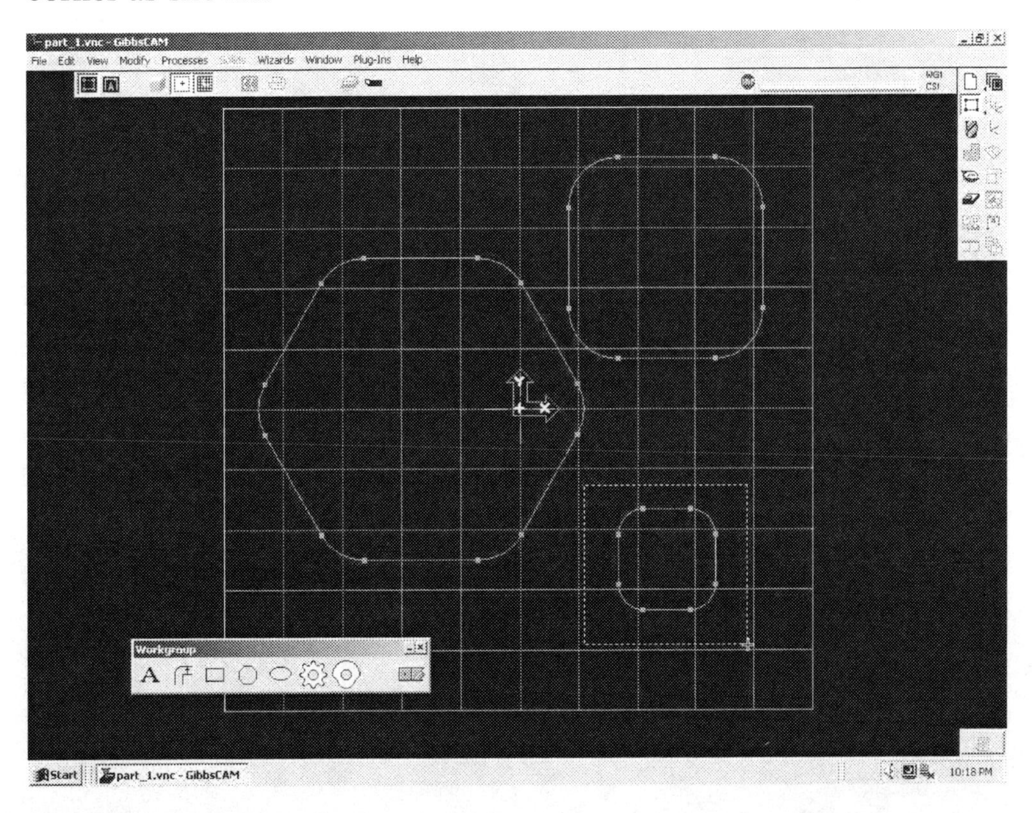

Hit the delete key. Your screen should look like the next graphic.

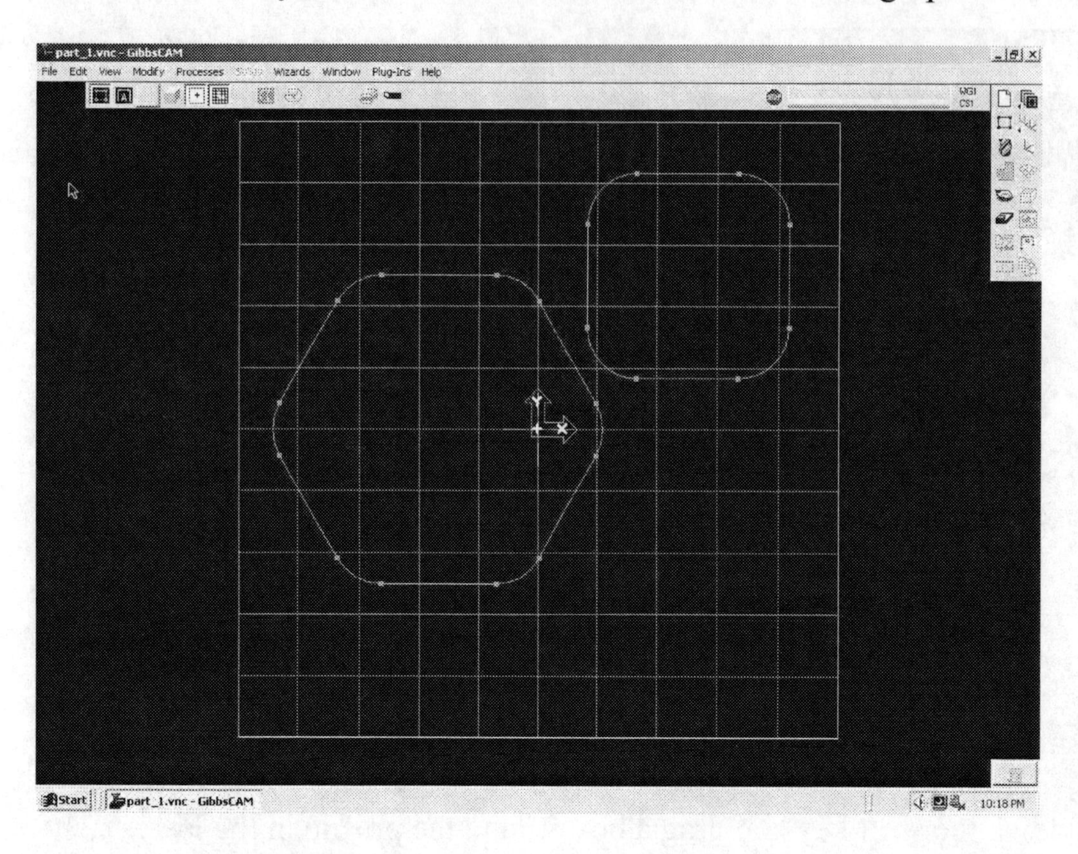


Close the file without saving. Choose *File-Close* from the *Main Menu*.

This completes the exercise.

## Exercise 2 – Creating Points for a Hole Pattern

In this exercise you will use the CAD tools to plot points that can be used to drill a hole pattern.

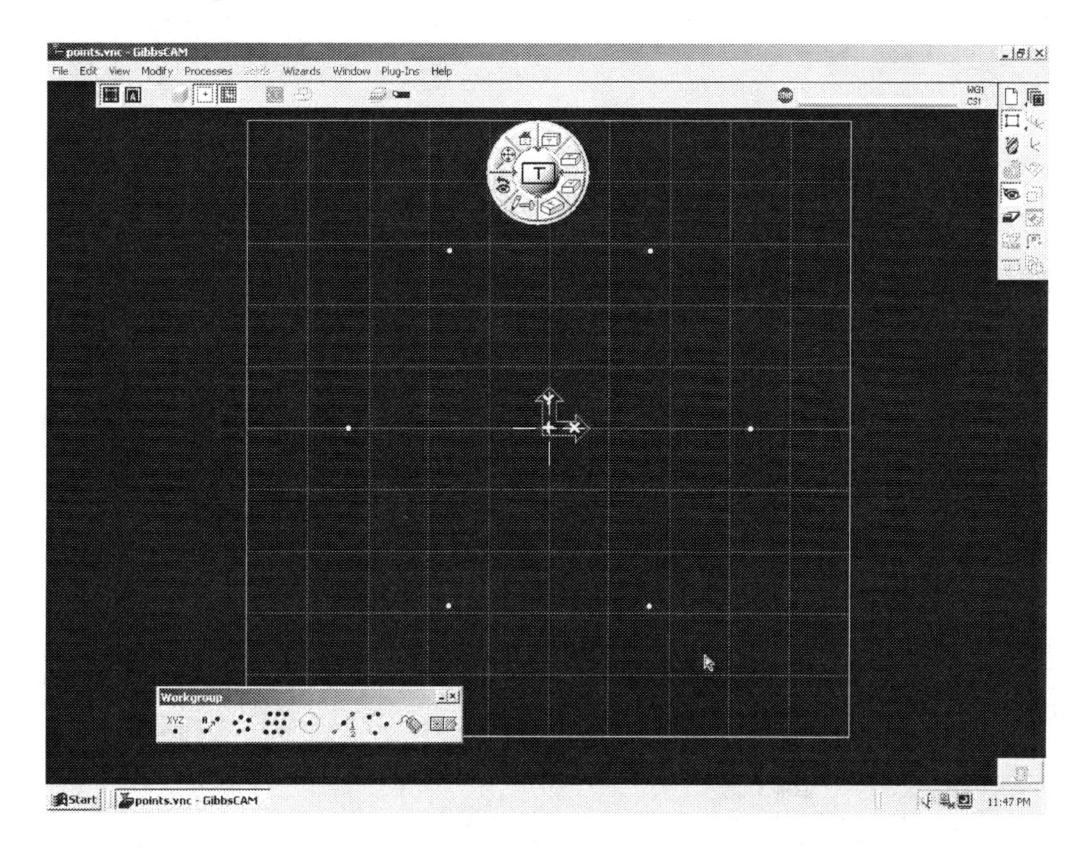


From the *Main Menu* select *File – New*.

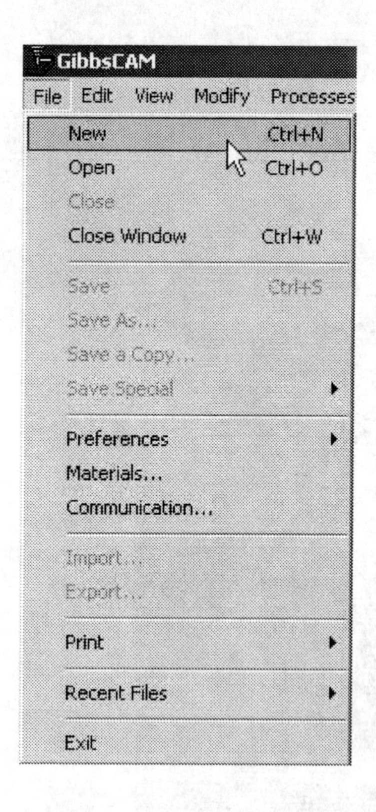


The *Save New Part File Dialog Box* will appear. In the Save in box, navigate to the storage location for your files. Enter POINTS in the *File Name* box and click on Save.

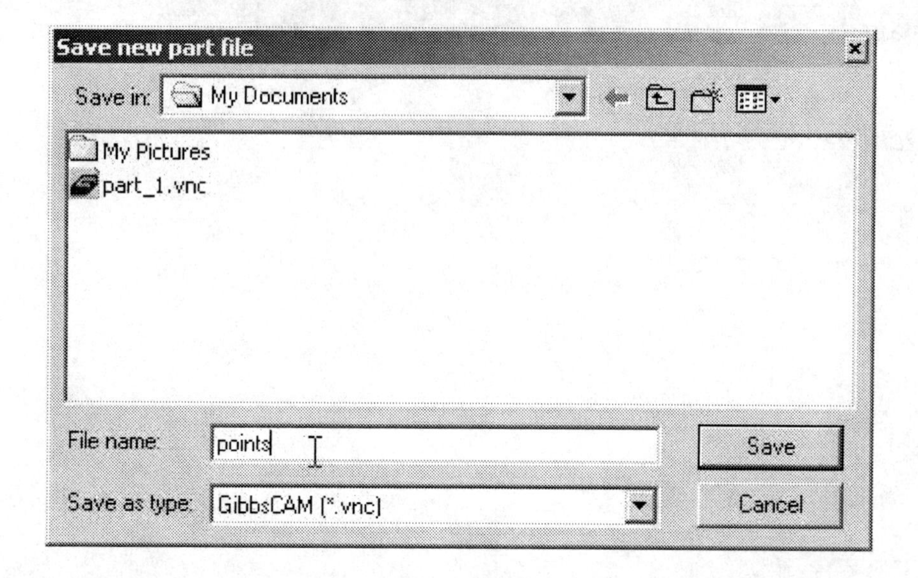


Select the Documents icon from the Top Level Palette.

Set the machine type to 3 axis vertical mill and the units to inch. Enter the values as shown for the stock size.

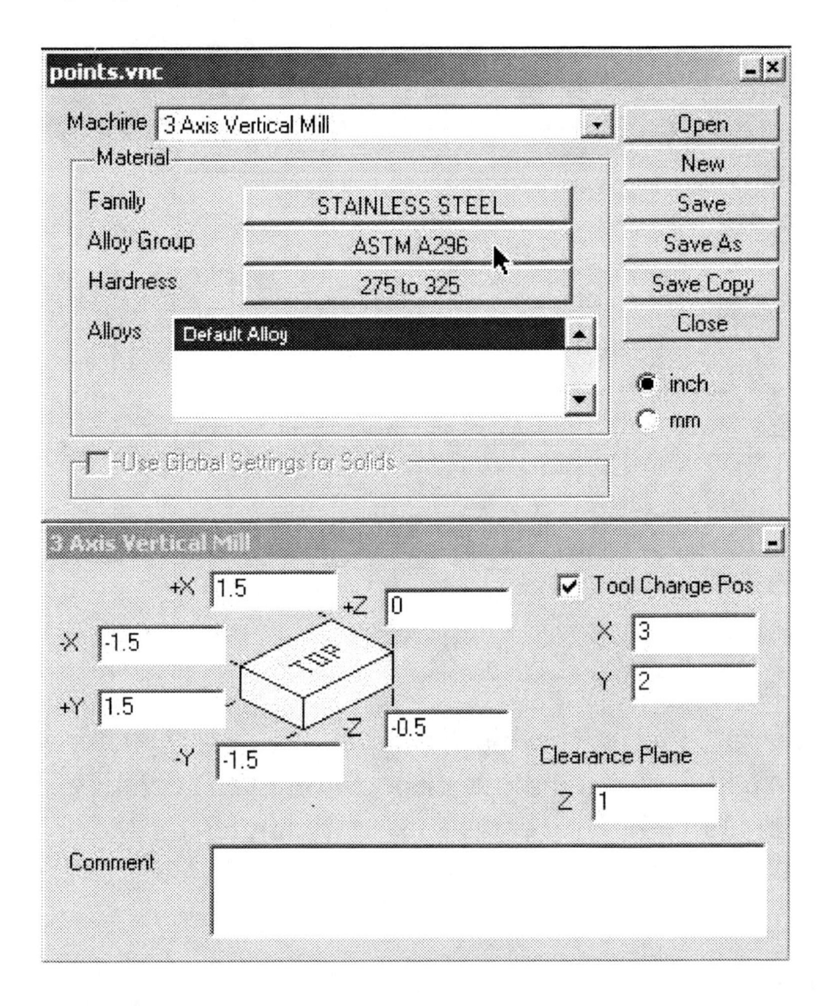


Close the dialog box.

Select the Geometry icon from the Top Level Palette.

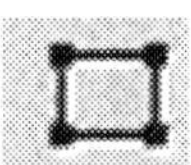

The Geometry Palette will appear.

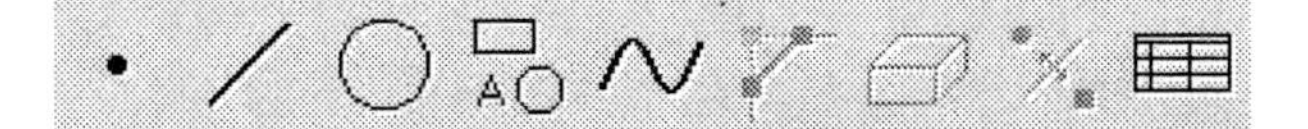

Select the Point icon from the Geometry Palette.

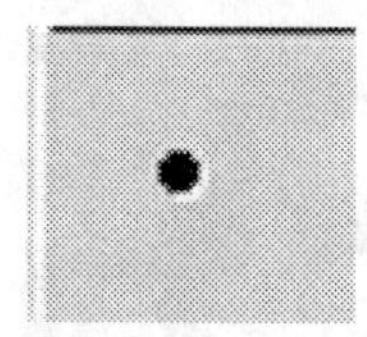

The Points Palette will appear.

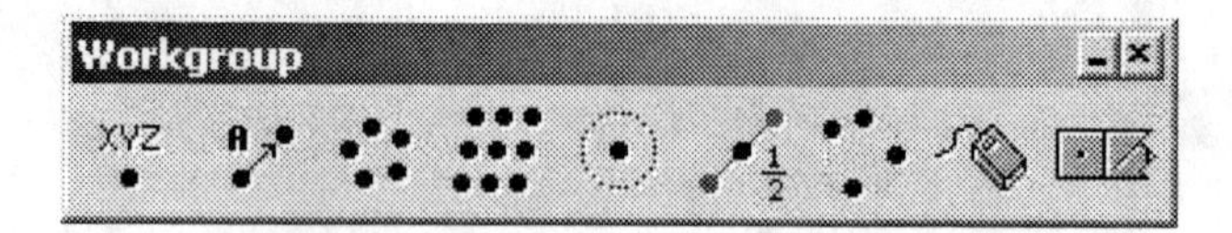


Select the Bolt Circle icon from the Points Palette.

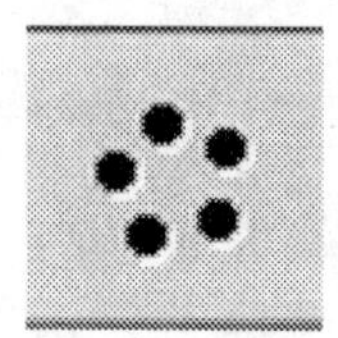

The Bolt Circle dialog box will appear. In this dialog box you will specify the radius of the bolt circle and determine the XYZ location of the center point of the bolt circle from the PRZ. You can also specify angles for the holes and direction (clockwise or counter-clockwise). Enter the values as shown and click on Do It.

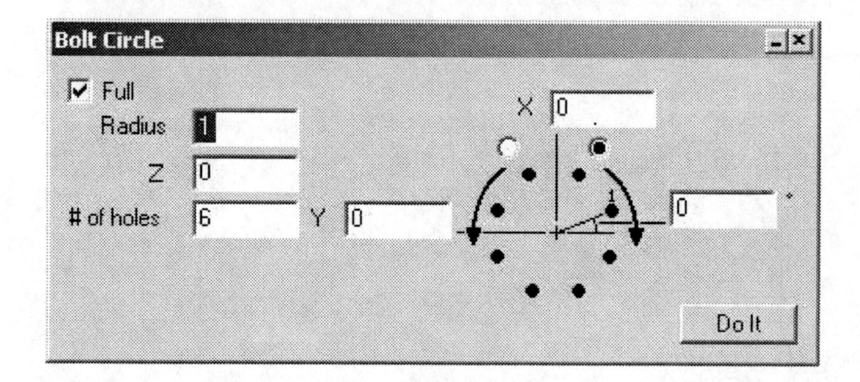


Close the Bolt Circle dialog box. Your screen should look like the next graphic.

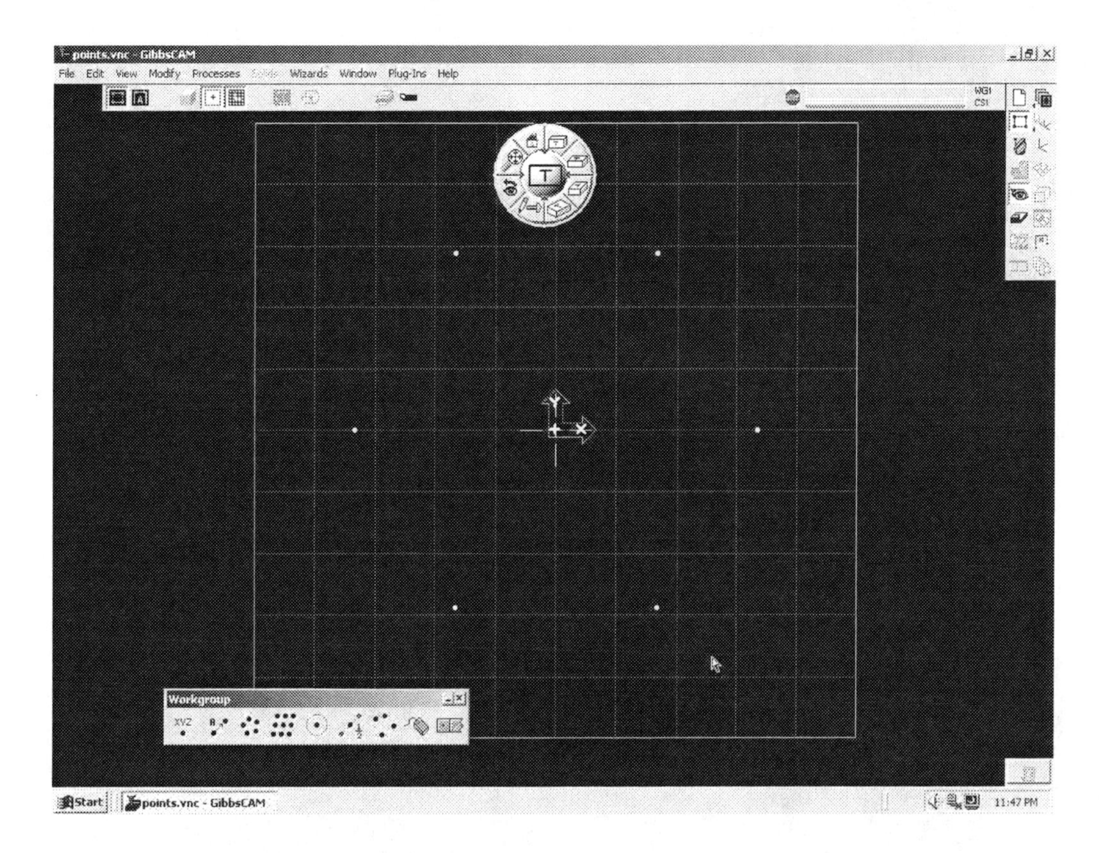

This completes the exercise. Save your file. Choose *File-Save* from the *Main Menu*. Close the file.

## Exercise 3 – Connectors

In this exercise you will use the CAD tools to create a square and a circle and then connect/trim the profiles to create the geometry shown below.

In GibbsCAM, blue geometry is connected and can be selected as a single profile in the machining process. Yellow geometry is not connected and can only be selected as a single line, point, or arc.

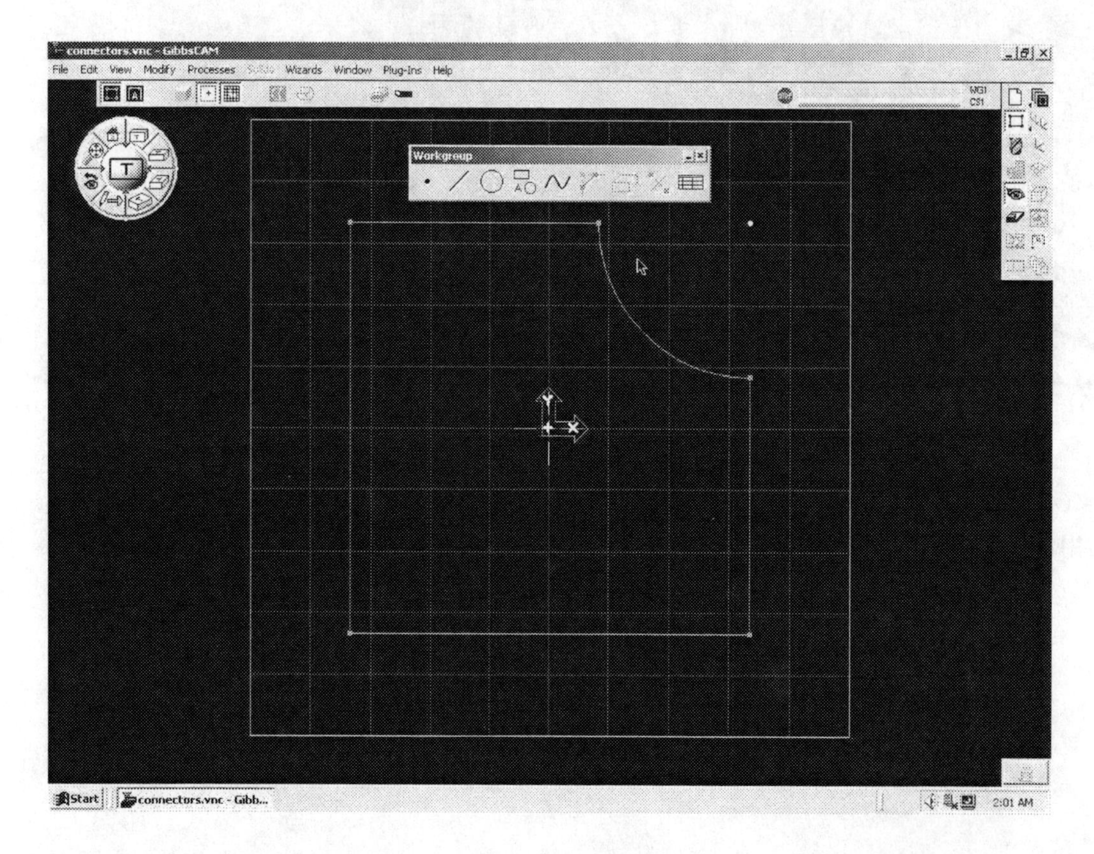


Create a new file named CONNECTORS. Set the machine type to 3 axis vertical mill and the units to inch. Enter the values in the next graphic for the stock size.

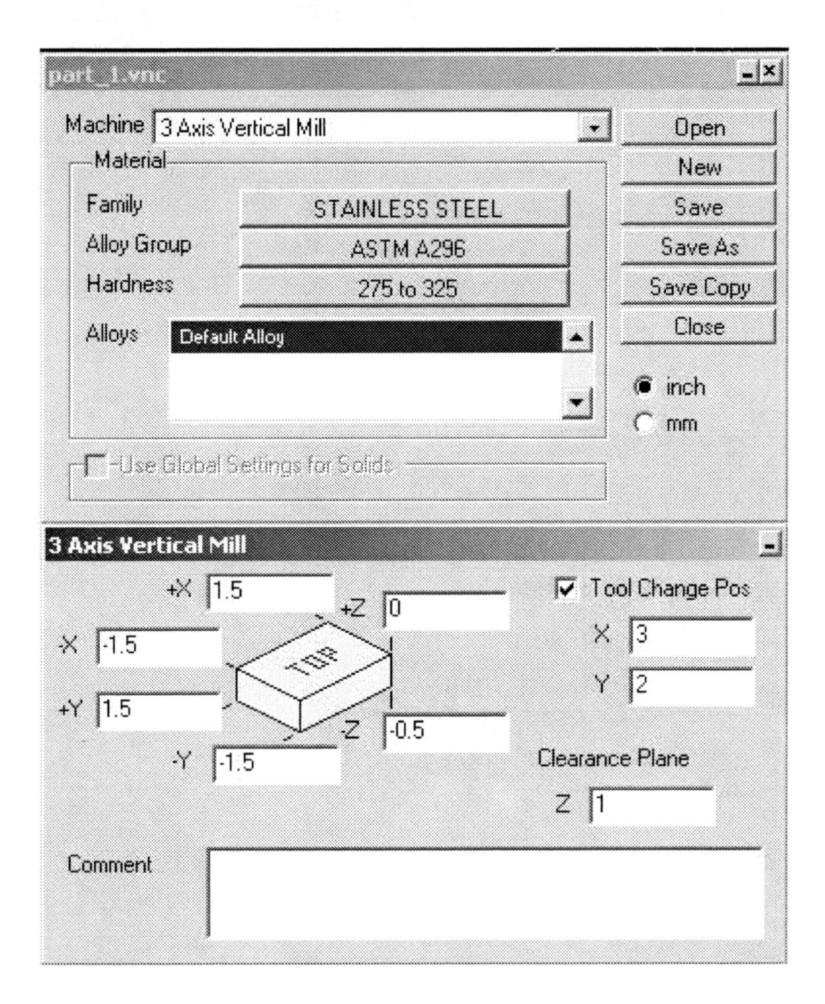


Select the Geometry icon from the Top Level Palette.

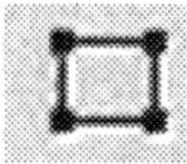

The Geometry Palette will appear.

Note: You may need to hit the return button to see the Geometry Palette if another set of geometry icons appear than the ones in the graphic above.

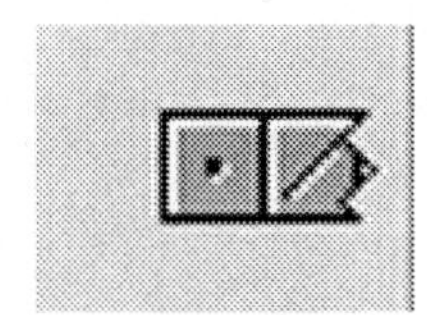

Select the Shapes icon.

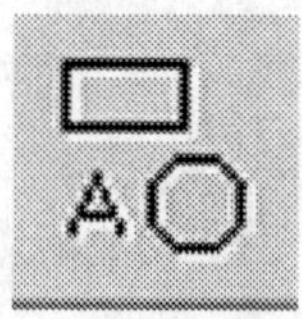

The Shapes Palette will appear.

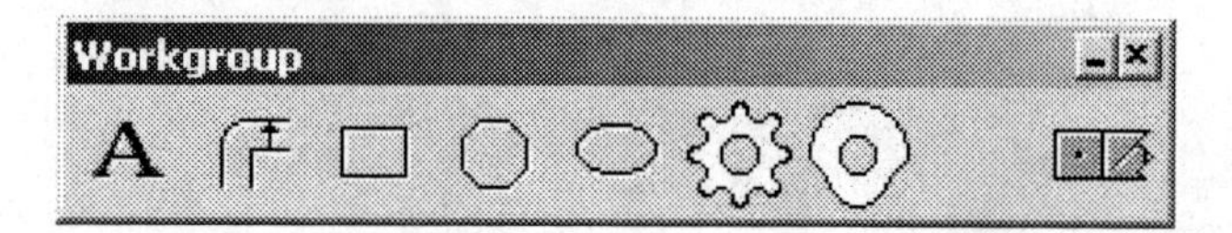


Select the Box icon from the Shapes Palette.

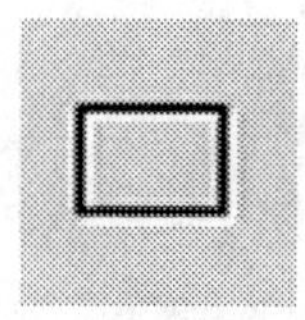

Enter the values below to create a square. Click on Do It.

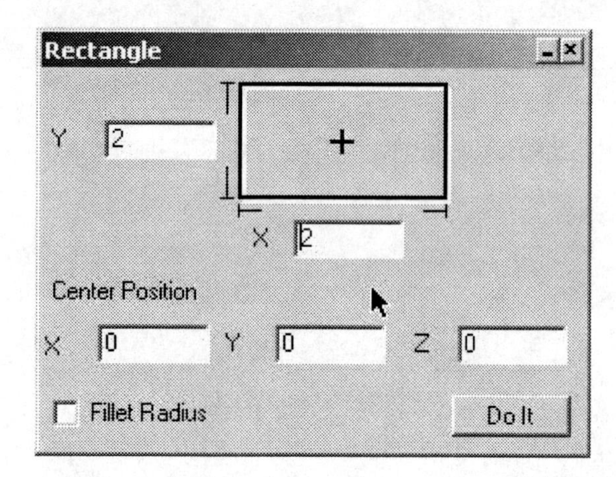

Close the Rectangle dialog box. Your screen should look like this.

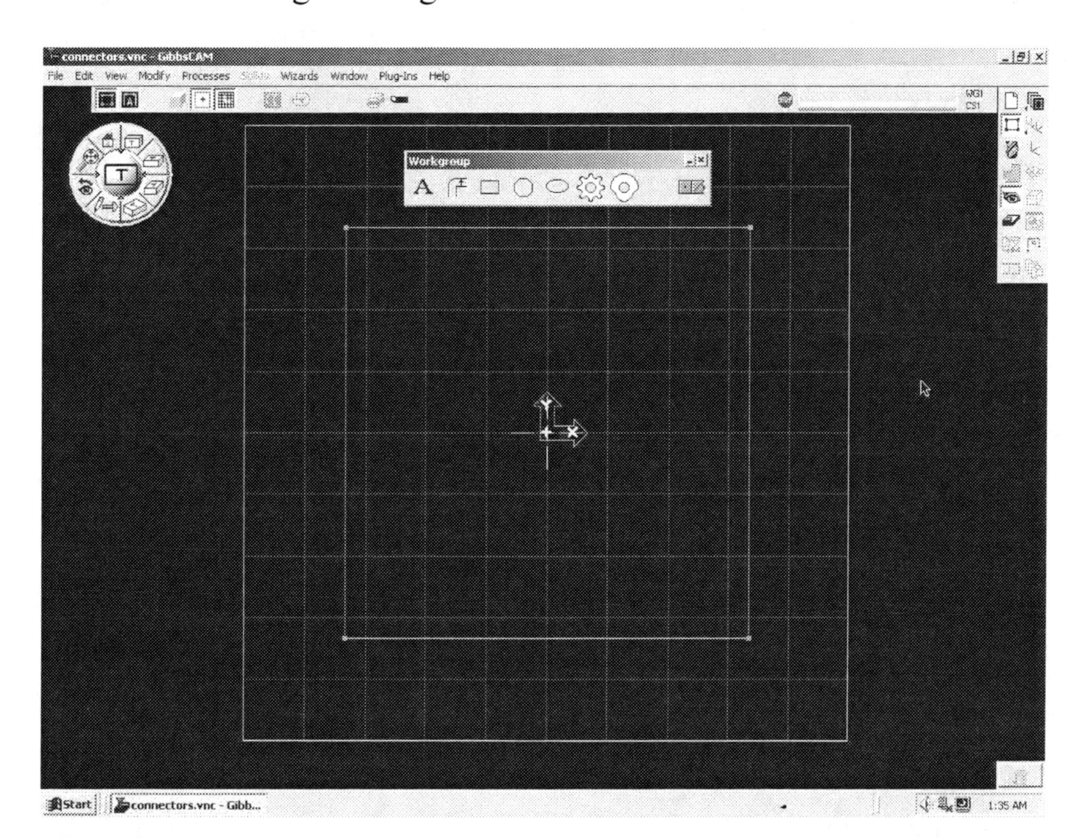


Notice that the endpoints of the lines of the square have blue points. These are *connectors*. Connectors allow the shape to be selected as one continuous entity later on in the machining process.

You will now add the arc to the upper-right hand corner of the square.

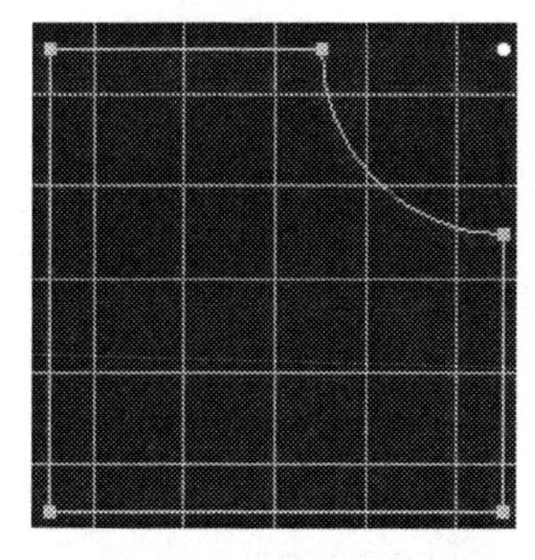

To do this you will create a circle and connect/trim it to the existing profile.

Select the Return icon to exit the Shapes Palette and return to the Geometry Palette.

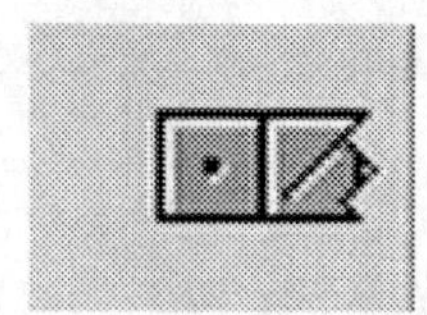

The Geometry Palette will appear.

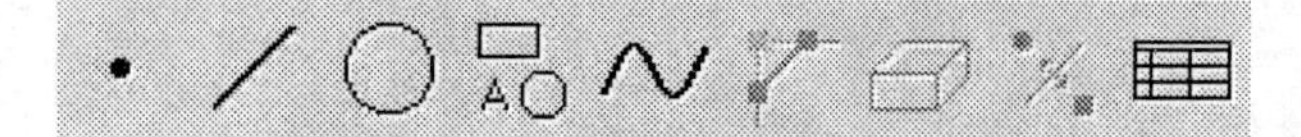


Click on the Circle icon in the Geometry Palette.

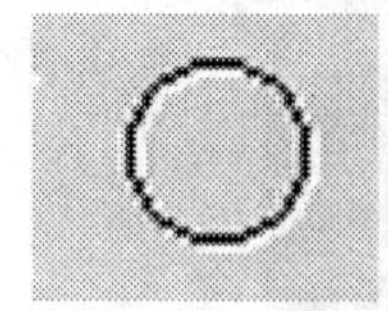

The Circle Palette will appear.

Select the Radius and Center Point icon.

Select the point in the upper right-hand corner of the square as shown. This will be the center point of the circle. Enter .75 for the radius and click on the Single Circle icon.

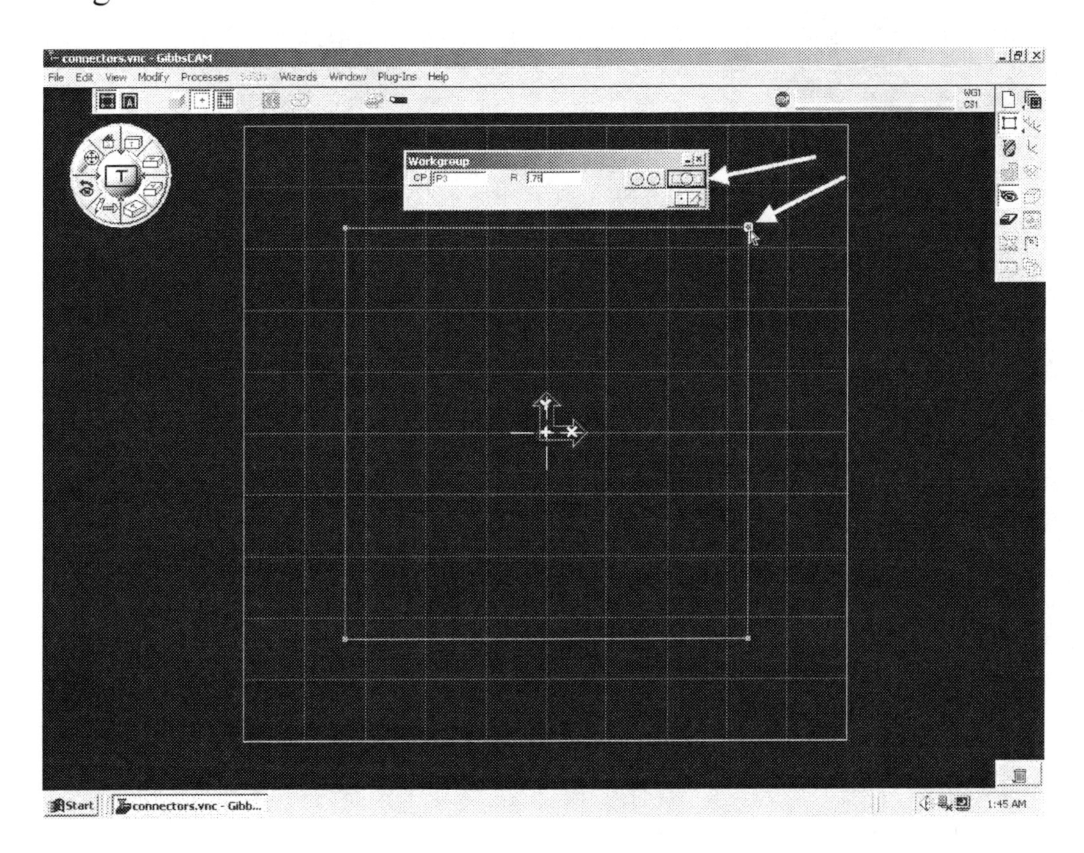

Your screen should look like this.

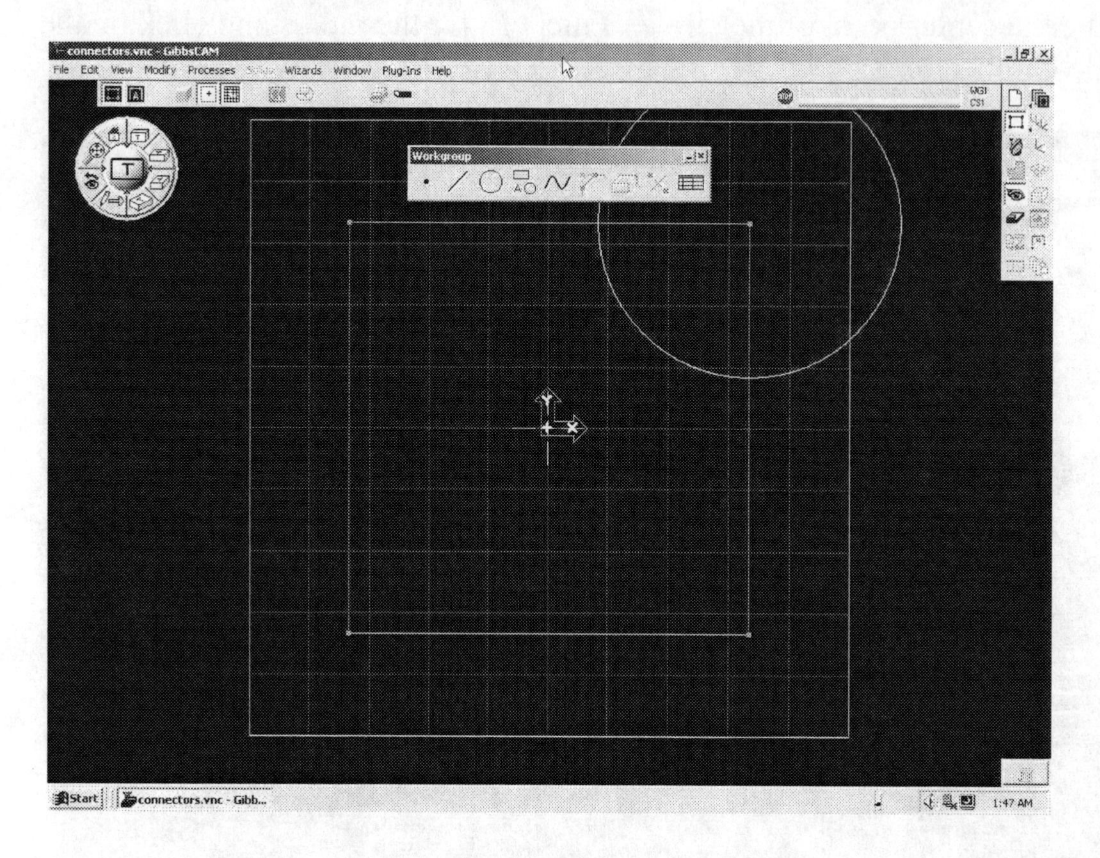


You will now connect the circle to the square. First, you must disconnect the point you used as the center point of the circle.

Select the point and click on the connect/disconnect icon in the Geometry Palette as shown in the next graphic.

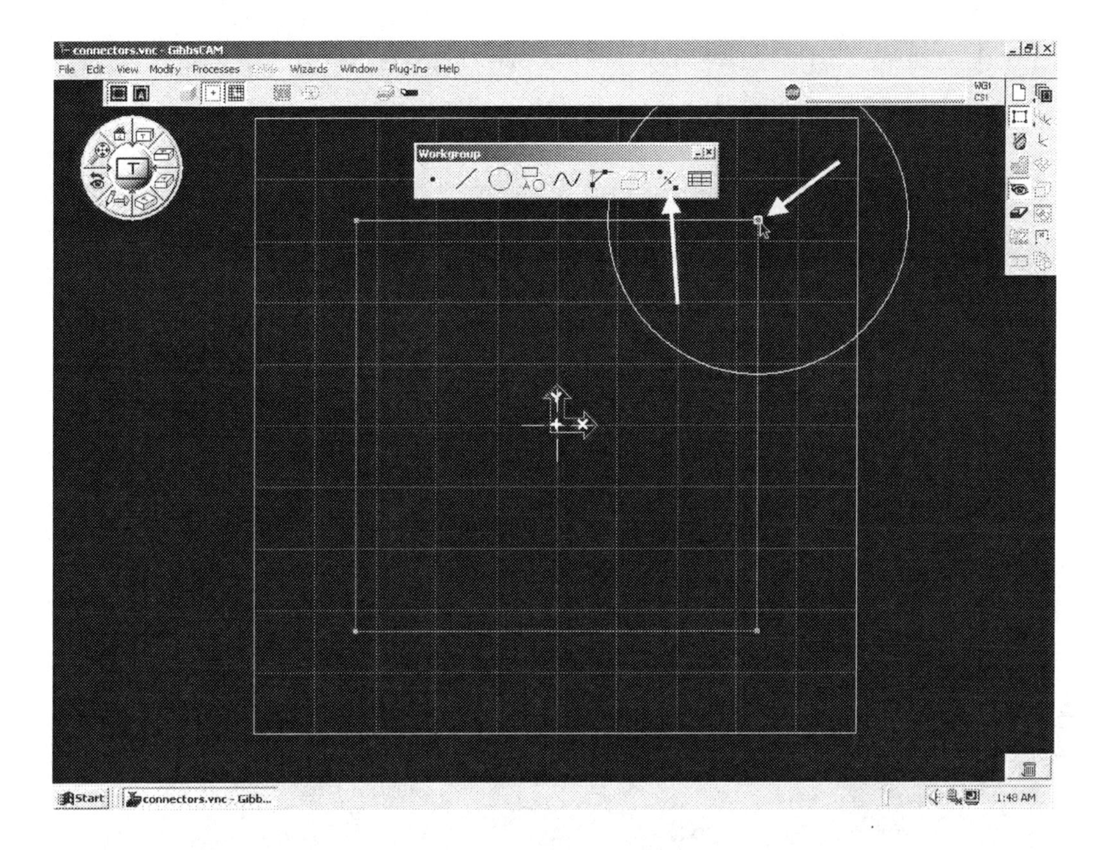


Your screen should look like this.

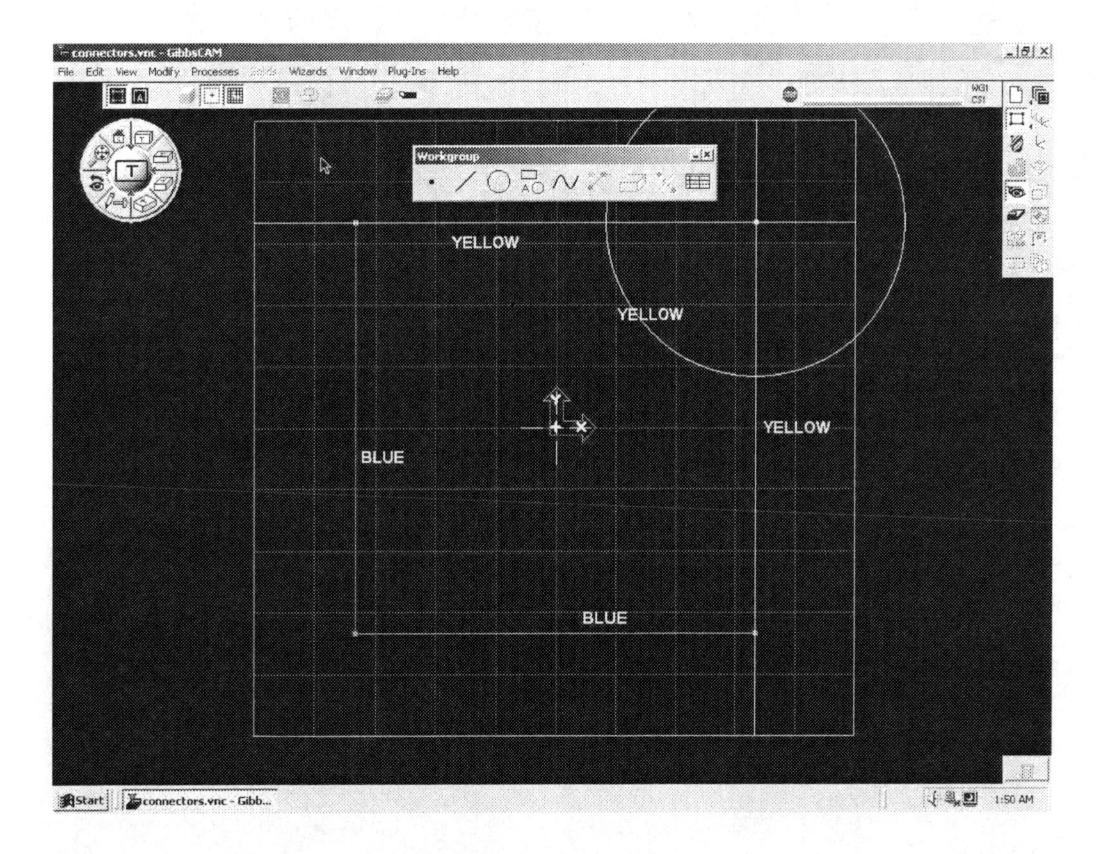

Notice that the 2 lines joined by the corner point are now disconnected and have turned yellow. You will now connect the arc to the lines.

Select the Point icon from the Geometry Palette.

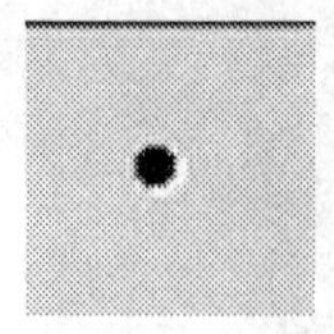

Hold down the control key and select the top horizontal line of the square and the circle. Click on the single point icon as shown in the next graphic.

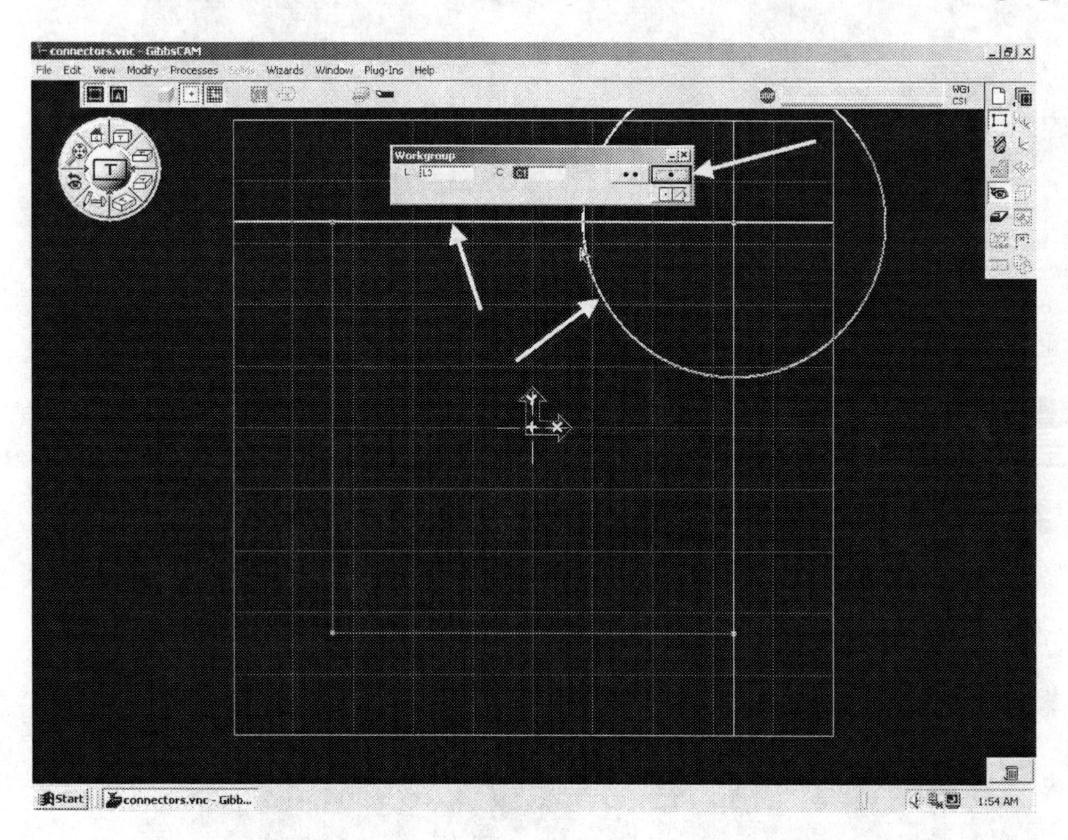

The system will find 2 points where the line intersects the circle. A dialog box will ask you which point you want to connect. Select the point on the left and click on OK as shown in the next graphic.

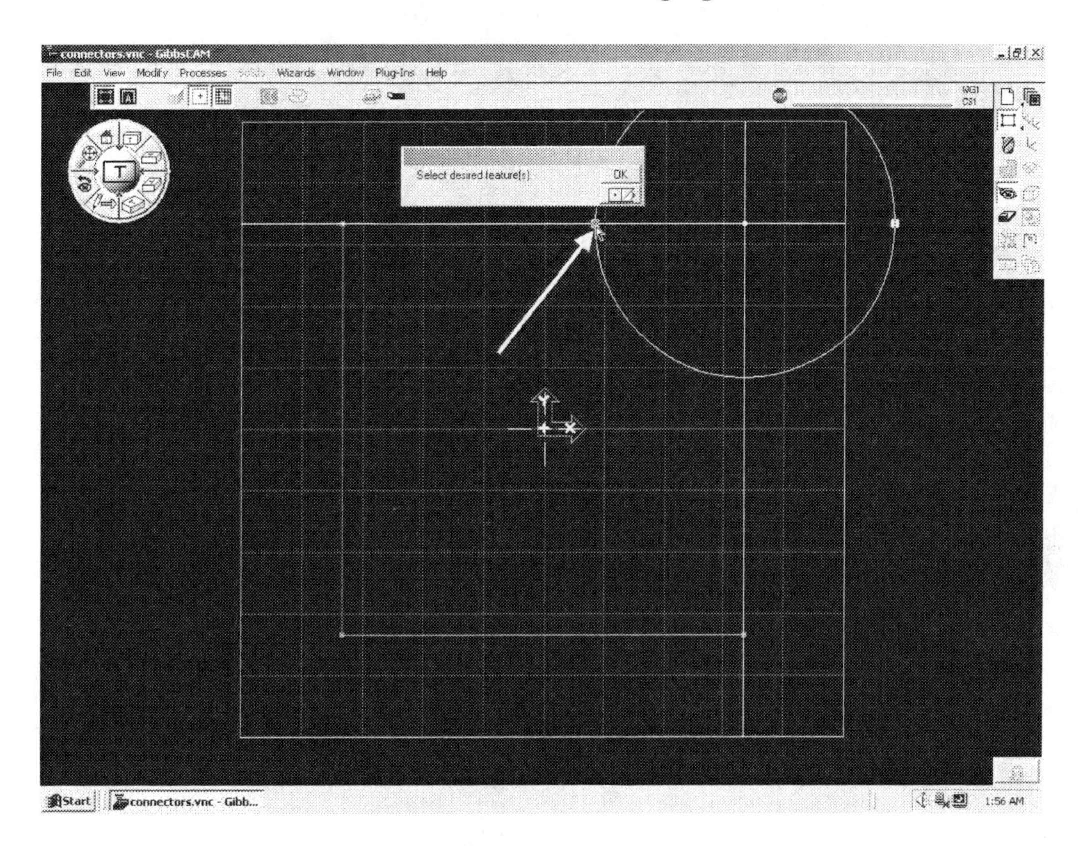

This will trim the line and place a blue connector point between the line and the circle as shown in the next graphic.

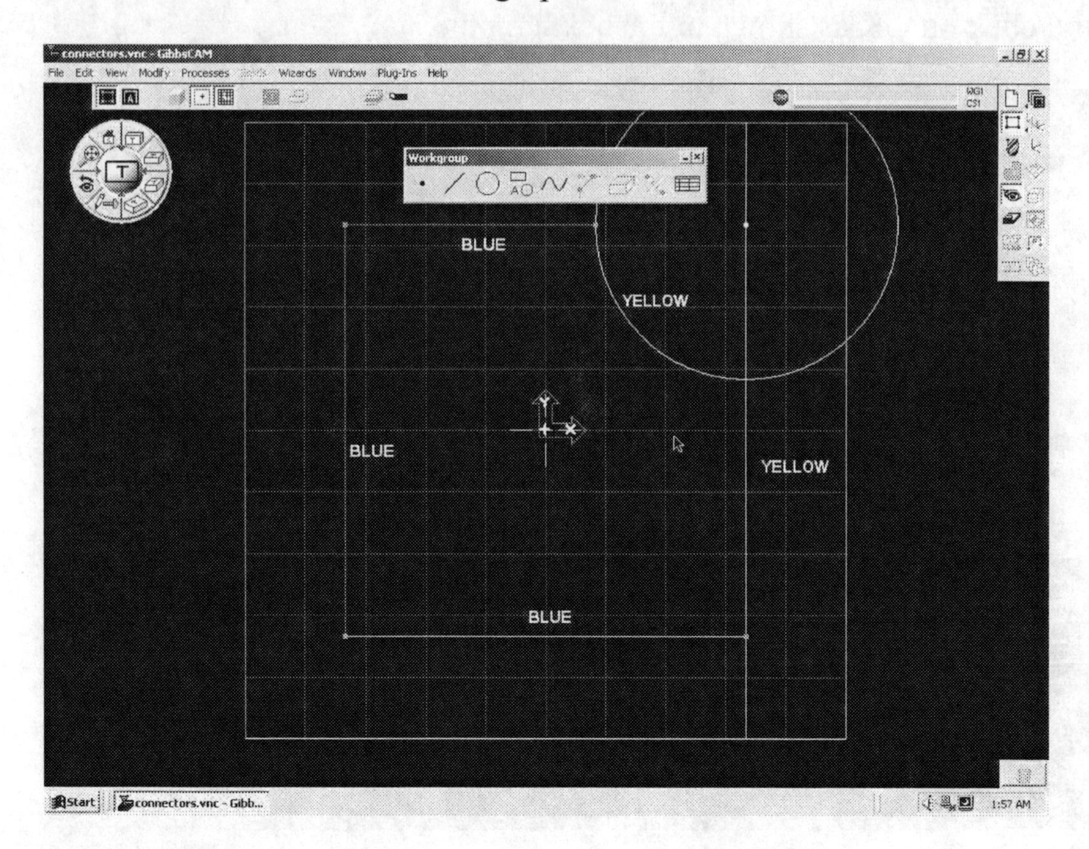


Select the Point icon from the Geometry Palette.

Hold down the control key and select the vertical line and the circle as shown. Click on the Single Point icon as shown in the next graphic.

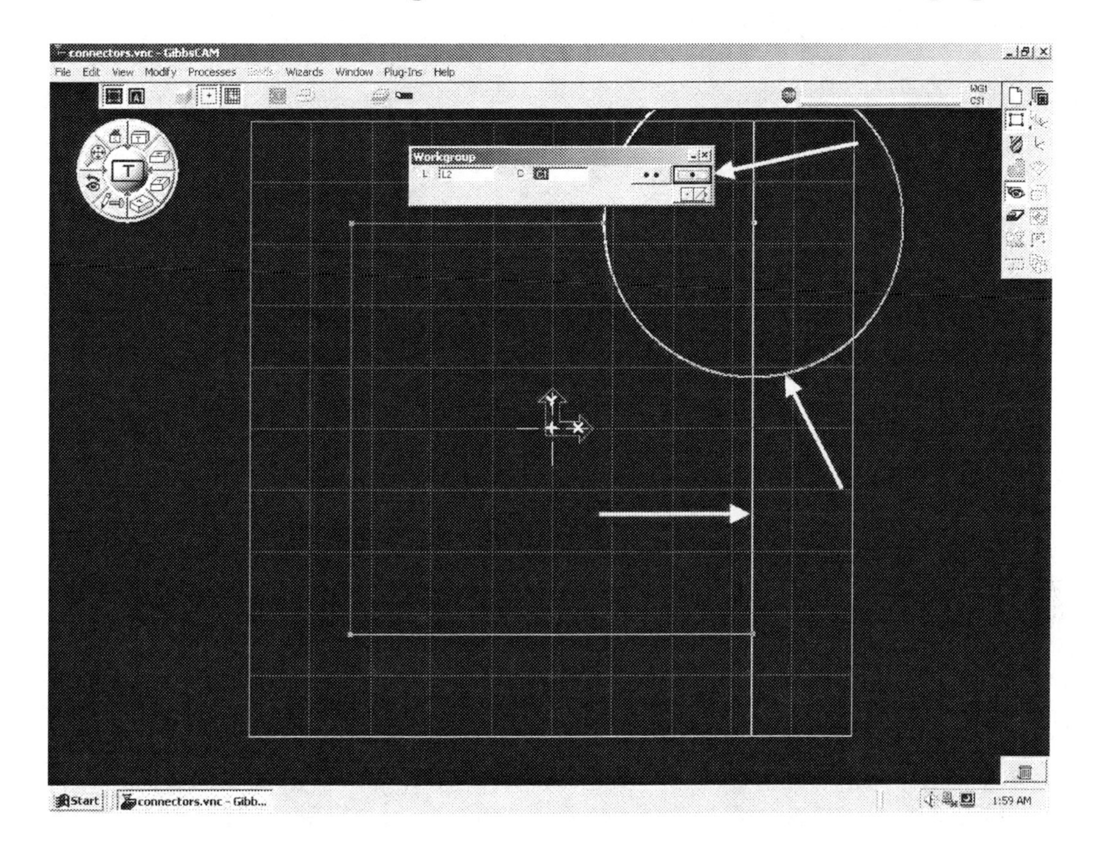

The system will find 2 points where the line intersects the circle. A dialog box will ask you which point you want to connect. Select the bottom point and click OK as shown in the next graphic.

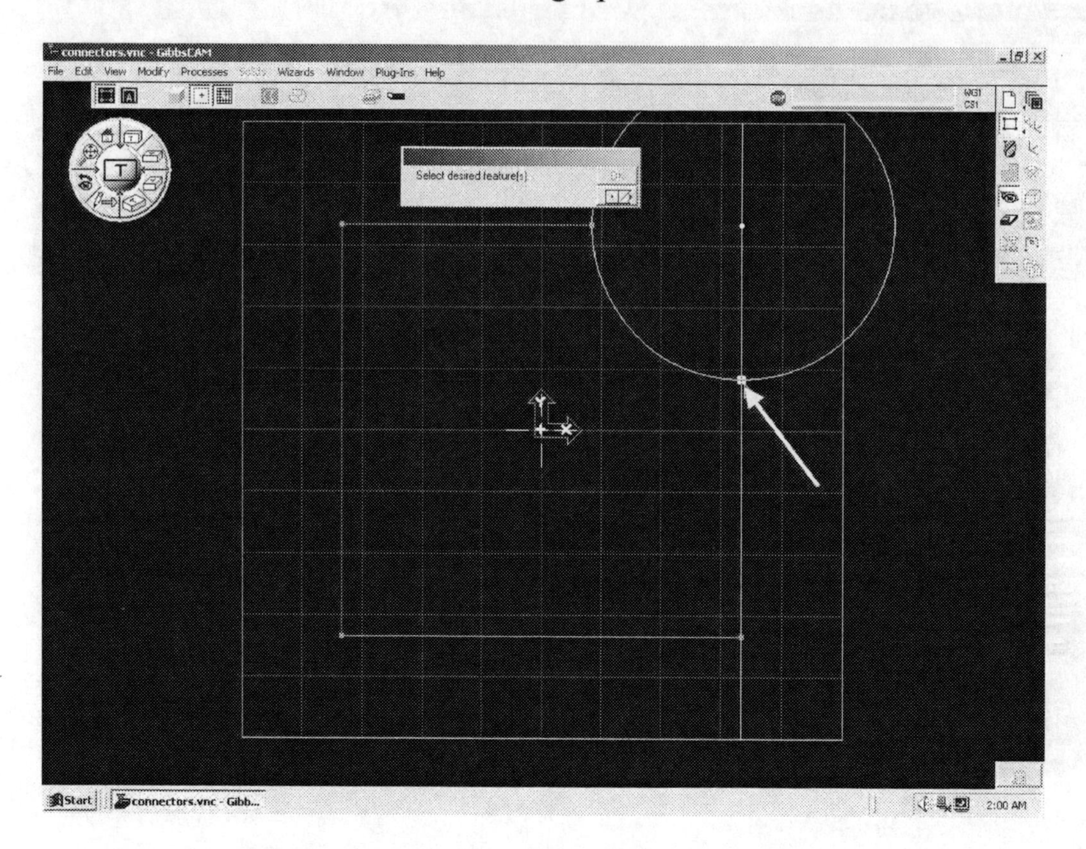

The geometry trims and the arc is now fully connected. Your screen should look like this.

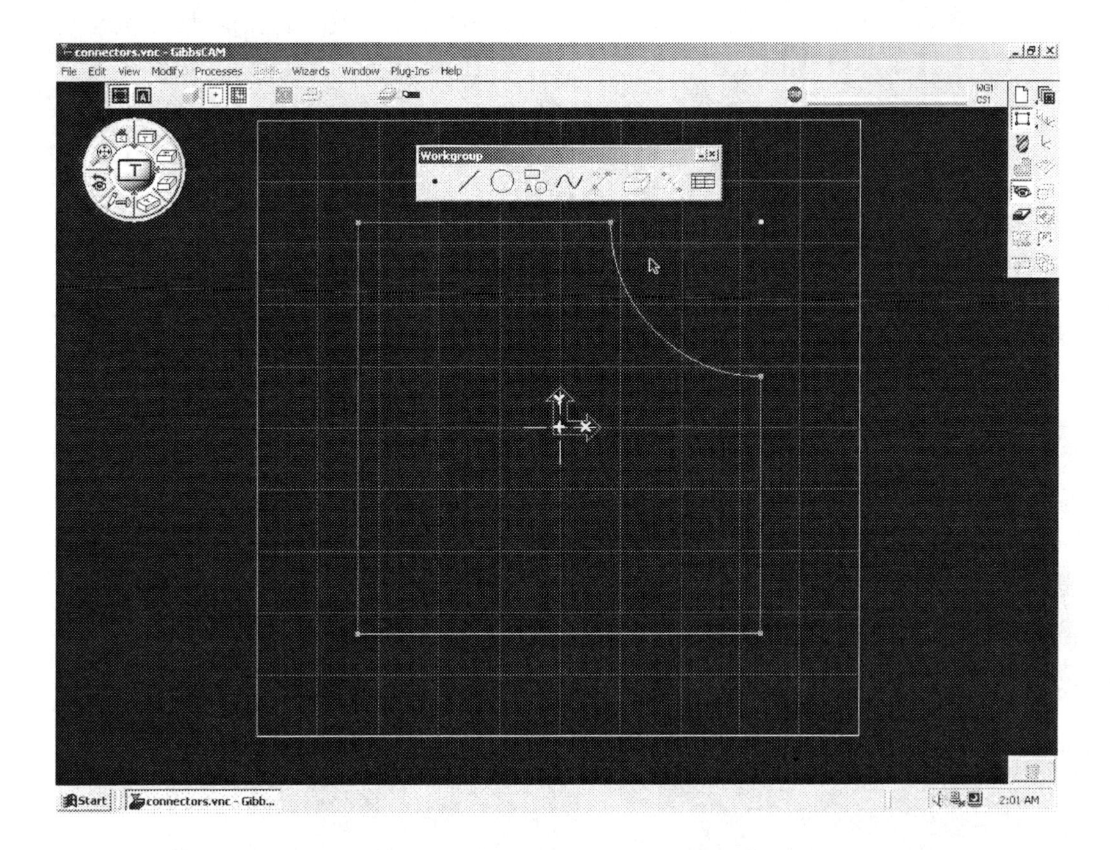


This completes the exercise. Save the file. Choose *File-Save* from the *Main Menu*. Close the file.

## Exercise 4 – Geometry Expert

In this exercise you will use the Geometry Expert to create the profile below. Geometry Expert is a good tool to use when you want to create a closed profile. You create profiles with Geometry Expert as if you were walking around the perimeter of the geometry. You will enter rows of information in the Geometry Expert as you create the profile in a clockwise direction.

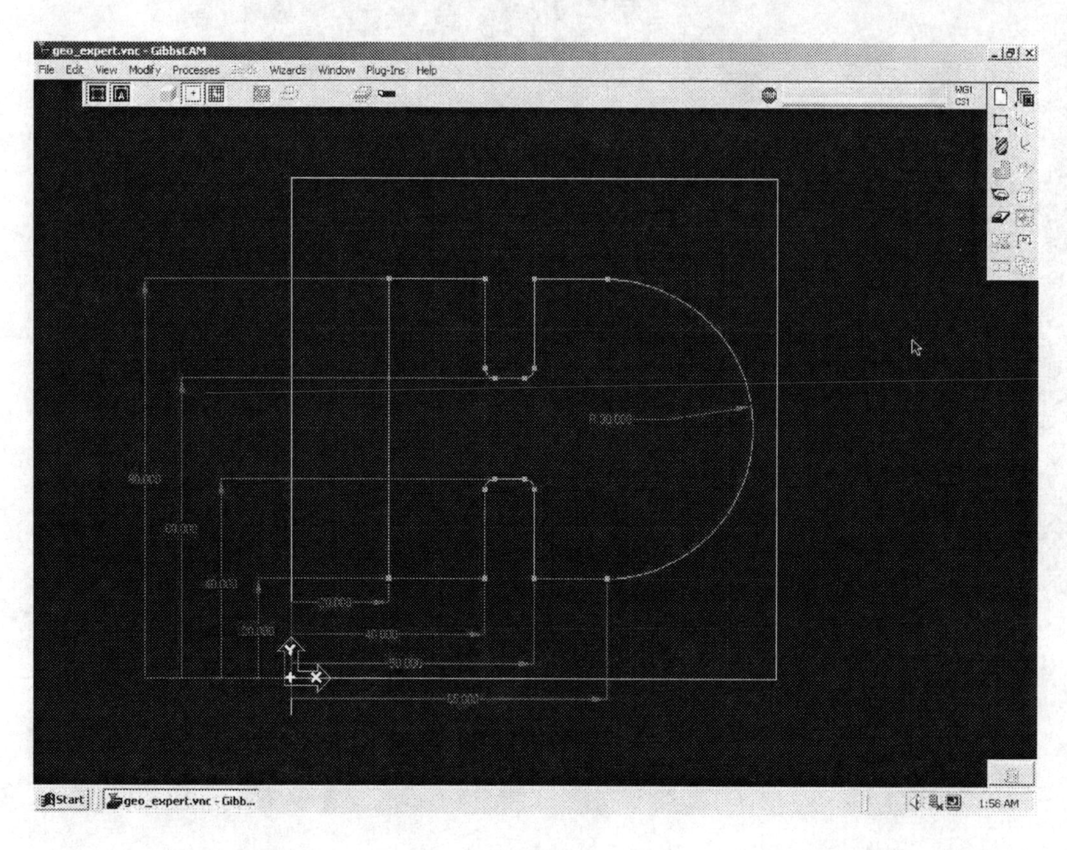


Create a new file named GEO_EXPERT. Set the machine type to 3 axis vertical mill and the units to mm. Enter the values in the next graphic for the stock size.

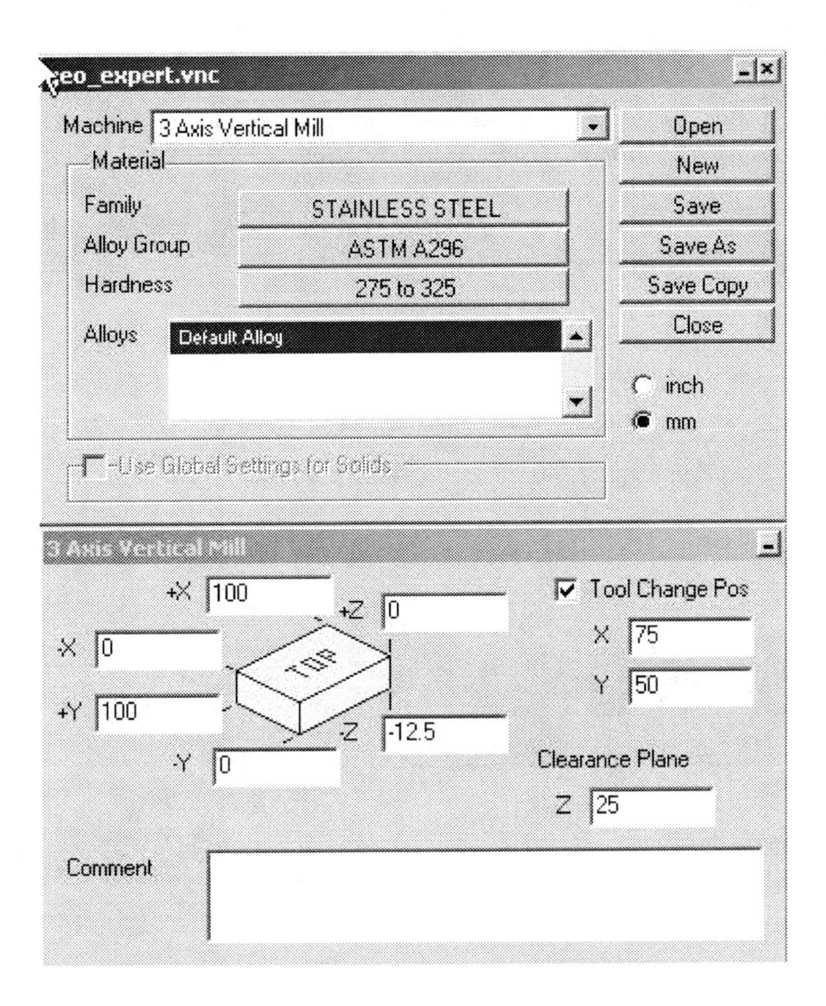


Select the Geometry icon from the Top Level Palette.

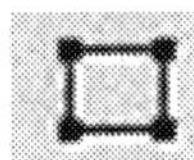

The Geometry Palette will appear.

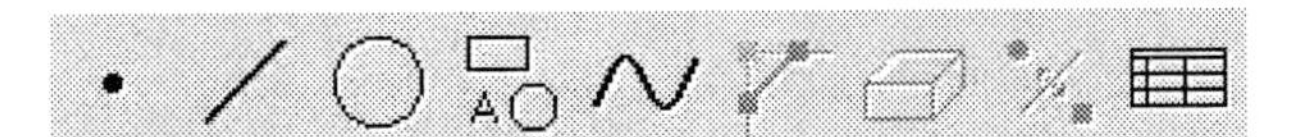

Select the Geometry Expert icon from the Geometry Palette.

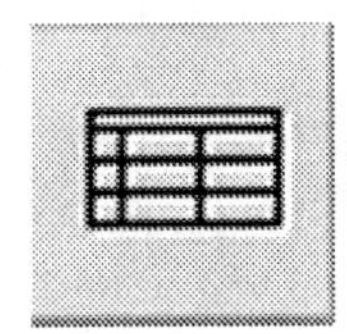

The Geometry Expert dialog box will appear.

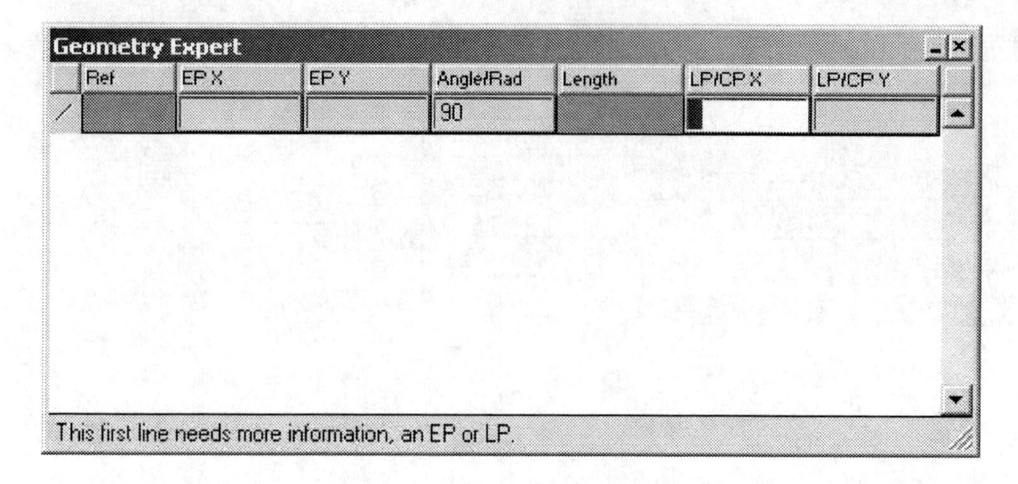


Each row of the Geometry Expert will have a Feature Type icon on the left side.

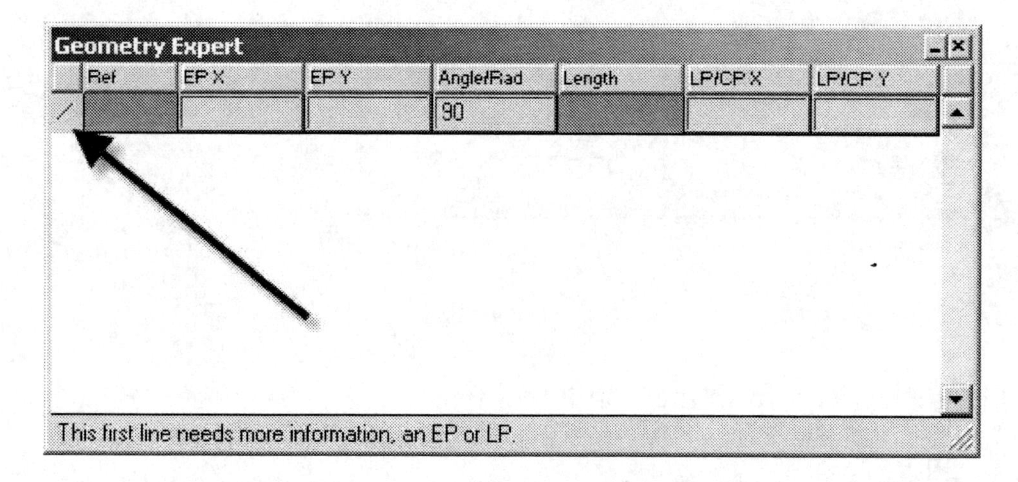


Click and hold the mouse down on this icon and the available feature types will appear. They are: Line, Chamfer, Fillet, Clockwise Arc, Counter-Clockwise Arc, Close Shape, and Macro.

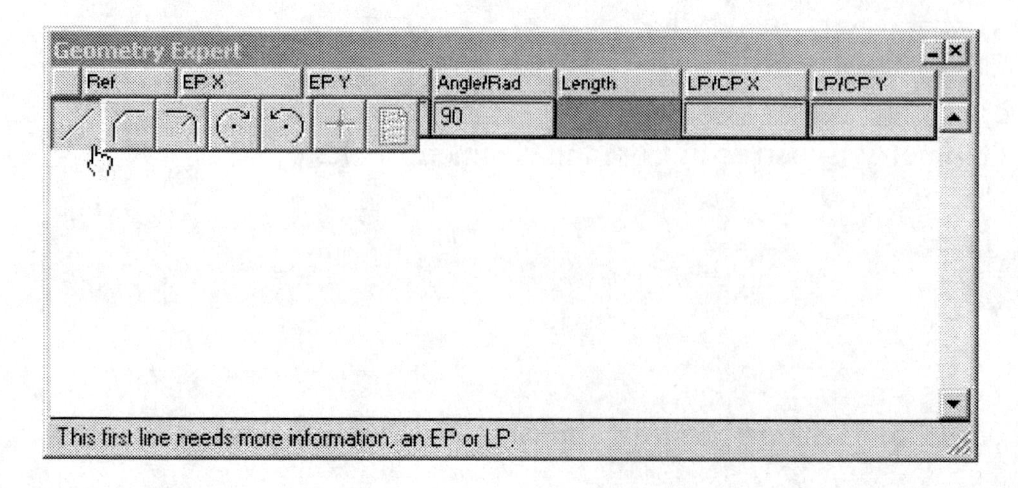

The Geometry Expert spreadsheet is set up as follows:

| Column | Information |
|---|---|
| Ref. | Reference information |
| EPX | X Value of Endpoint |
| EPY | Y Value of Endpoint |
| Angle/Rad | Line Angle or Arc Radius |
| Length | Line Length or Chamfer Length |
| LP/CPX | X Value of Line Point or Center point |
| LP/CPY | Y Value of Line Point or Center point |

The Geometry Expert also has a prompt line at the bottom. This prompt line will tell you when you have a fully defined line, an invalid feature, or when you need to add more information.

First, you will create a vertical line that passes thru X20. Make sure the feature type shows the Line icon. Enter 90 in the angle box and enter 20 in the LP/CPX box. Notice the prompt line at the bottom of the Geometry Expert tells you that you have a fully defined line. Your spreadsheet should look like the following image.

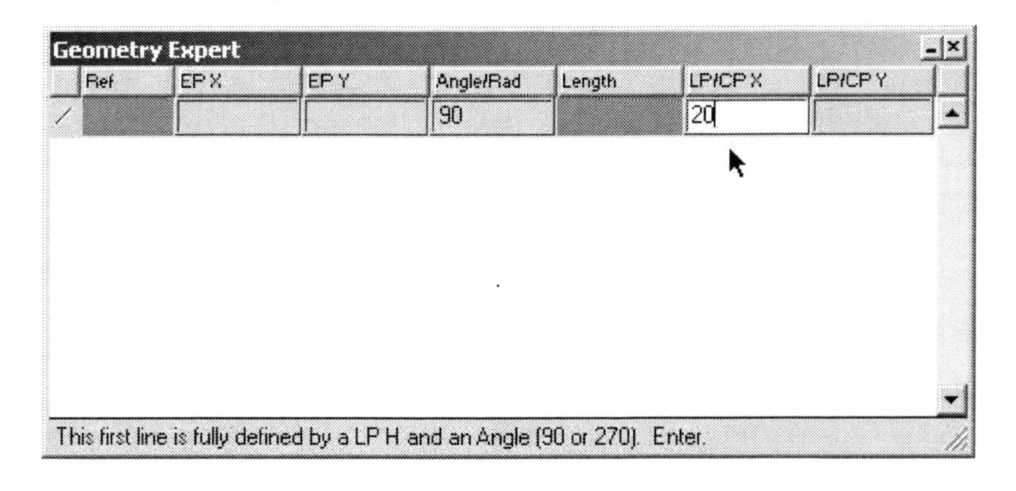


Press enter. A vertical line will appear.

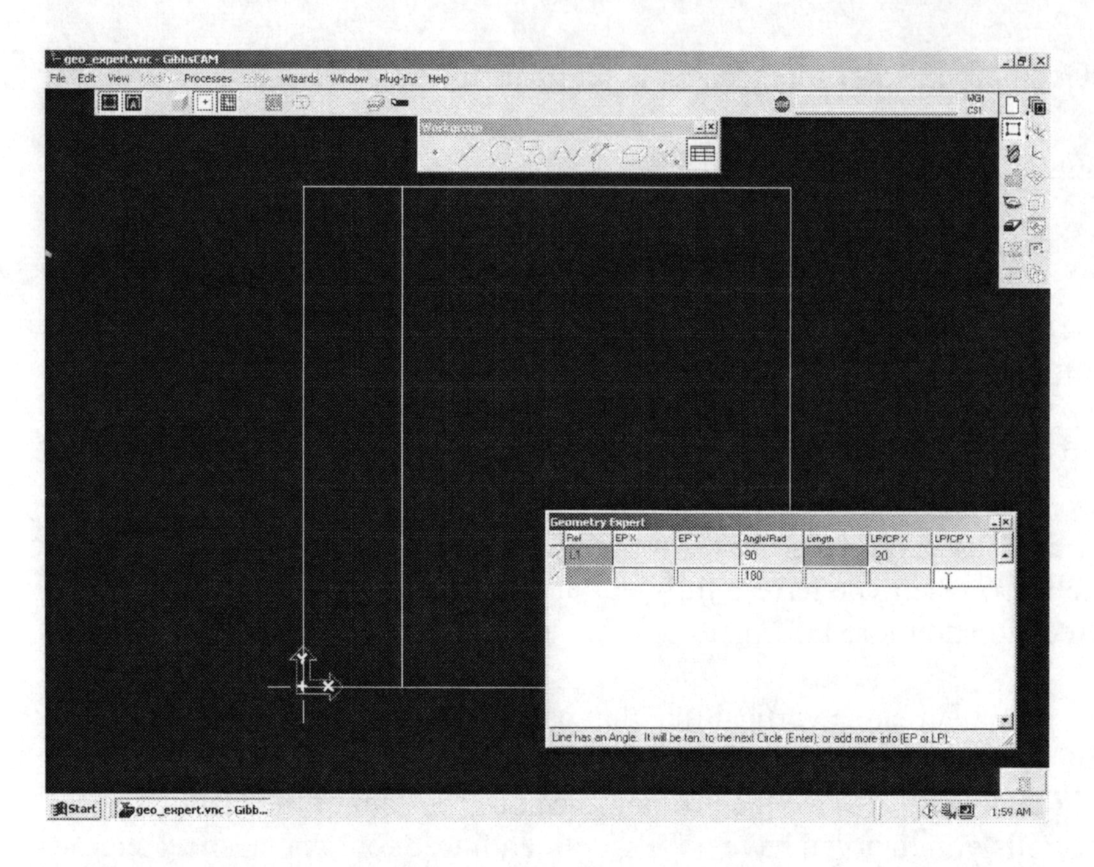

A new row will be added to the spreadsheet. Enter the values shown for line 2. This will create a horizontal line thru Y80.

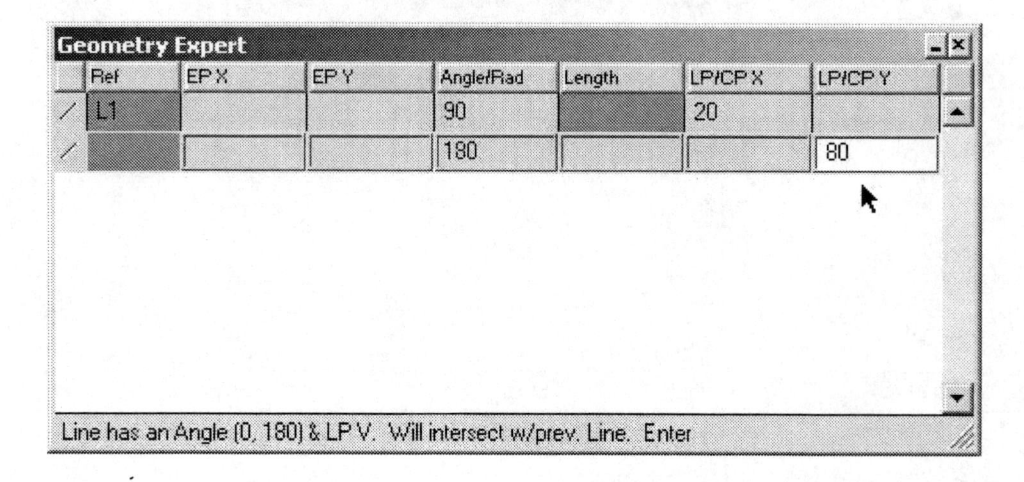


Press enter. You will notice a blue connector dot appears on the screen.

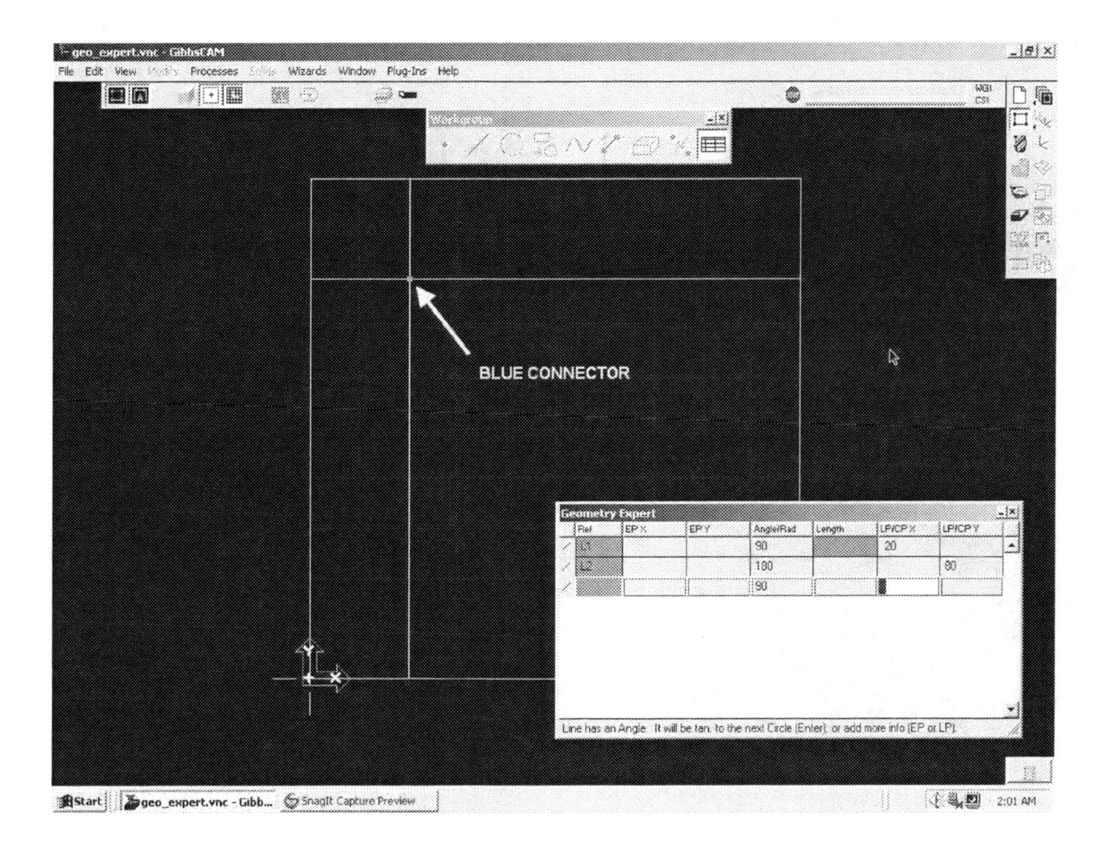


Enter the values shown for line 3.

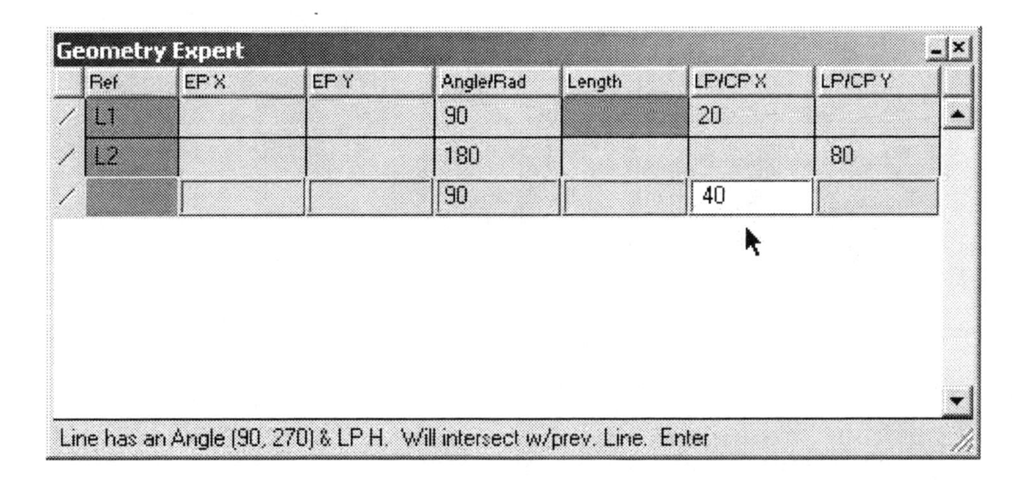


Press enter. The horizontal line you created passing thru Y80 will be trimmed.

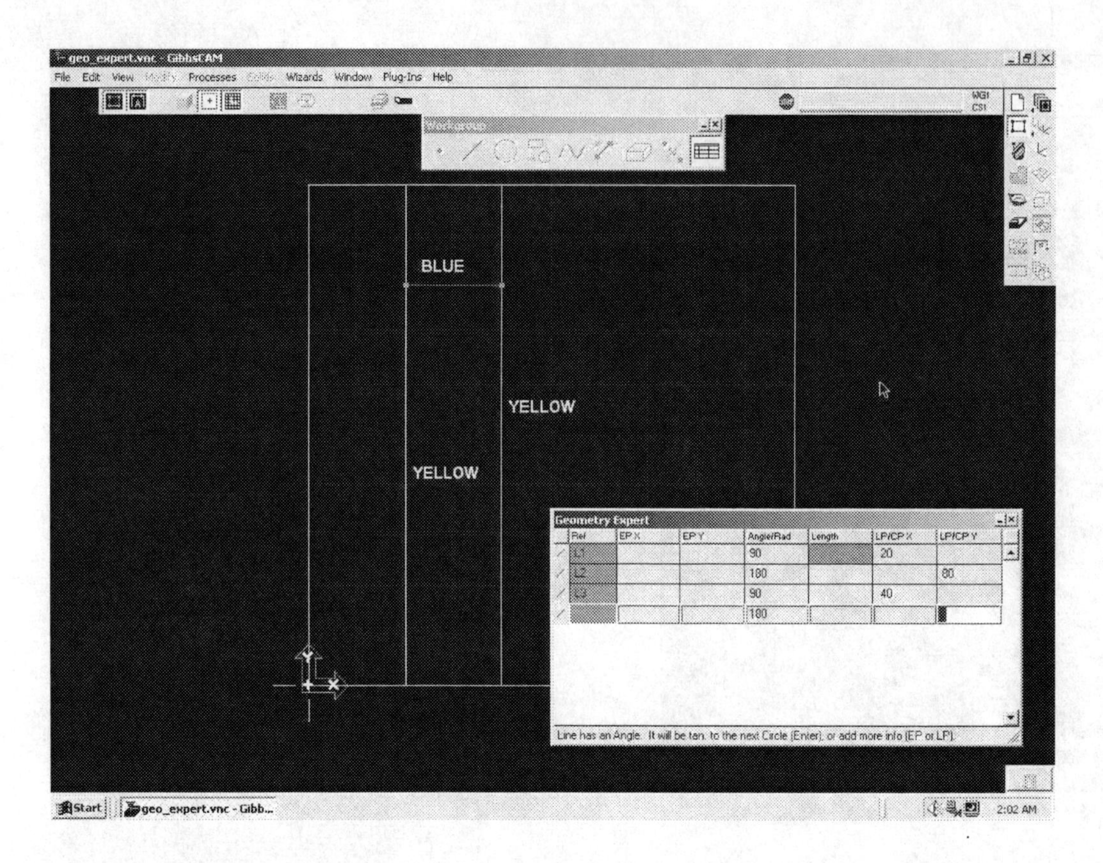


Enter the values in line 4. This will create a horizontal line thru Y60.

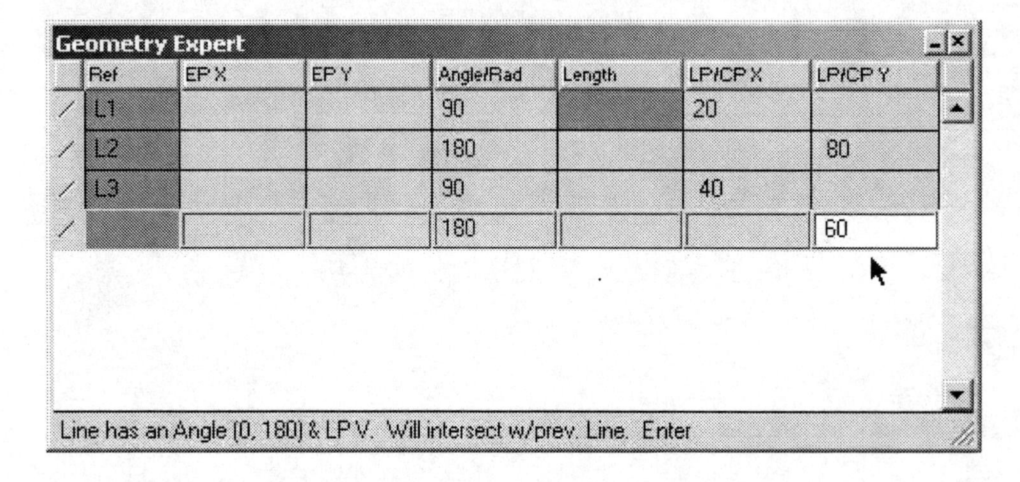


Press enter. Notice the lines trim again. The blue lines represent geometry that is fully connected.

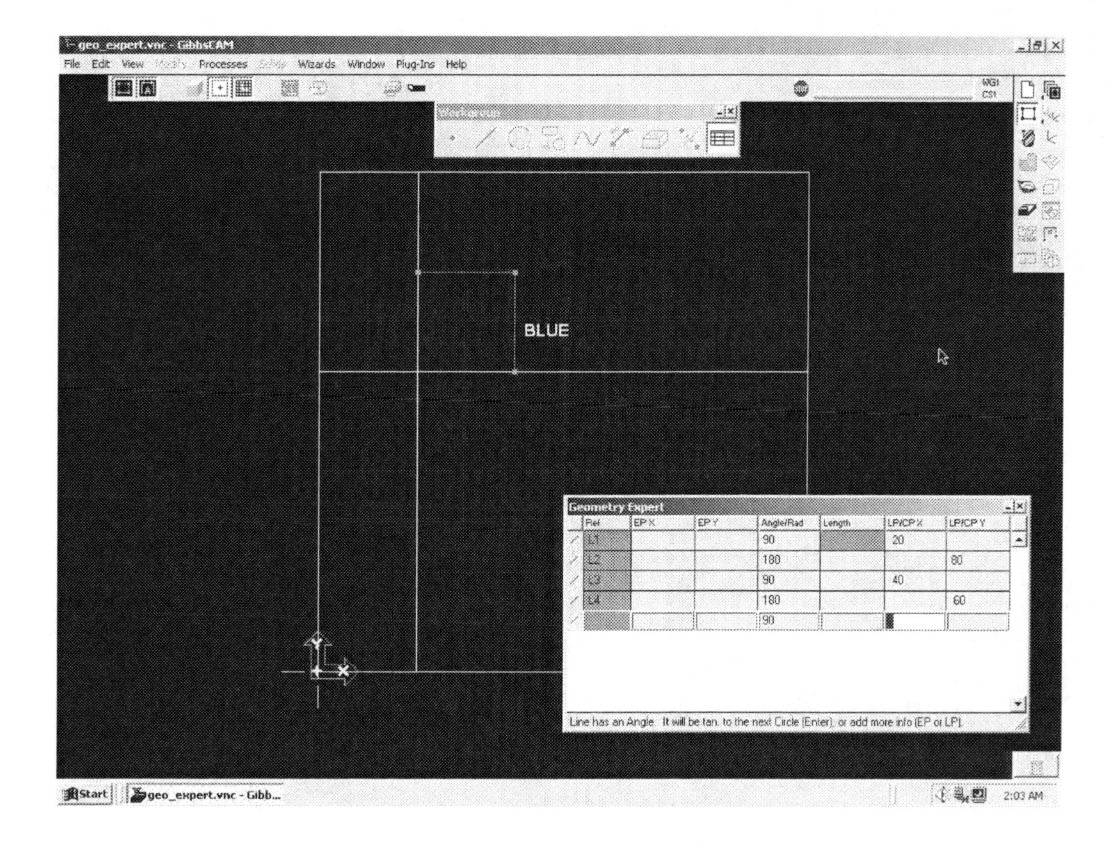


Enter the values shown in line 5. This will create a vertical line thru X80.

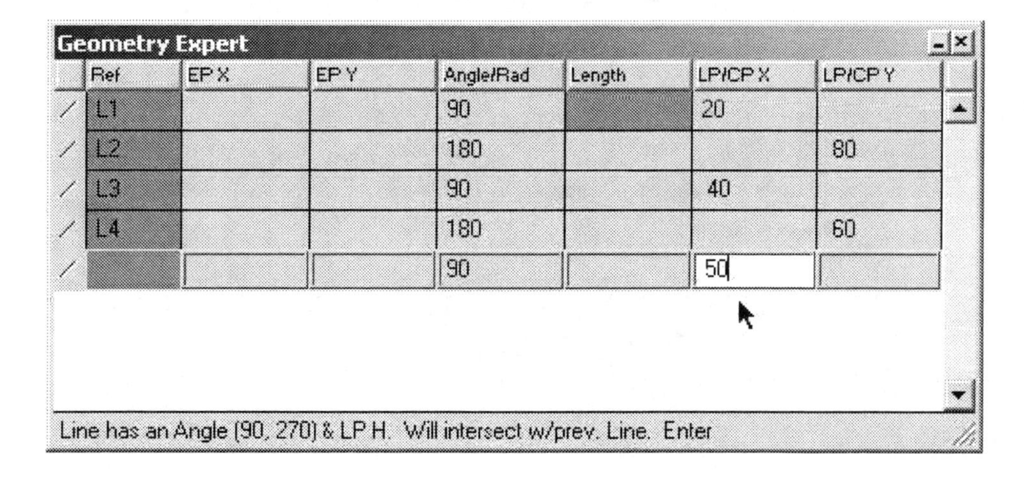
Geometry Expert

| | Ref | EP X | EP Y | Angle/Rad | Length | LP/CP X | LP/CP Y |
|---|---|---|---|---|---|---|---|
| / | L1 | | | 90 | | 20 | |
| / | L2 | | | 180 | | | 80 |
| / | L3 | | | 90 | | 40 | |
| / | L4 | | | 180 | | | 60 |
| / | | | | 90 | | 50 | |

Line has an Angle (90, 270) & LP H. Will intersect w/prev. Line. Enter

Press enter.

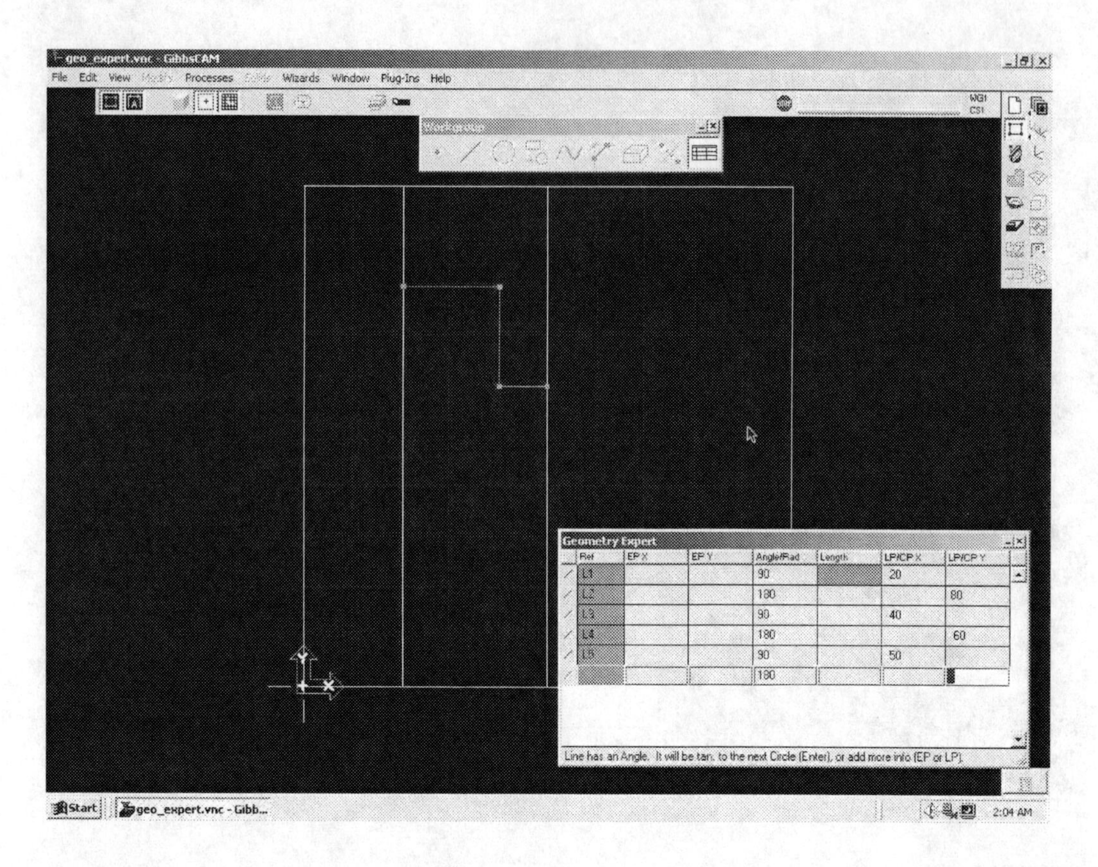


Next, you will create a horizontal line from the current location that has an endpoint of X65 Y80. You will use this endpoint to start a clockwise arc in the next step.

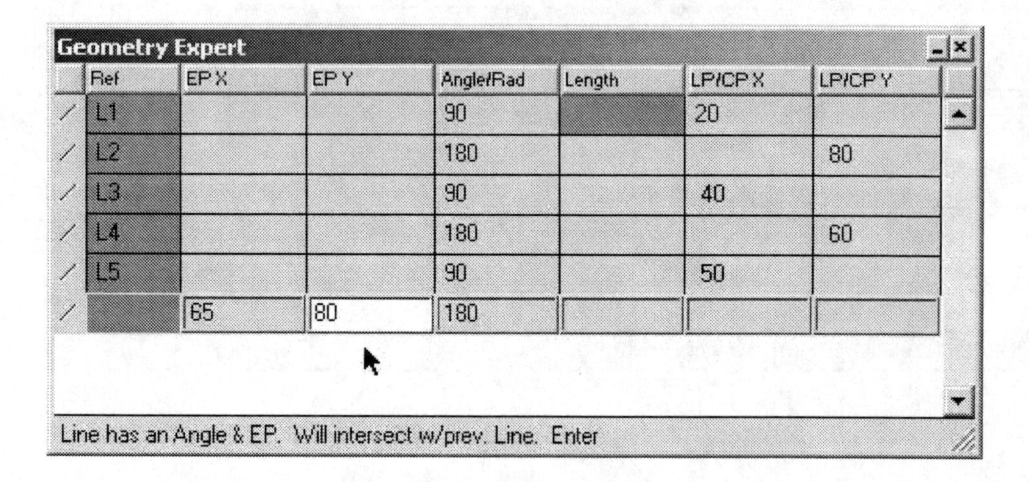


Press enter.

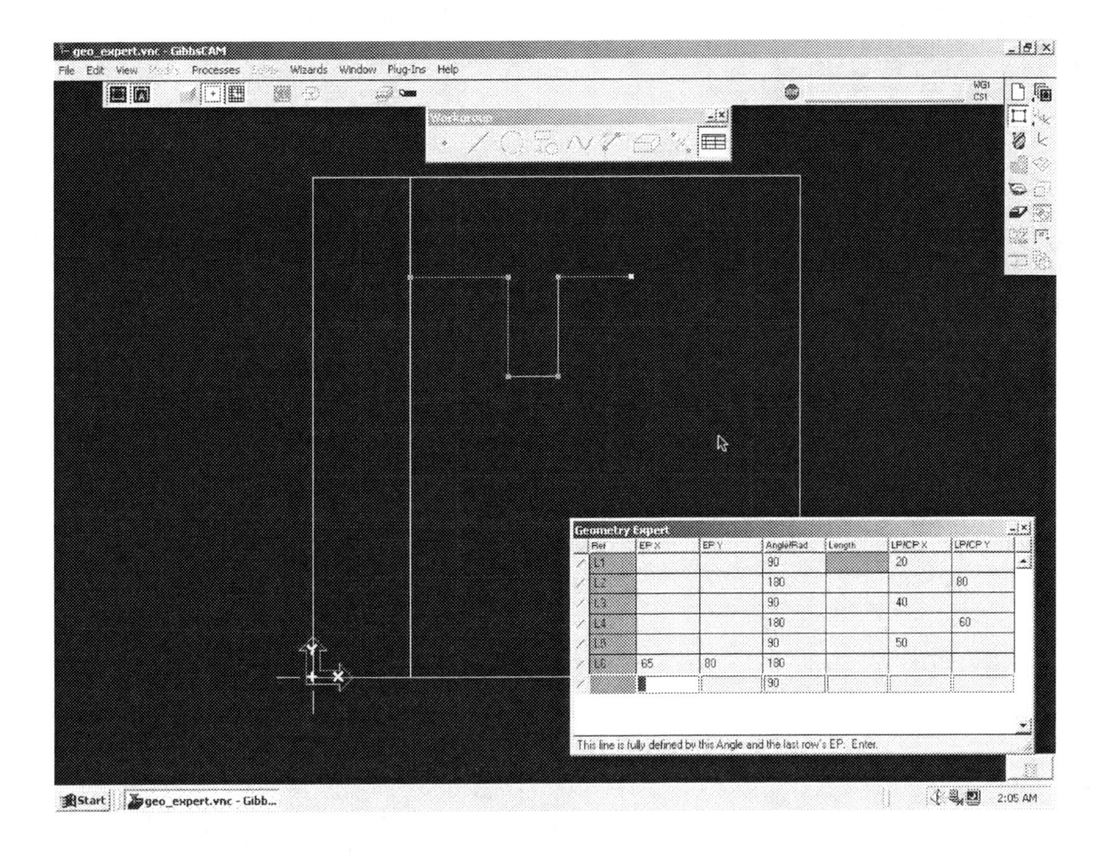


Now you will begin the arc. Click and hold the mouse on the Feature Type icon on line 7. Select the Clockwise Arc icon from the feature types and enter 30 for the radius.

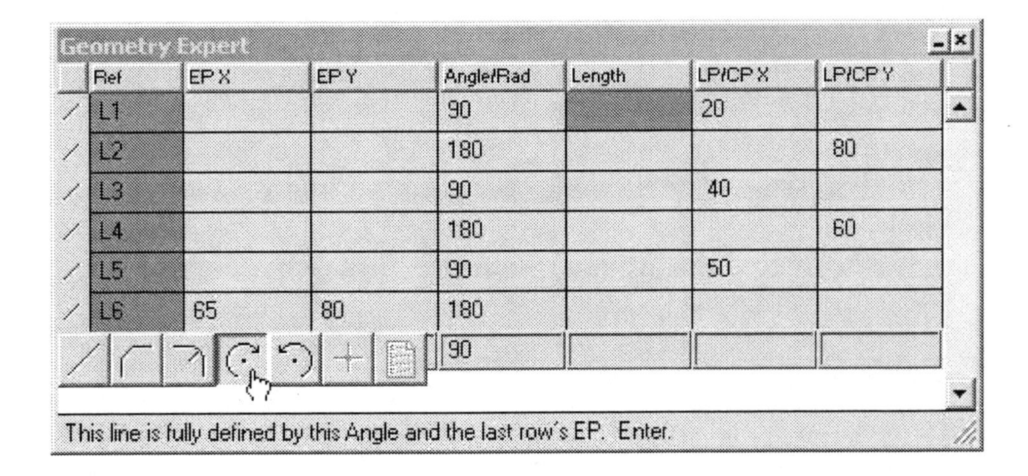

Geometry Expert

| Ref | EP X | EP Y | Angle/Rad | Length | LP/CP X | LP/CP Y |
|---|---|---|---|---|---|---|
| L1 | | | 90 | | 20 | |
| L2 | | | 180 | | | 80 |
| L3 | | | 90 | | 40 | |
| L4 | | | 180 | | | 60 |
| L5 | | | 90 | | 50 | |
| L6 | 65 | 80 | 180 | | | |
| | | | 90 | | | |

This line is fully defined by this Angle and the last row's EP. Enter.

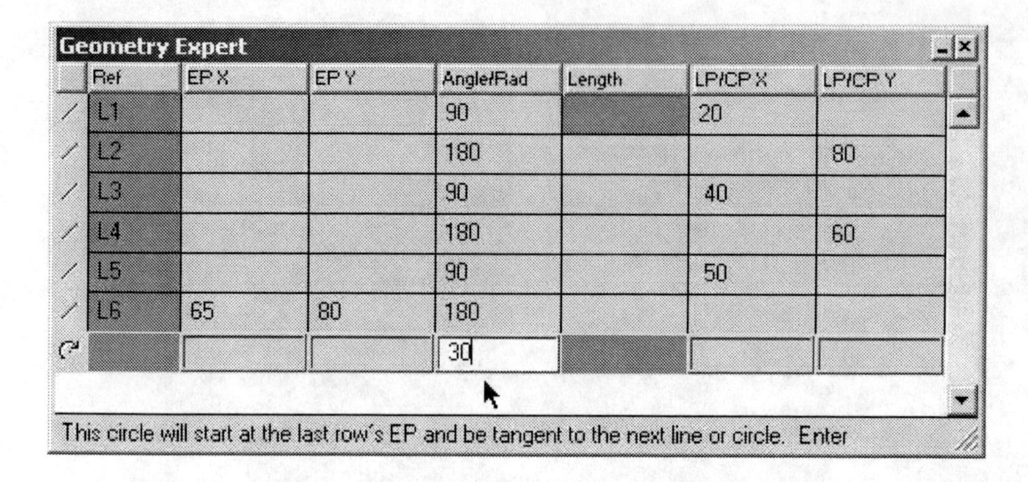

Geometry Expert

| Ref | EP X | EP Y | Angle/Rad | Length | LP/CP X | LP/CP Y |
|---|---|---|---|---|---|---|
| L1 | | | 90 | | 20 | |
| L2 | | | 180 | | | 80 |
| L3 | | | 90 | | 40 | |
| L4 | | | 180 | | | 60 |
| L5 | | | 90 | | 50 | |
| L6 | 65 | 80 | 180 | | | |
| | | | 30 | | | |

This circle will start at the last row's EP and be tangent to the next line or circle. Enter

Press enter. You will not see the arc yet because it is dependent on the next line.

The next line will be a horizontal line that passes thru Y20. Enter 180 for the Angle/Rad.

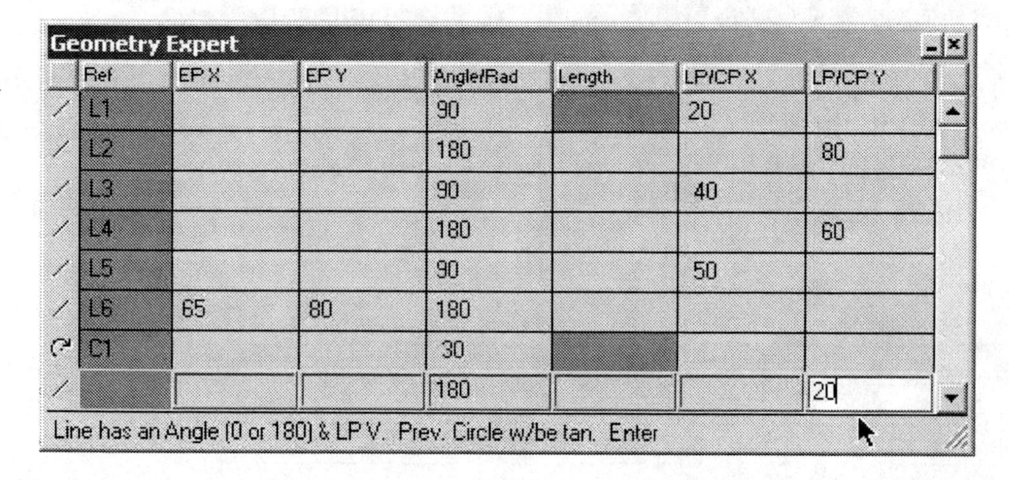

Geometry Expert

| Ref | EP X | EP Y | Angle/Rad | Length | LP/CP X | LP/CP Y |
|---|---|---|---|---|---|---|
| L1 | | | 90 | | 20 | |
| L2 | | | 180 | | | 80 |
| L3 | | | 90 | | 40 | |
| L4 | | | 180 | | | 60 |
| L5 | | | 90 | | 50 | |
| L6 | 65 | 80 | 180 | | | |
| C1 | | | 30 | | | |
| | | | 180 | | | 20 |

Line has an Angle (0 or 180) & LP V. Prev. Circle w/be tan. Enter

Press enter. You will now see the arc and the line.

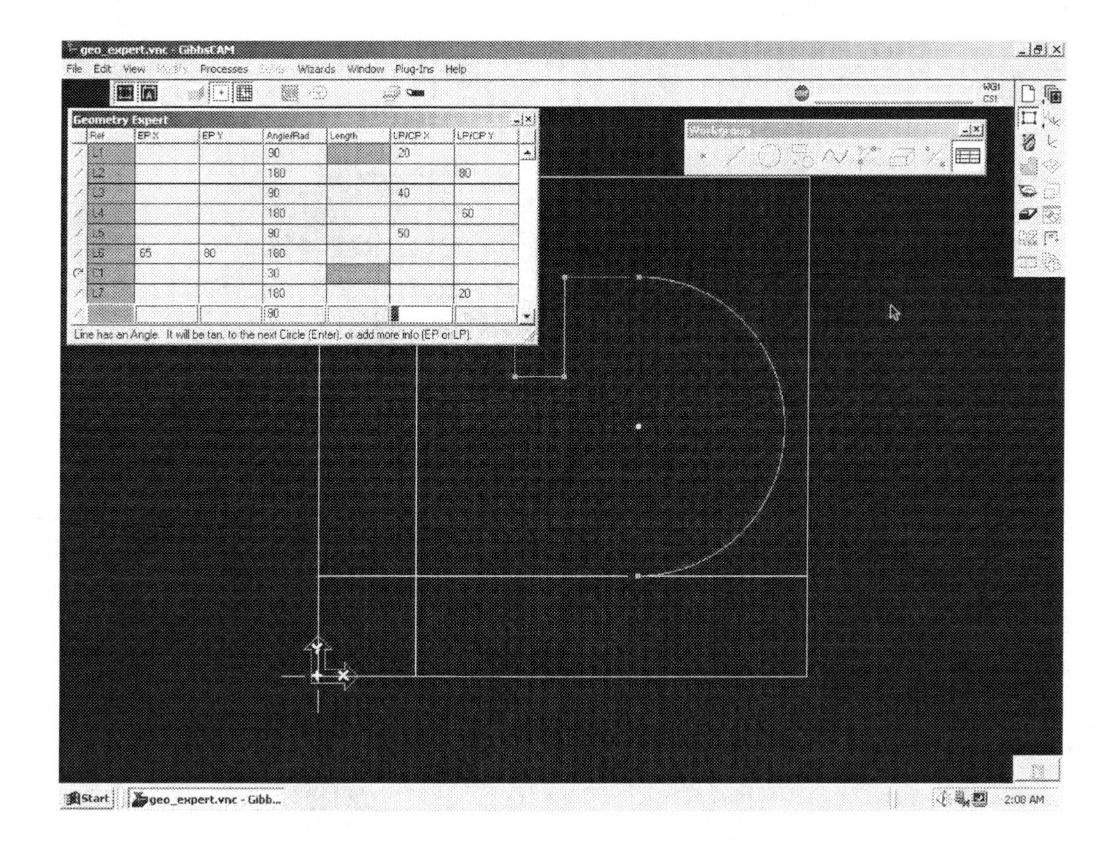


The next line will be a vertical line that will pass thru X50.

Geometry Expert

| | Ref | EP X | EP Y | Angle/Rad | Length | LP/CP X | LP/CP Y |
|---|---|---|---|---|---|---|---|
| / | L2 | | | 180 | | | 80 |
| / | L3 | | | 90 | | 40 | |
| / | L4 | | | 180 | | | 60 |
| / | L5 | | | 90 | | 50 | |
| / | L6 | 65 | 80 | 180 | | | |
| ↻ | C1 | | | 30 | | | |
| / | L7 | | | 180 | | | 20 |
| / | | | | 90 | | 50 | |

Line has an Angle (90, 270) & LP H. Will intersect w/prev. Line. Enter

Press enter.

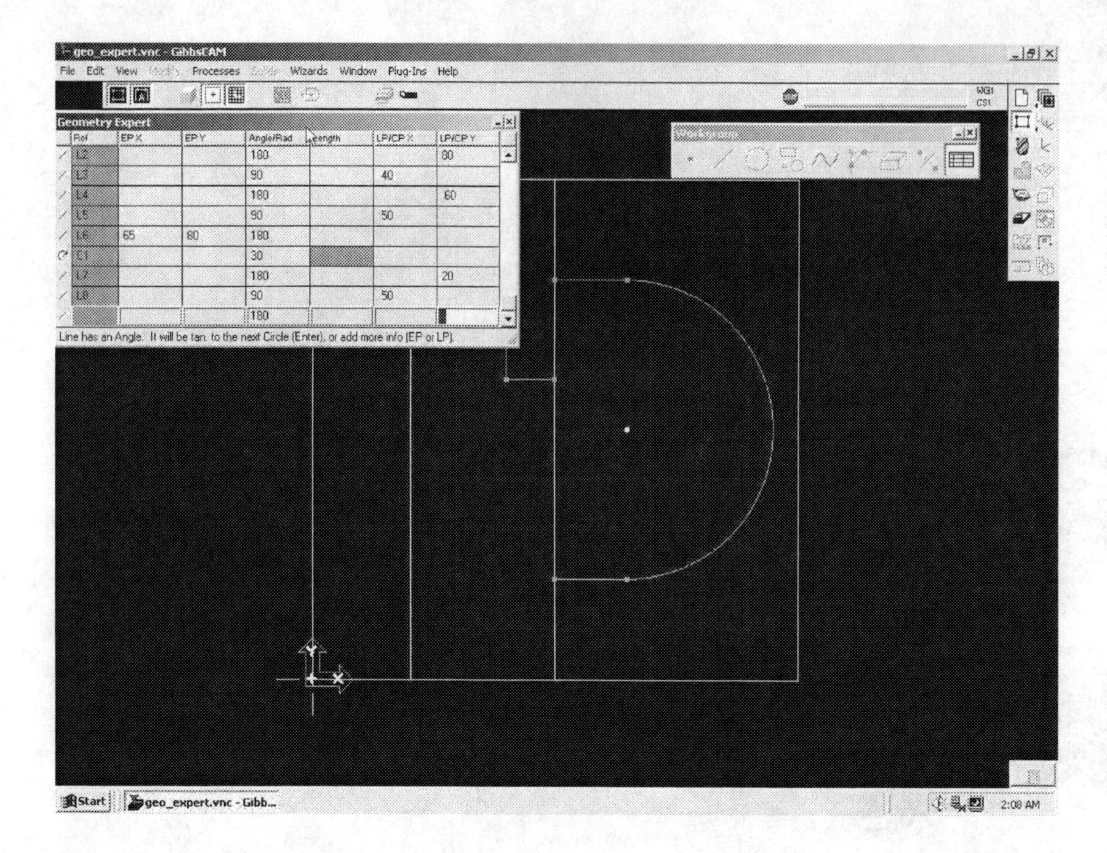


The next line will be a horizontal line that passes thru Y40.

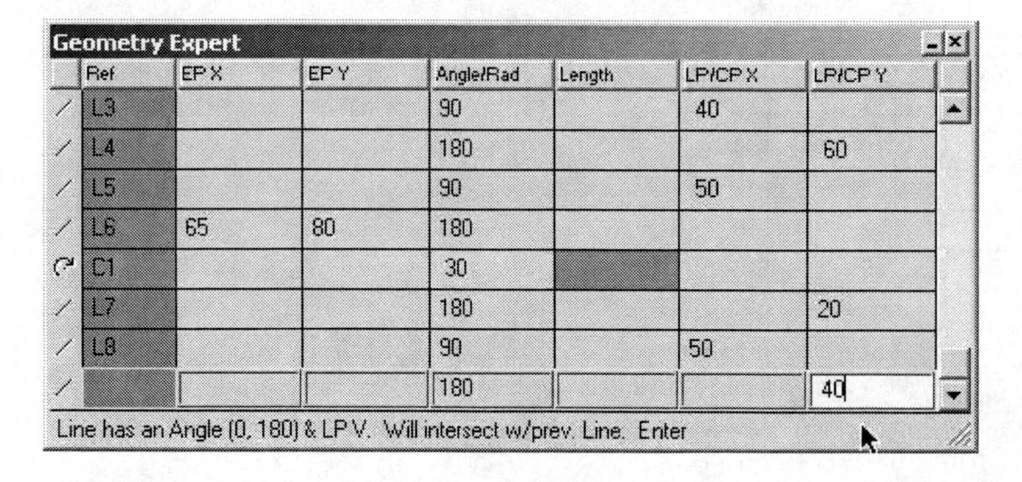


Press enter.

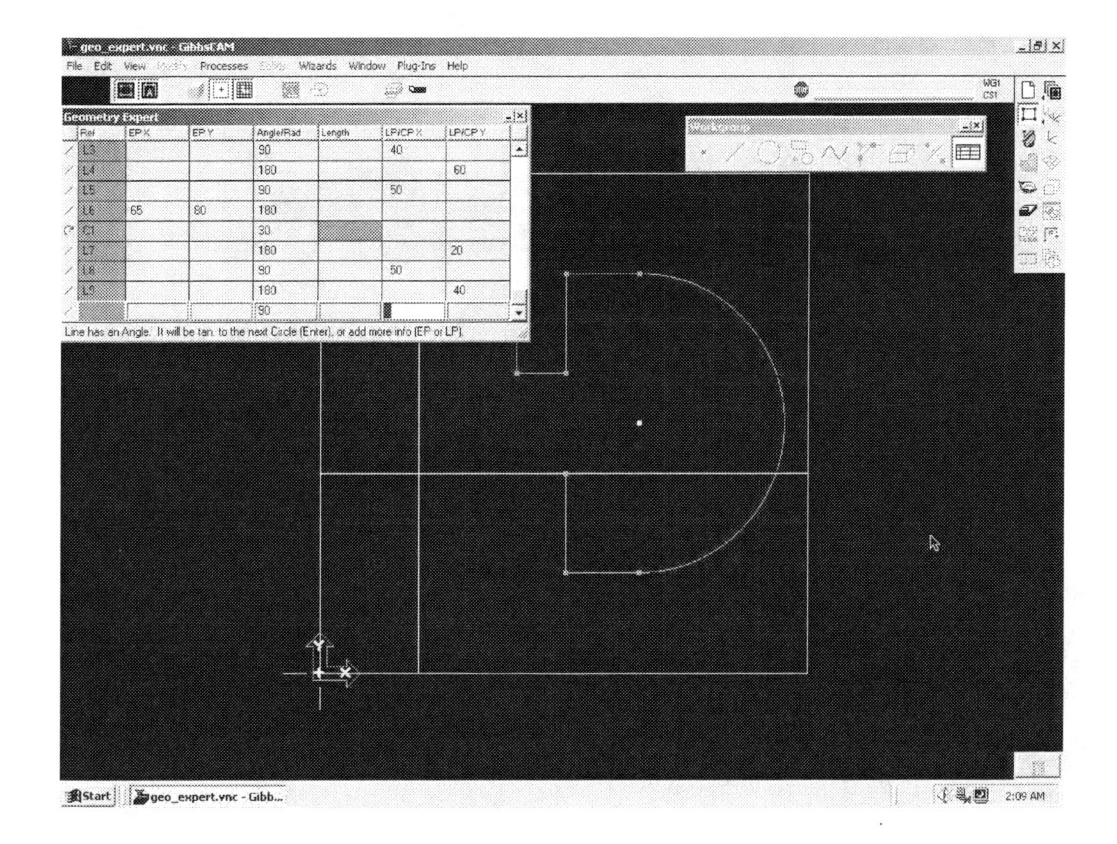


The next line will be a vertical line that passes thru X40.

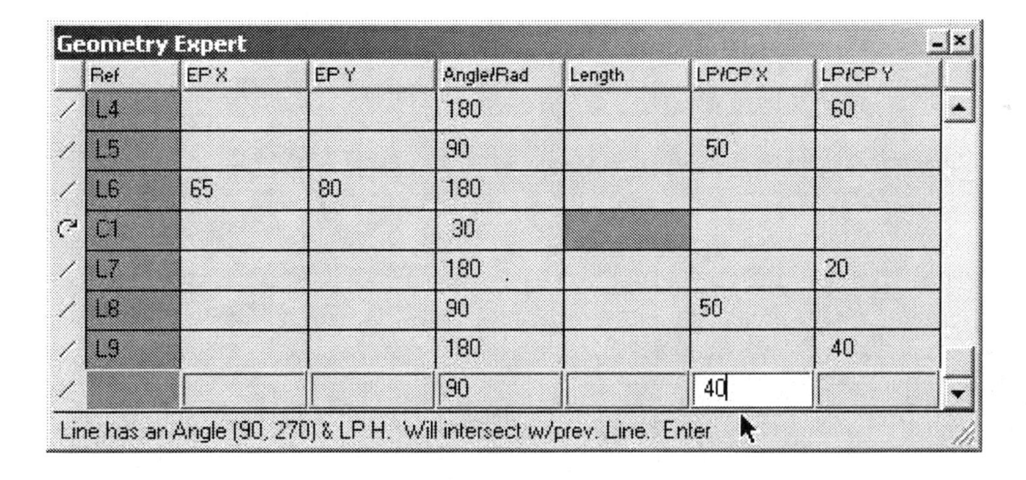


Press enter.

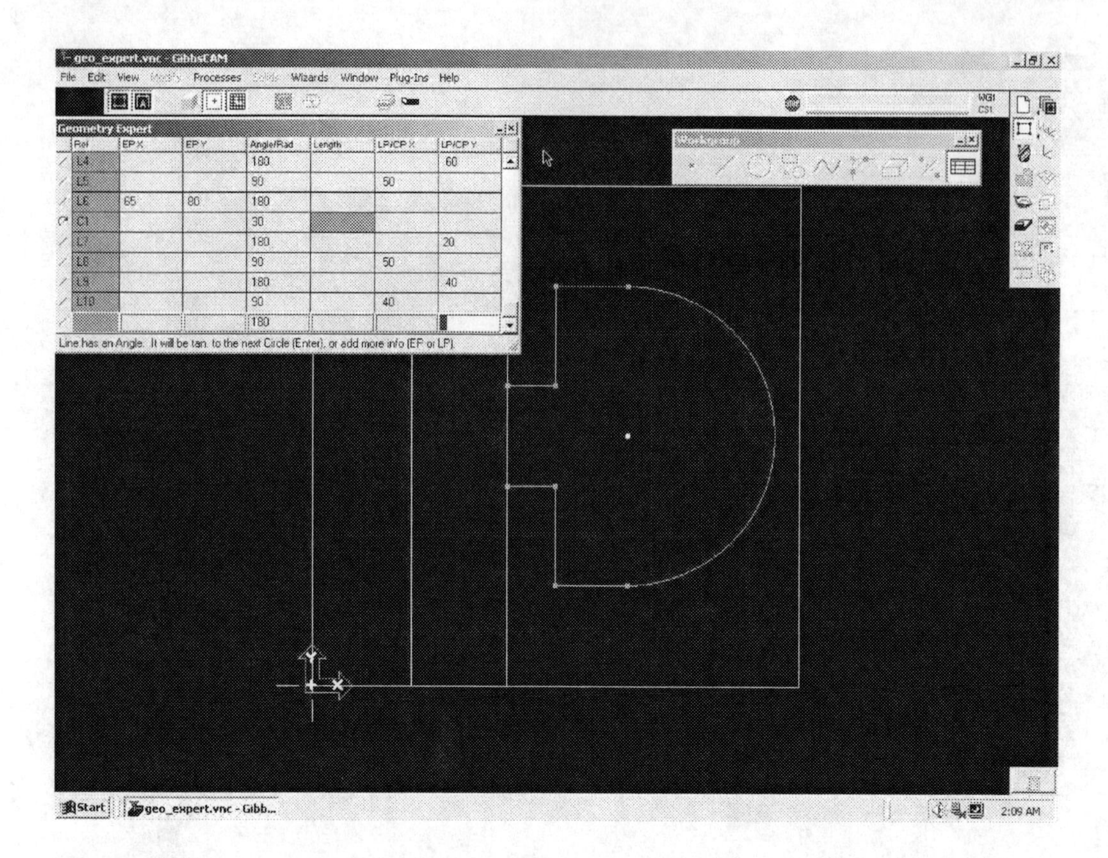


The next line will be a horizontal line thru Y20.

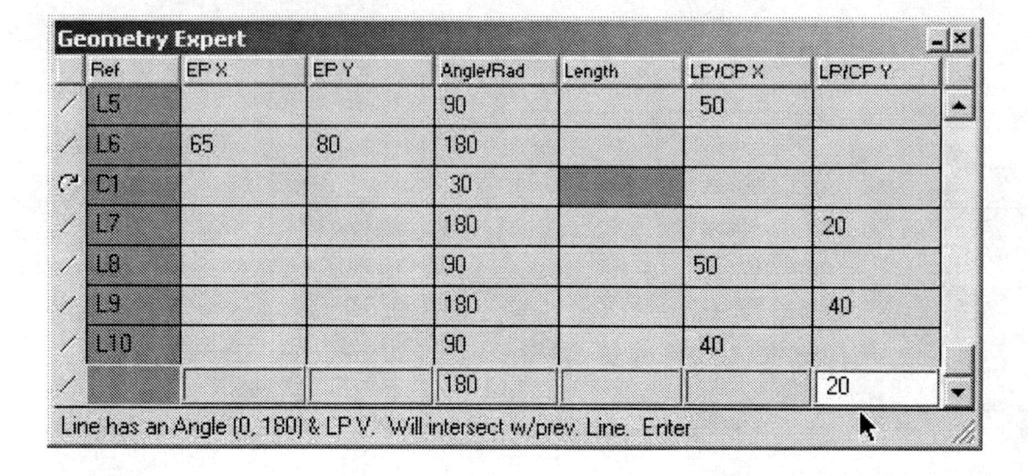


Press enter.

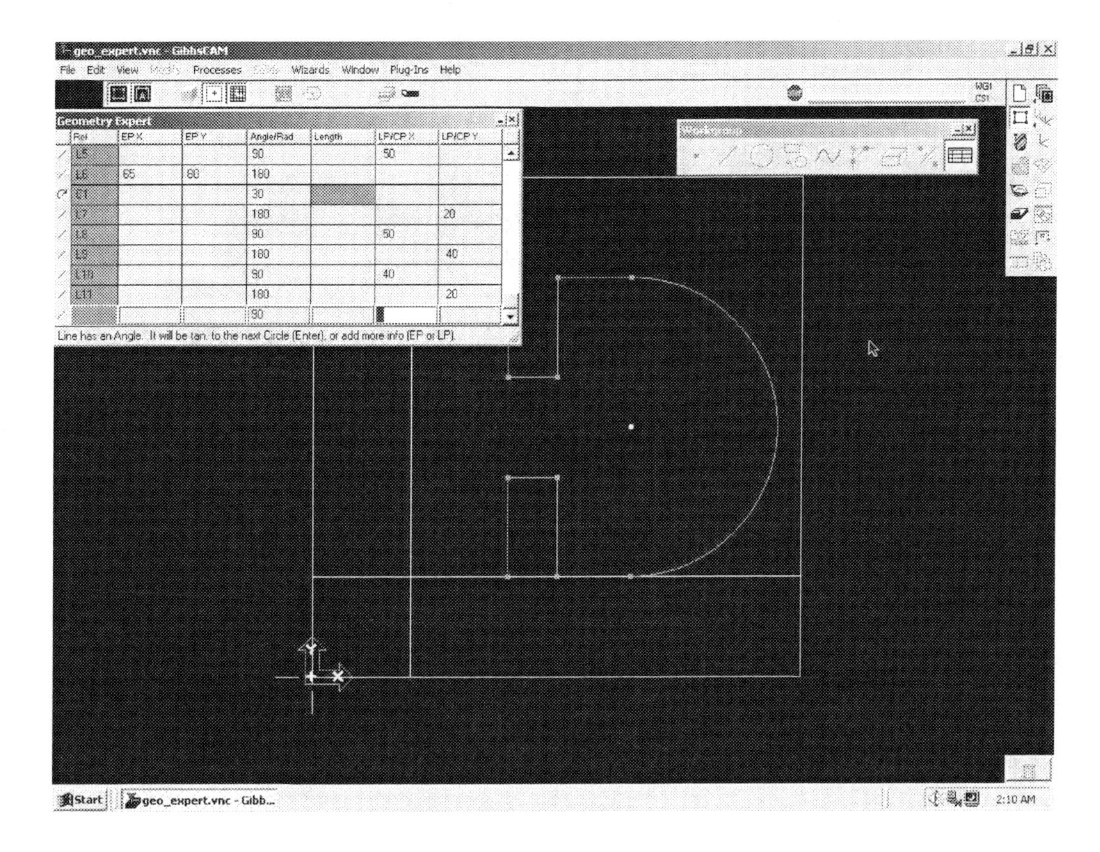

You are now ready to connect the last line drawn to the first line drawn. Select the Connect icon accessed from the Feature Type icon.

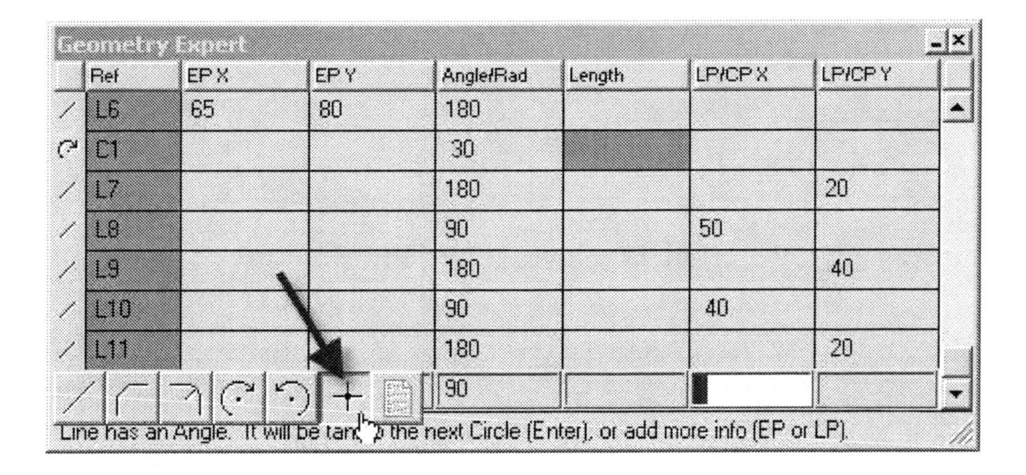


Press enter.

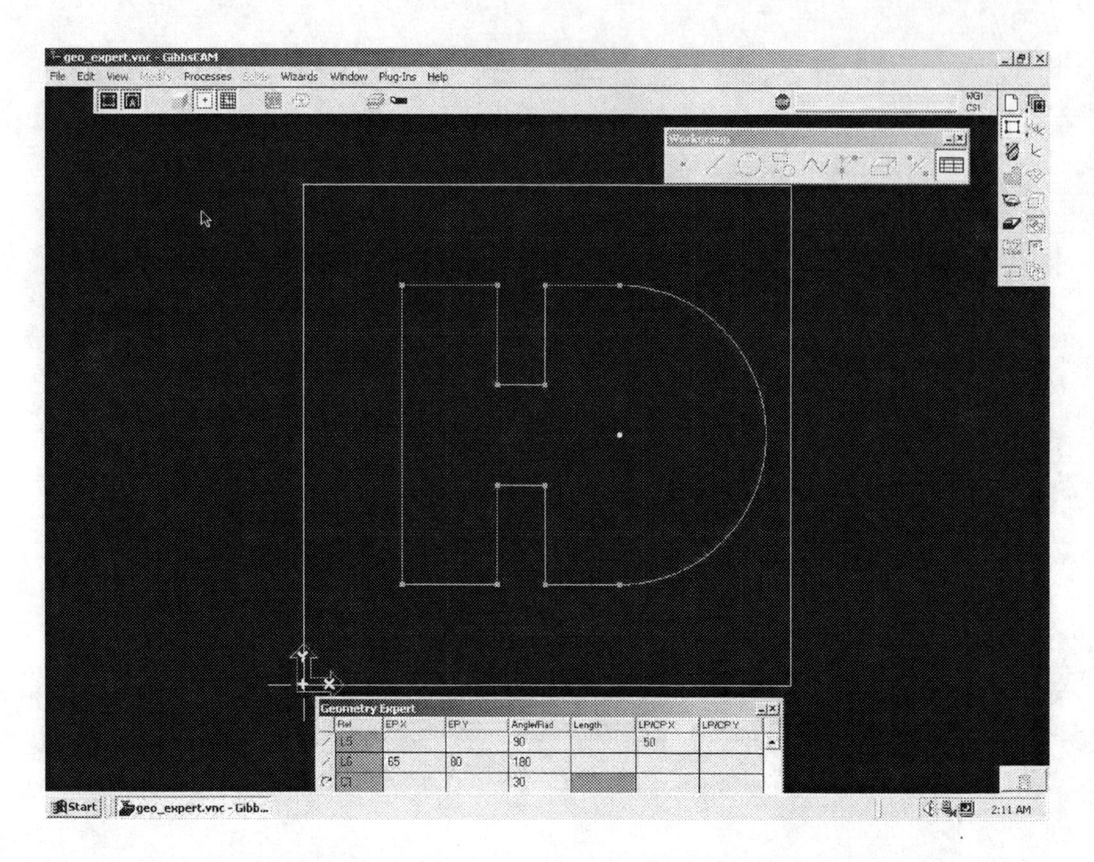


You will now add some fillets. This can be done inside the Geometry Expert, but you will use the CAD tool instead because it allows you to do multiple fillets at once. With Geometry Expert you must enter a line for each fillet.

Exit the Geometry Expert by clicking on the X located at the upper-right hand corner of the spreadsheet.

Hold down the control key and select the 4 corner points as shown in the next graphic.

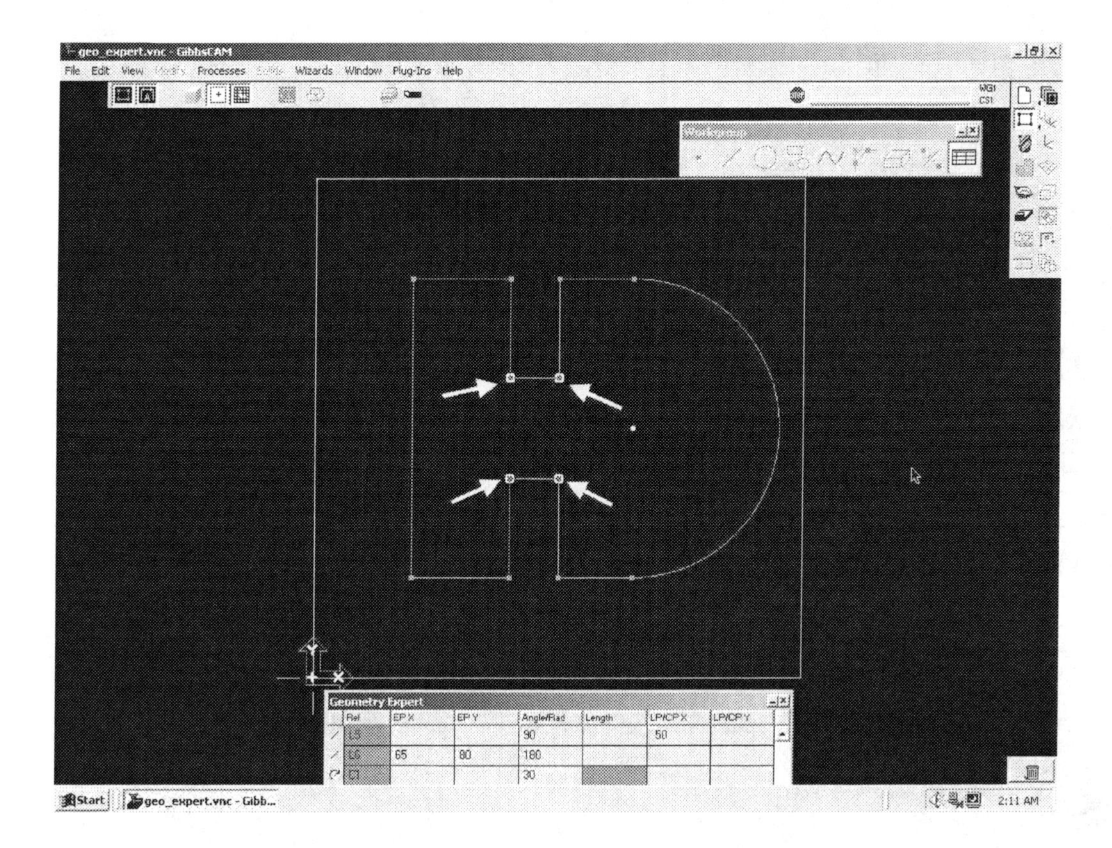


Select the Chamfer icon from the Geometry Palette.

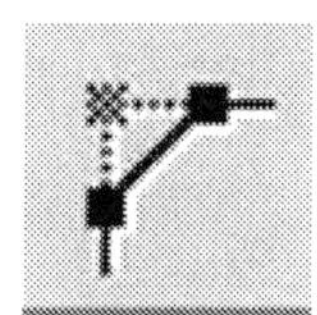

The Fillet-Chamfer Sub-Palette will appear. Select the radius icon as shown and enter 2 for the radius value. Click on the Circle icon in the upper-right corner of this box.

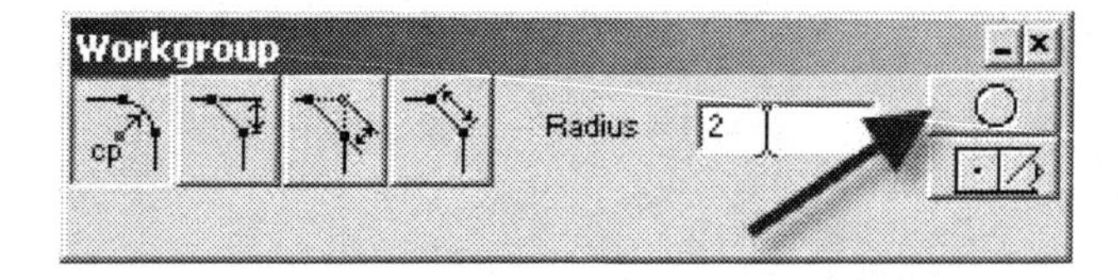

The fillets are now added as shown in the following image.

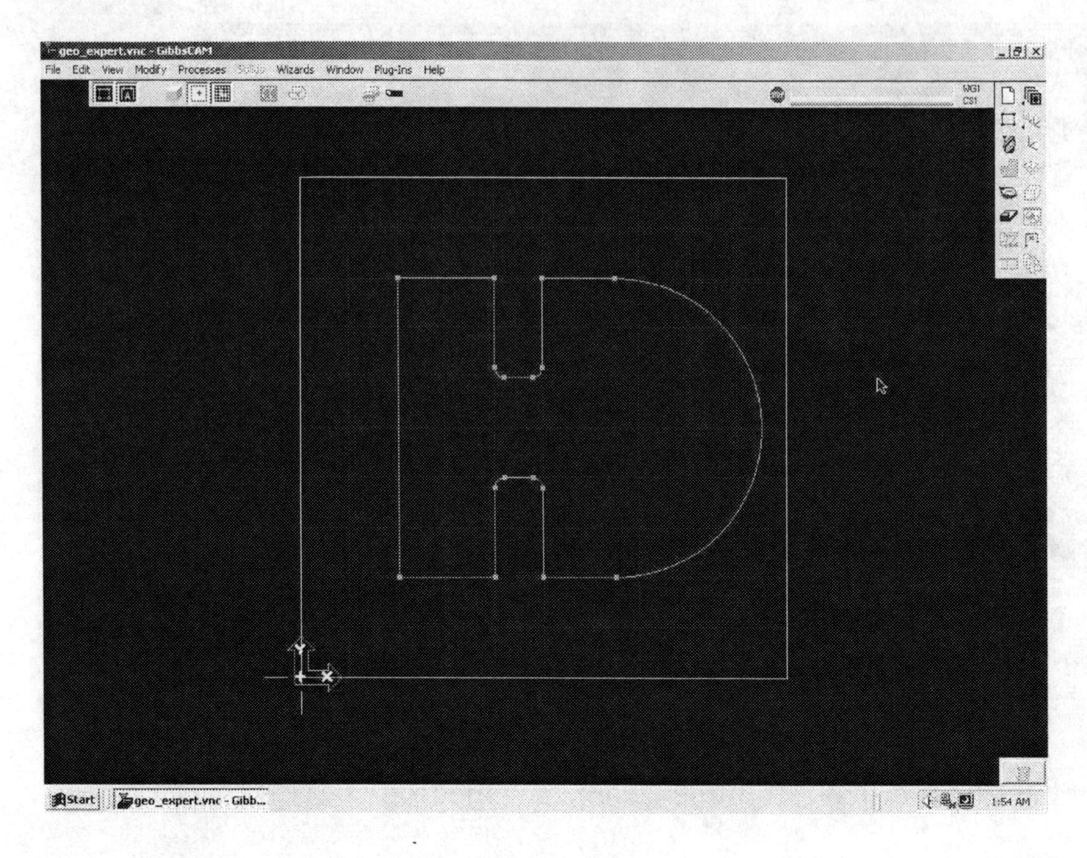


This completes the exercise. Save the file. Choose *File-Save* from the *Main Menu*. Close the file.

## Exercise 5 – Editing Geometry with Geometry Expert

In this exercise you will modify part geometry using Geometry Expert.

Open the file EDIT.vnc. *This file can be downloaded at www.schroff1.com.*

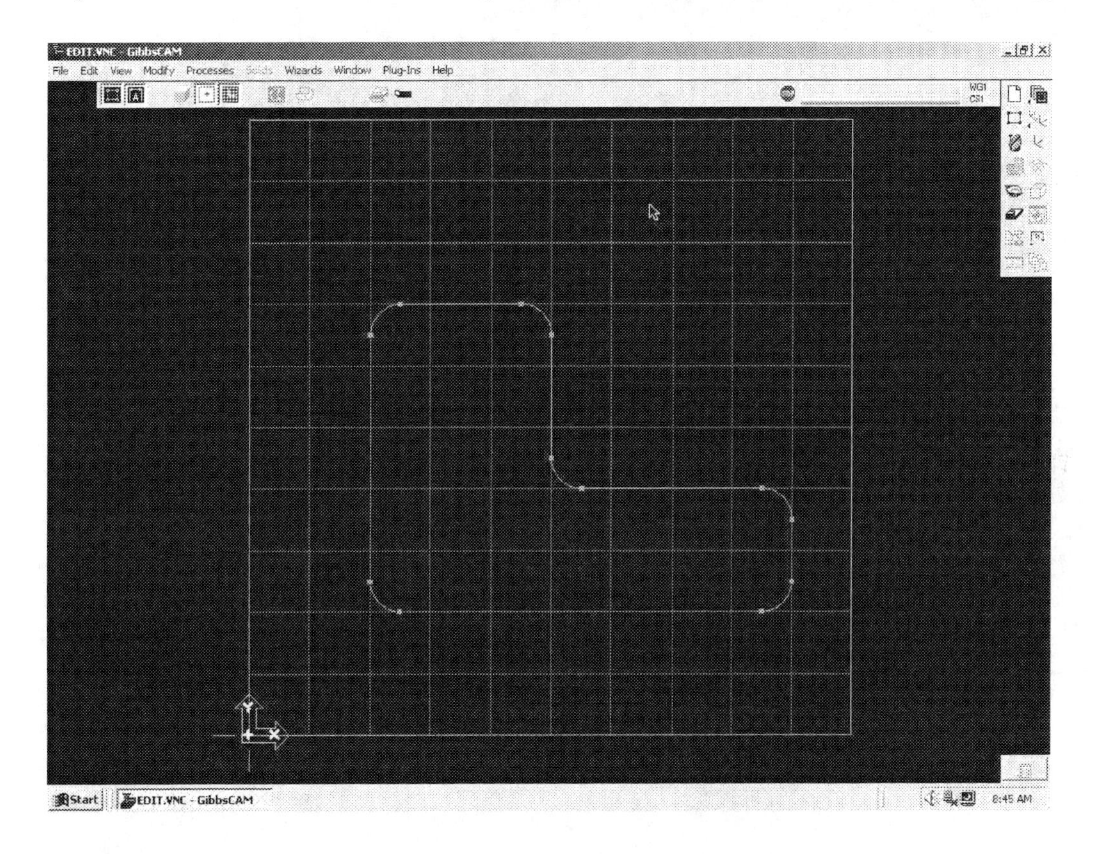

Select the geometry icon from the Top Level Palette.

The Geometry Palette will appear.

Select the Geometry Expert icon.

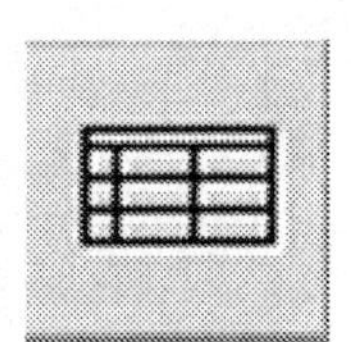

Double click on the bottom horizontal line to load the profile into the Geometry Expert.

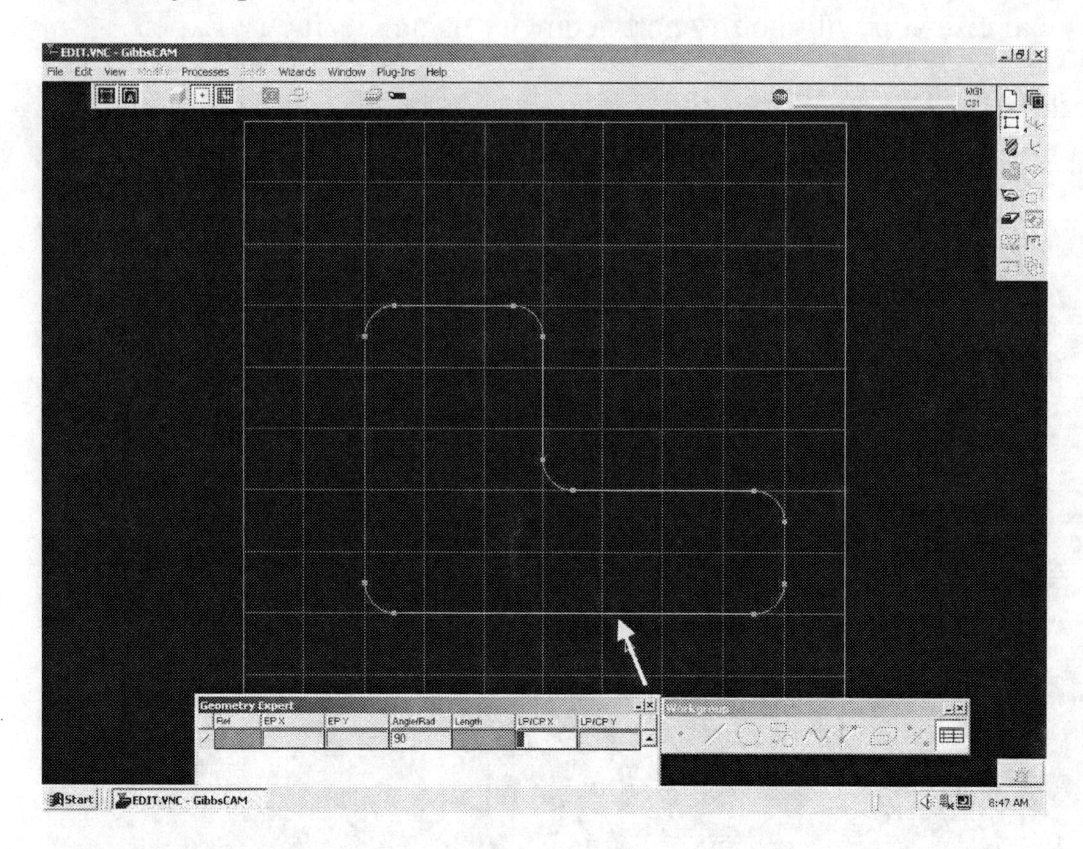


The geometric information for the profile will be loaded into the Geometry Expert.

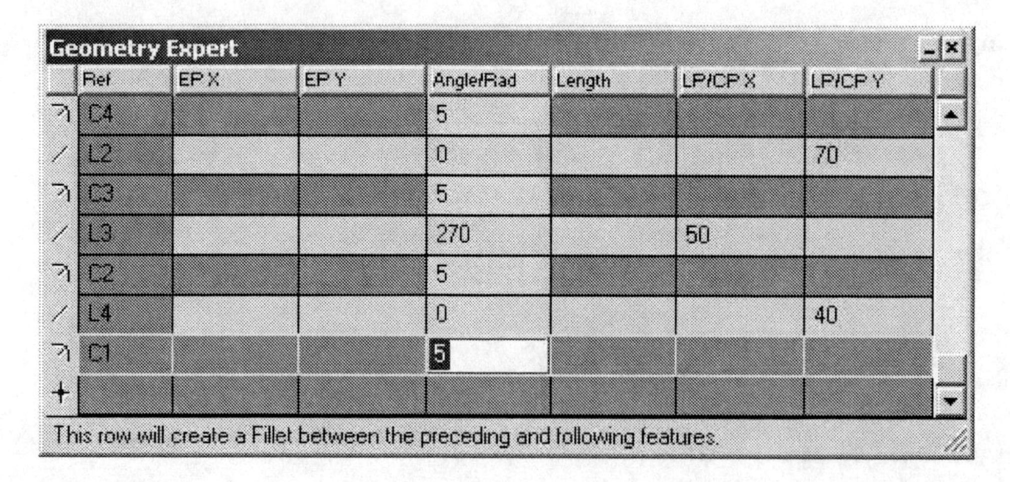

Geometry Expert

| Ref | EP X | EP Y | Angle/Rad | Length | LP/CP X | LP/CP Y |
|---|---|---|---|---|---|---|
| C4 | | | 5 | | | |
| L2 | | | 0 | | | 70 |
| C3 | | | 5 | | | |
| L3 | | | 270 | | 50 | |
| C2 | | | 5 | | | |
| L4 | | | 0 | | | 40 |
| C1 | | | 5 | | | |
| | | | | | | |

This row will create a Fillet between the preceding and following features.

Select L2 of the Geometry Expert. Notice the selected line will highlight on the screen.

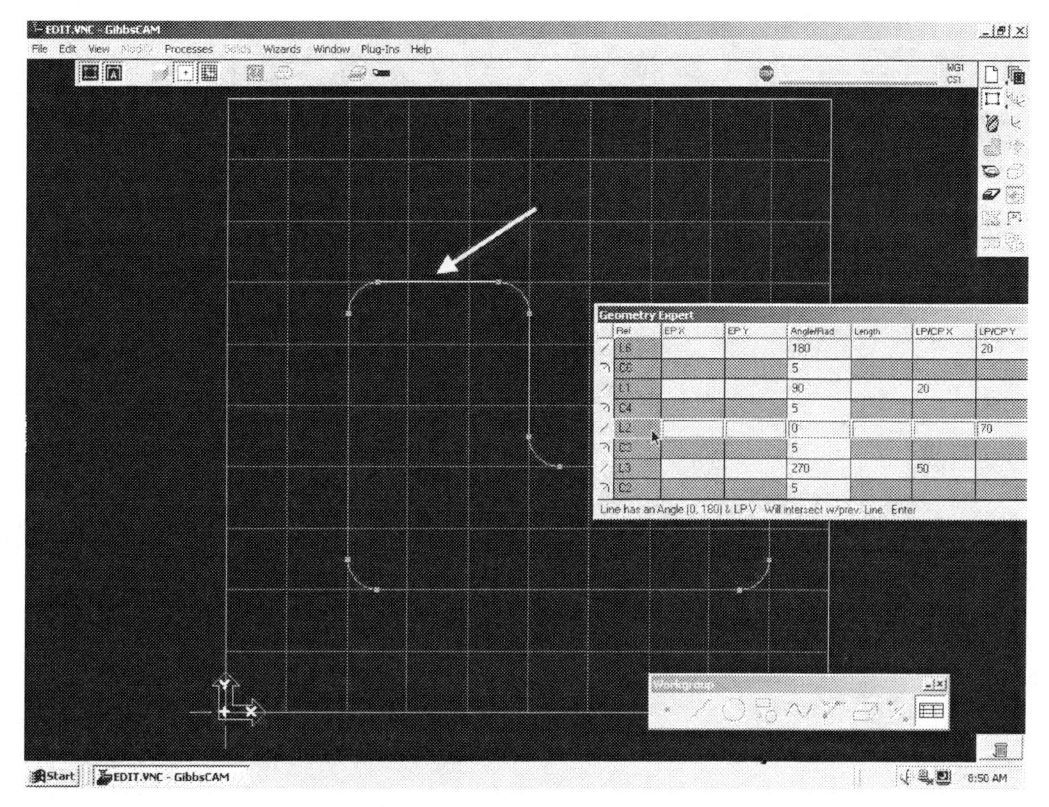


Change the LP/CPY value to 90.

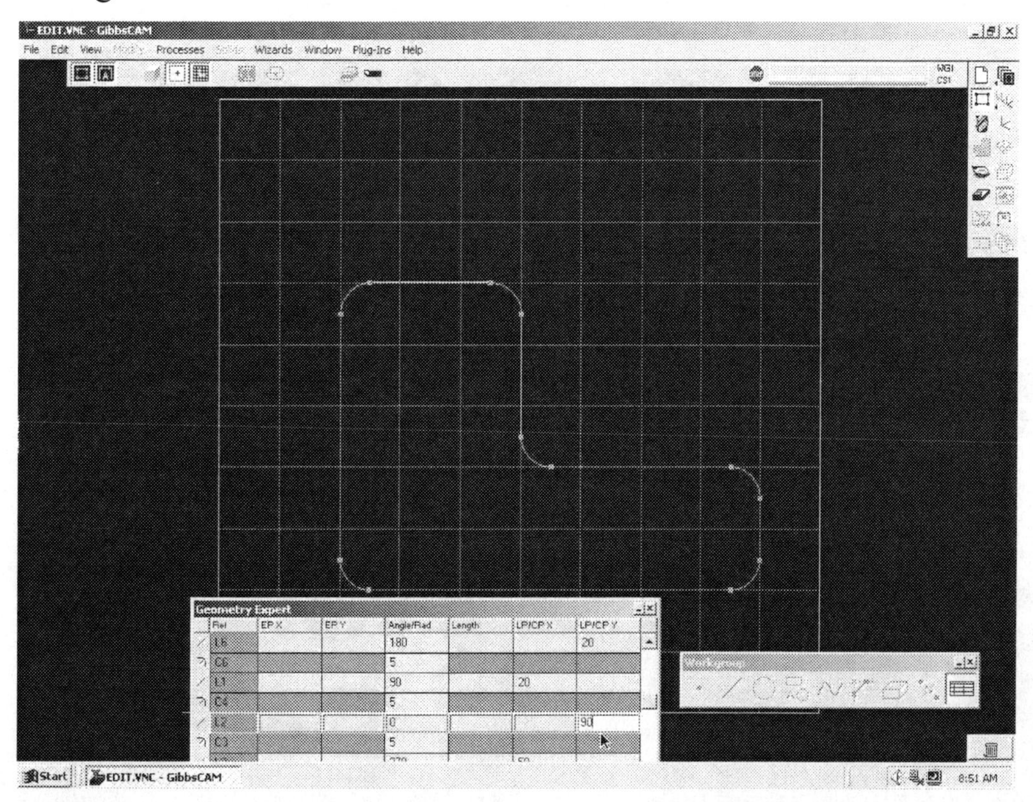

Hit enter. The geometry will update.

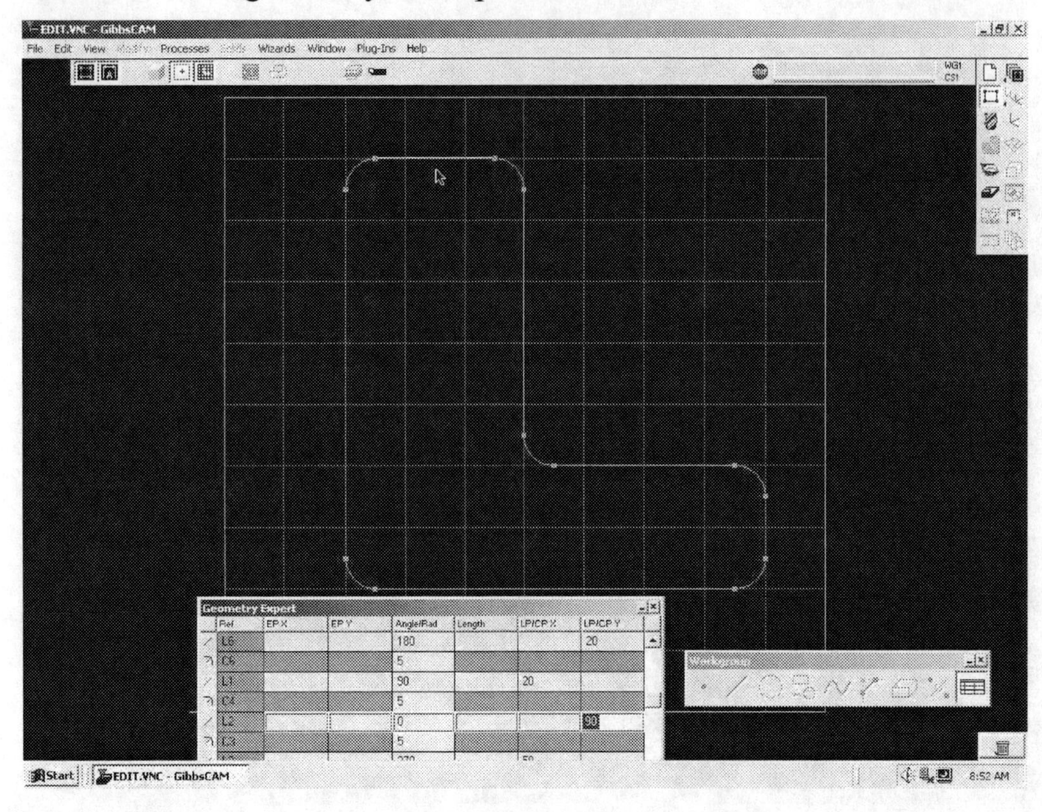


You will now remove a fillet. Select row C9.

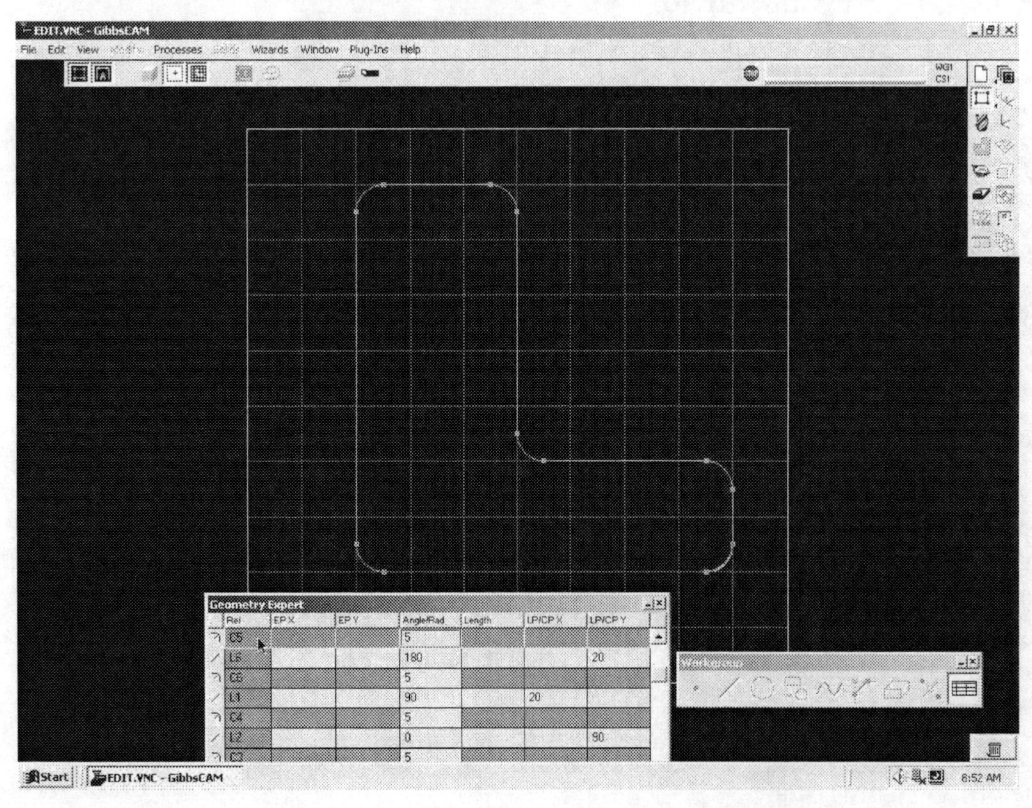

Select *Edit-Delete Row* from the *Main Menu*. This will remove the fillet.

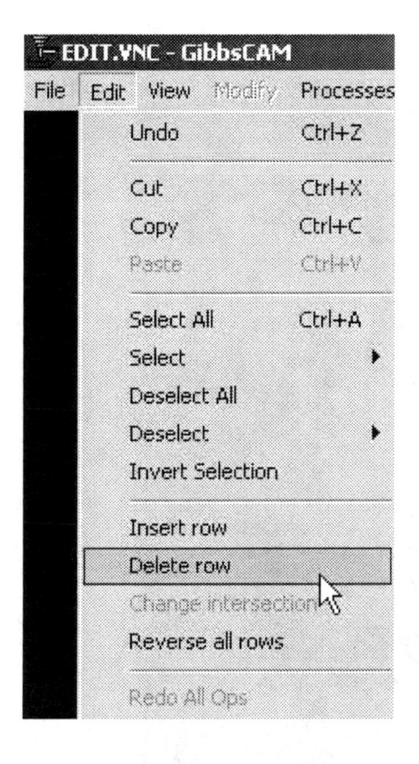


The geometry will update.

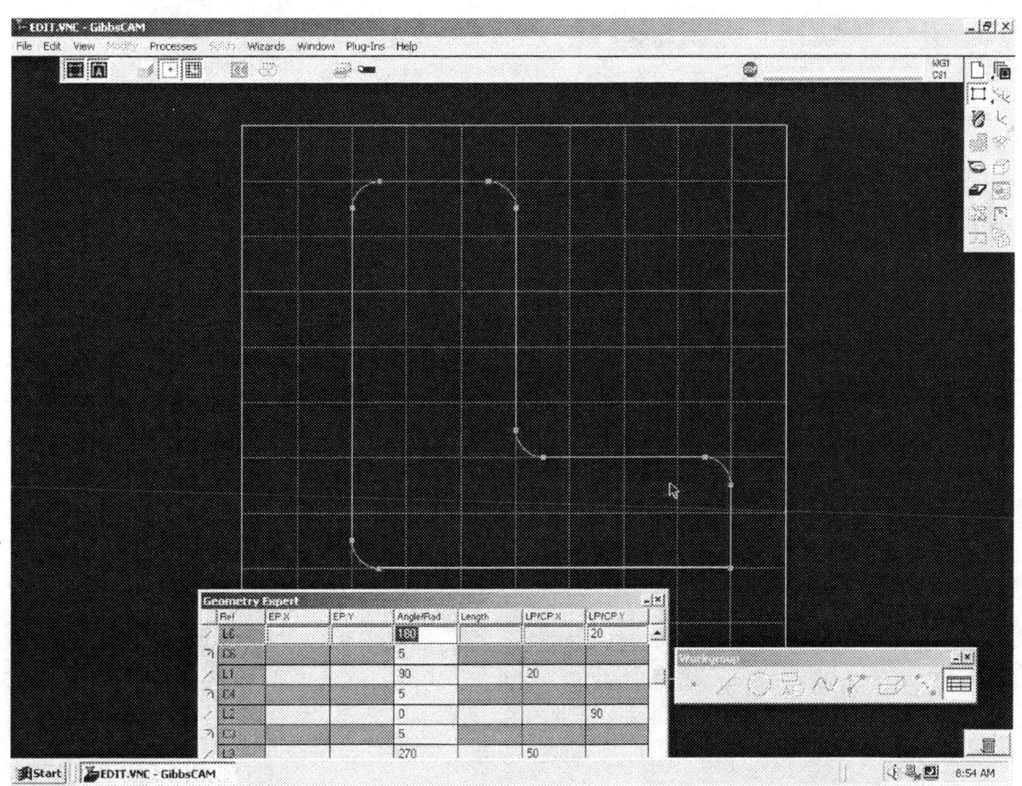


This completes the exercise. Save the file and then close it.

## Exercise 6 – CAD

In this exercise you will create 2 profiles using the CAD tools.

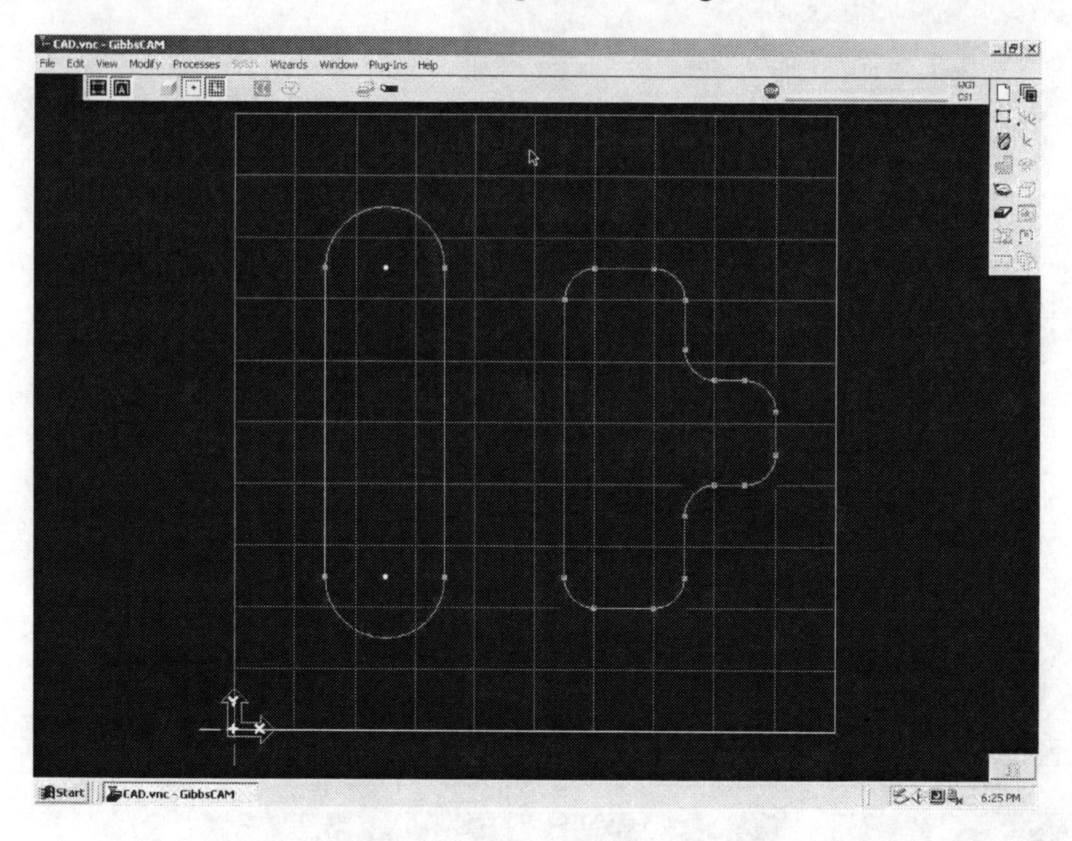


Create a new file named CAD. Set the machine type to 3 axis vertical mill and the units to mm. Enter the values as shown in the next graphic for the stock size.

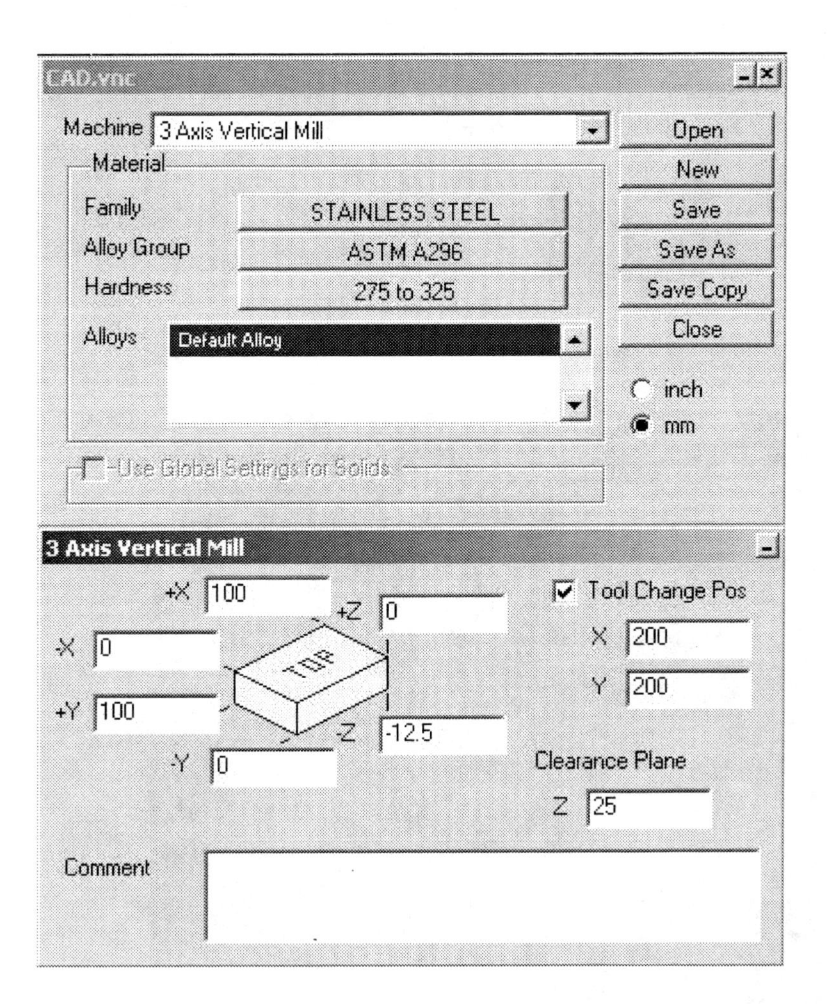


Select the Geometry icon from the Top Level Palette.

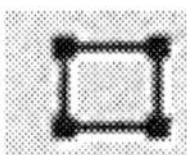

Select the Point icon from the Geometry Palette.

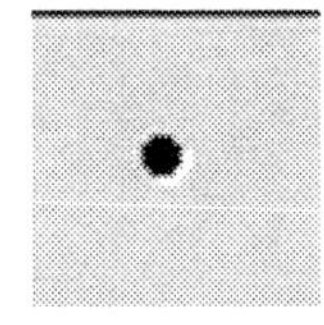

The Points Palette will appear. Select the Explicit Point icon.

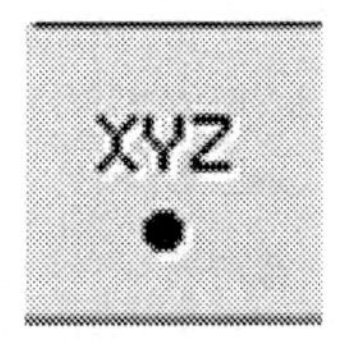

You will insert the first point at X25 Y75. Click on the Multiple Points icon to create the point and keep this dialog box open.

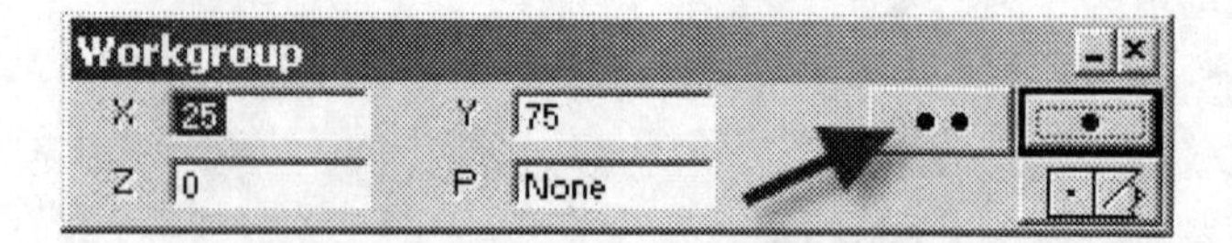


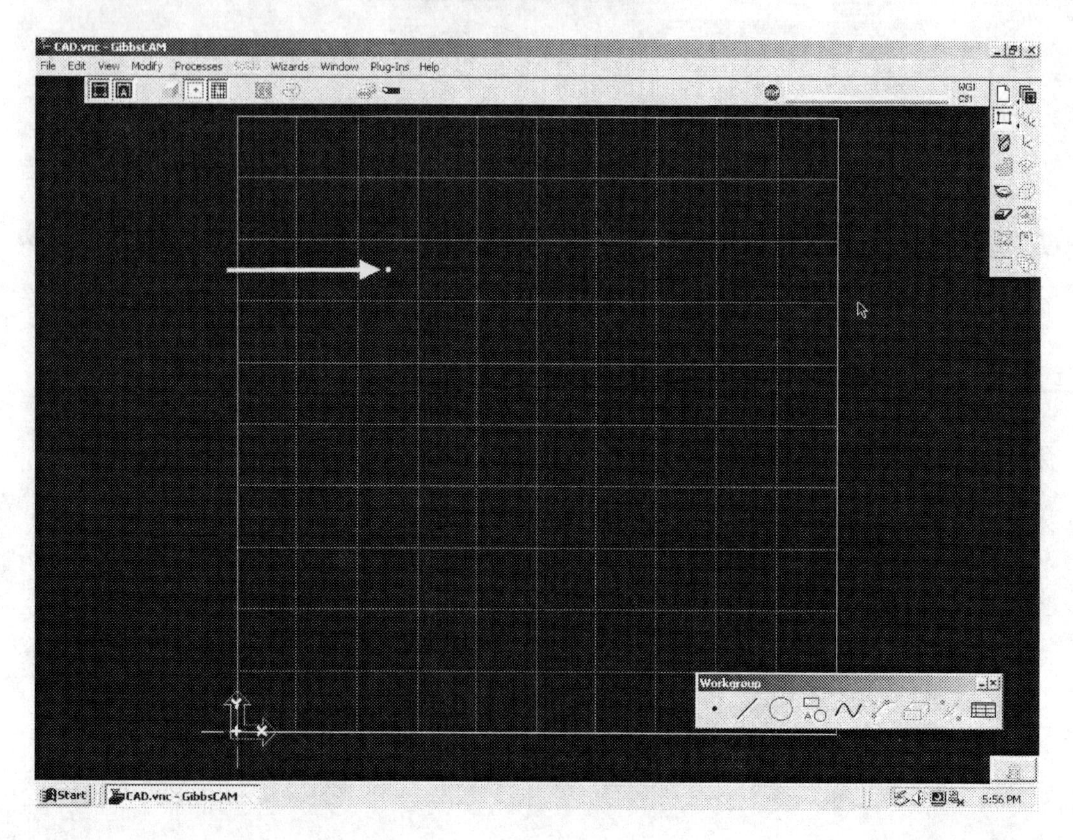


The next point is at X25 Y25. Enter the values as shown and click on the Single Point icon.

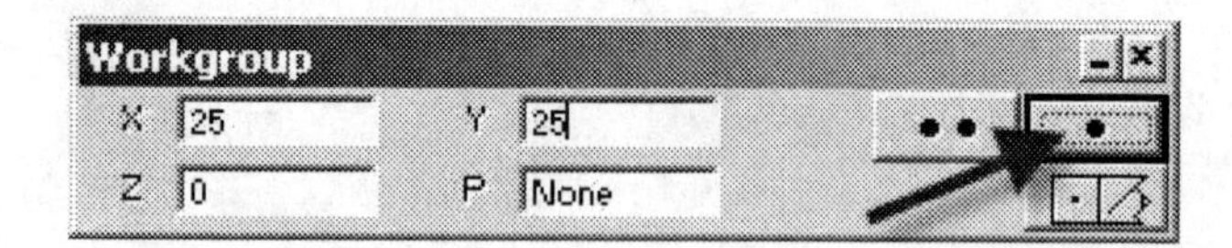


Your screen should look like as shown in the next graphic.

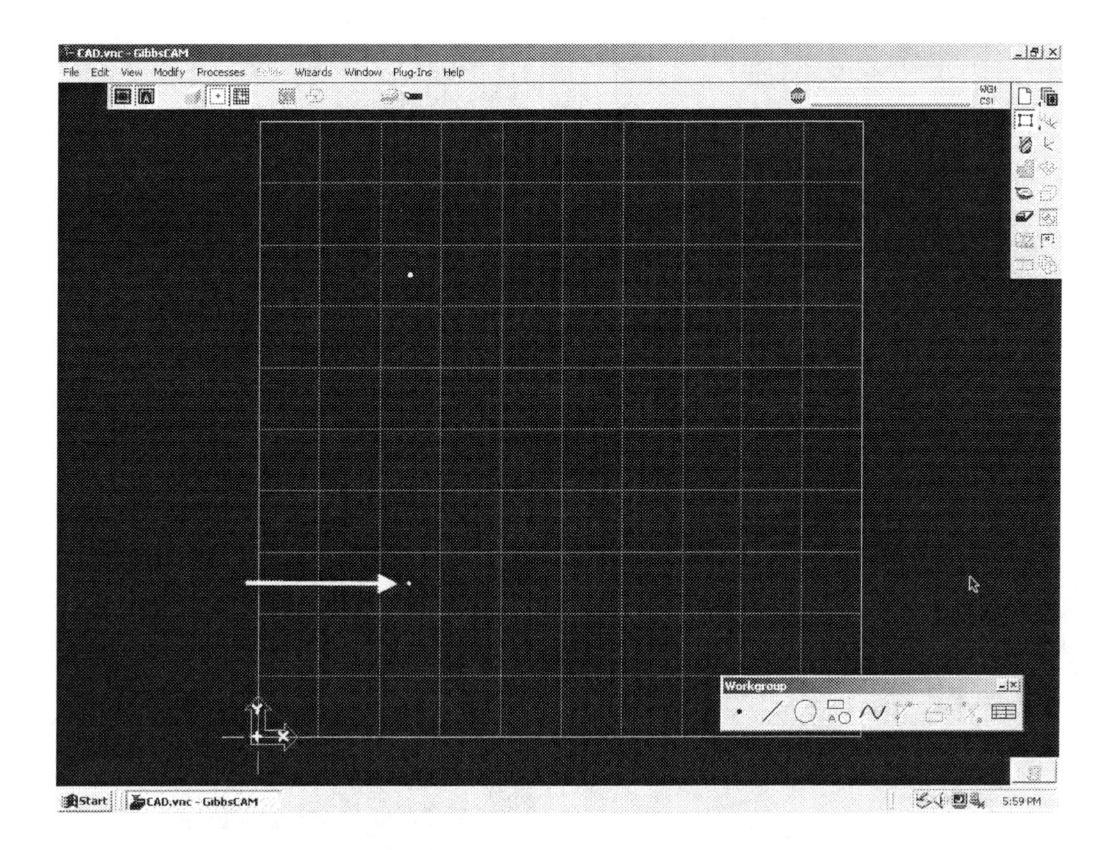


Now you will create circles at the center points. Select the circle icon on the Geometry Palette.

The Circle Palette will appear.

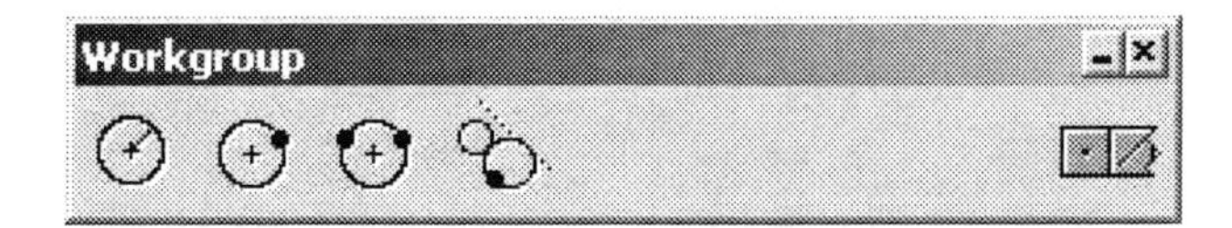


Select the Radius and Center Point icon.

Select the top point and enter 10 for the radius. Click on the Multiple Circles icon to create the top circle and leave the dialog box open.

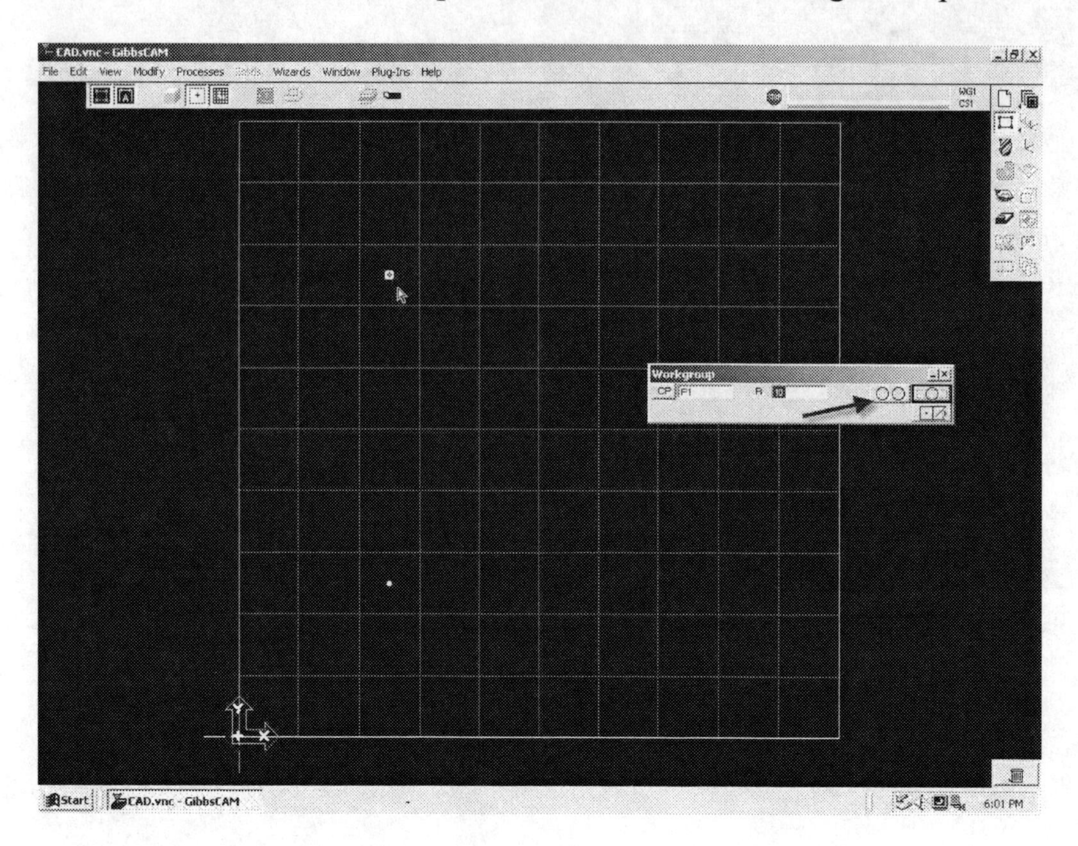

Select the bottom point and enter 10 for the radius. Click on the Single Circle icon. This will create the circle and return you to the Geometry Palette.

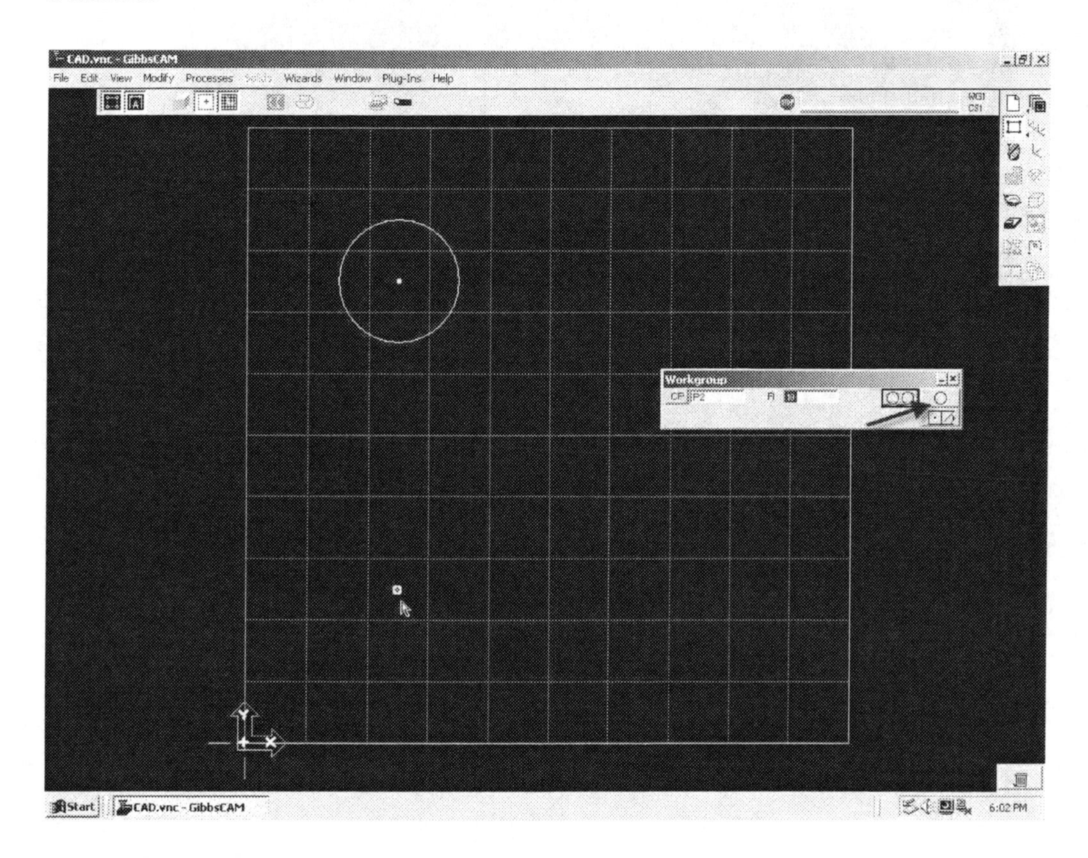

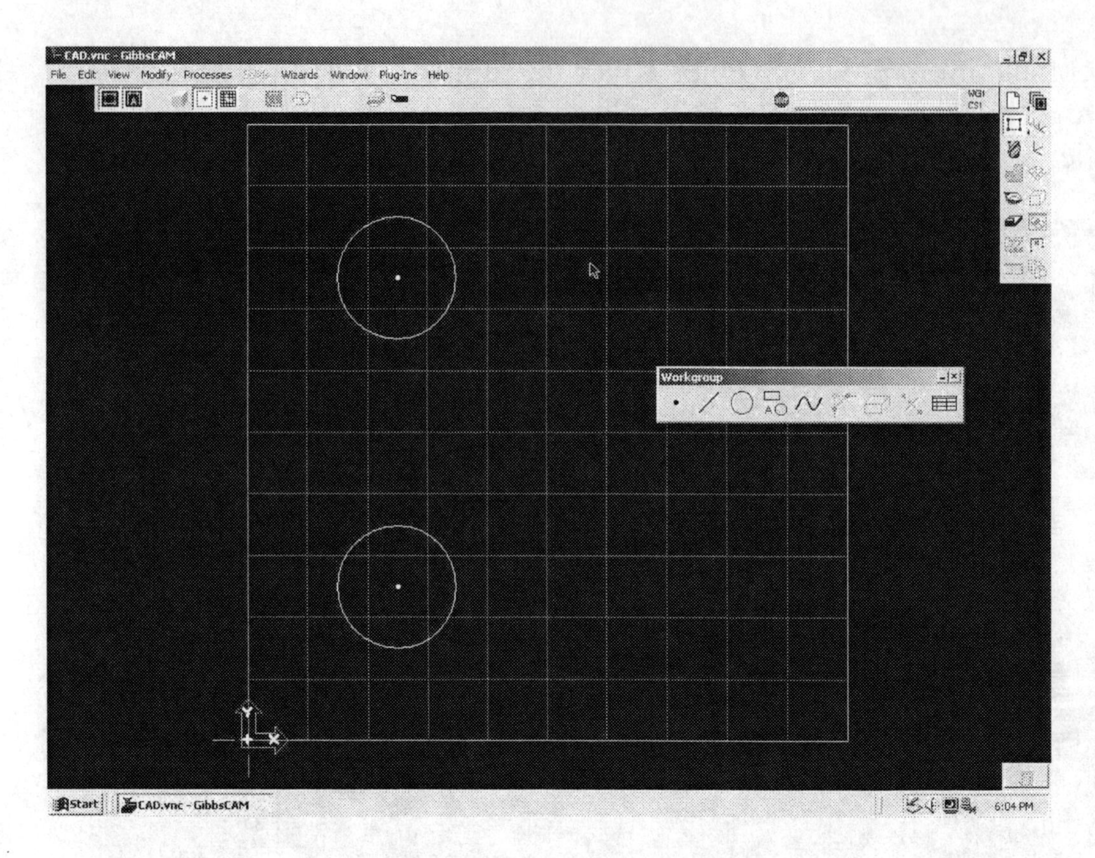


Select the Line icon on the Geometry Palette.

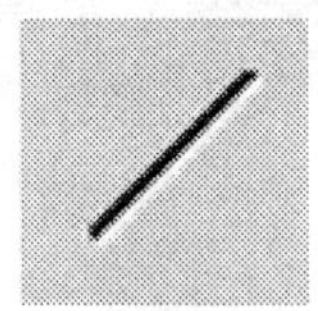

Hold down the control key and select the 2 circles to create tangent lines to the circles. Select the Single Line icon as shown.

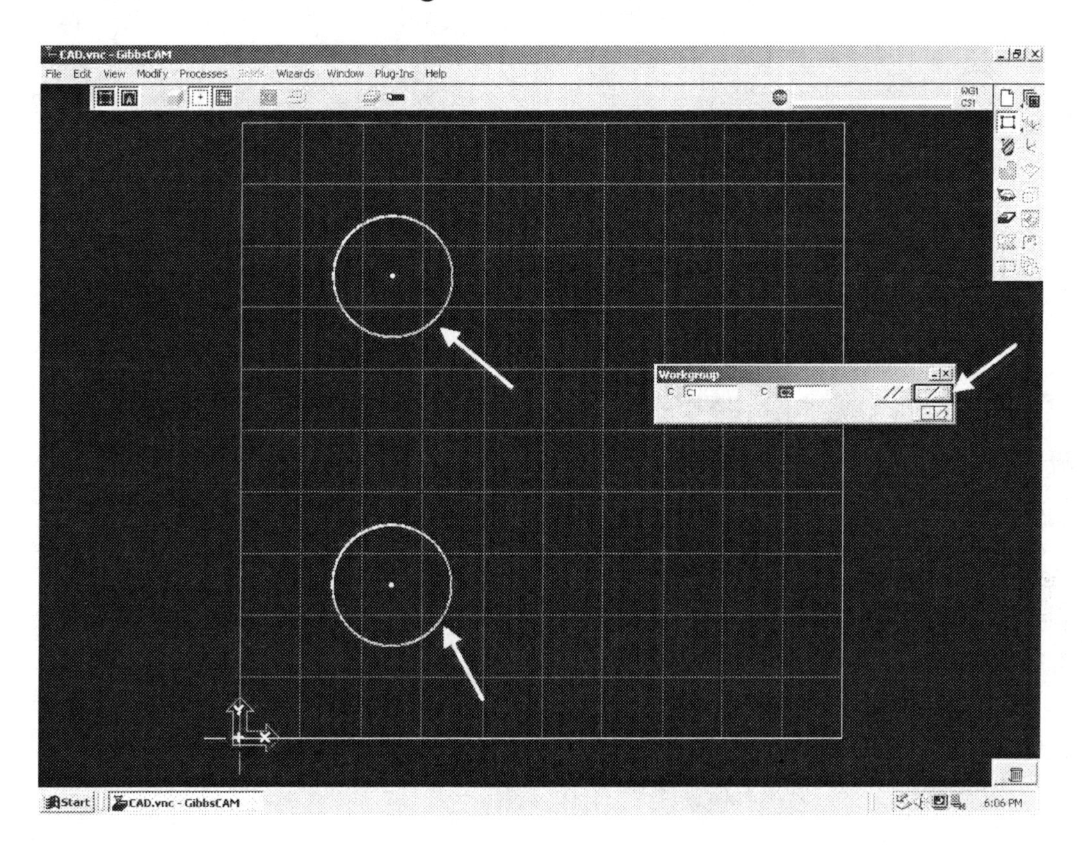

You will be prompted to select the desired feature(s). Select the 2 vertical lines and click on OK as shown.

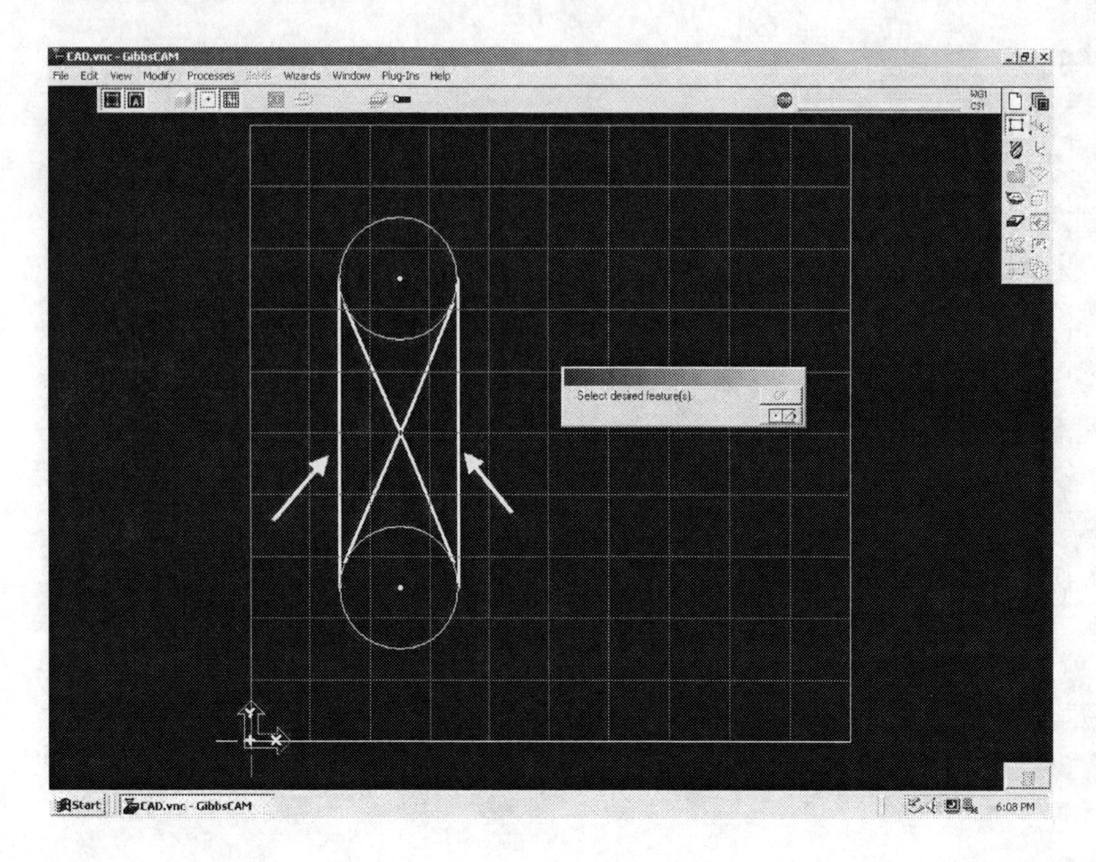


Your screen may look like the following image. Notice that the circles are trimmed and connected, but one of the arcs needs to be reversed.

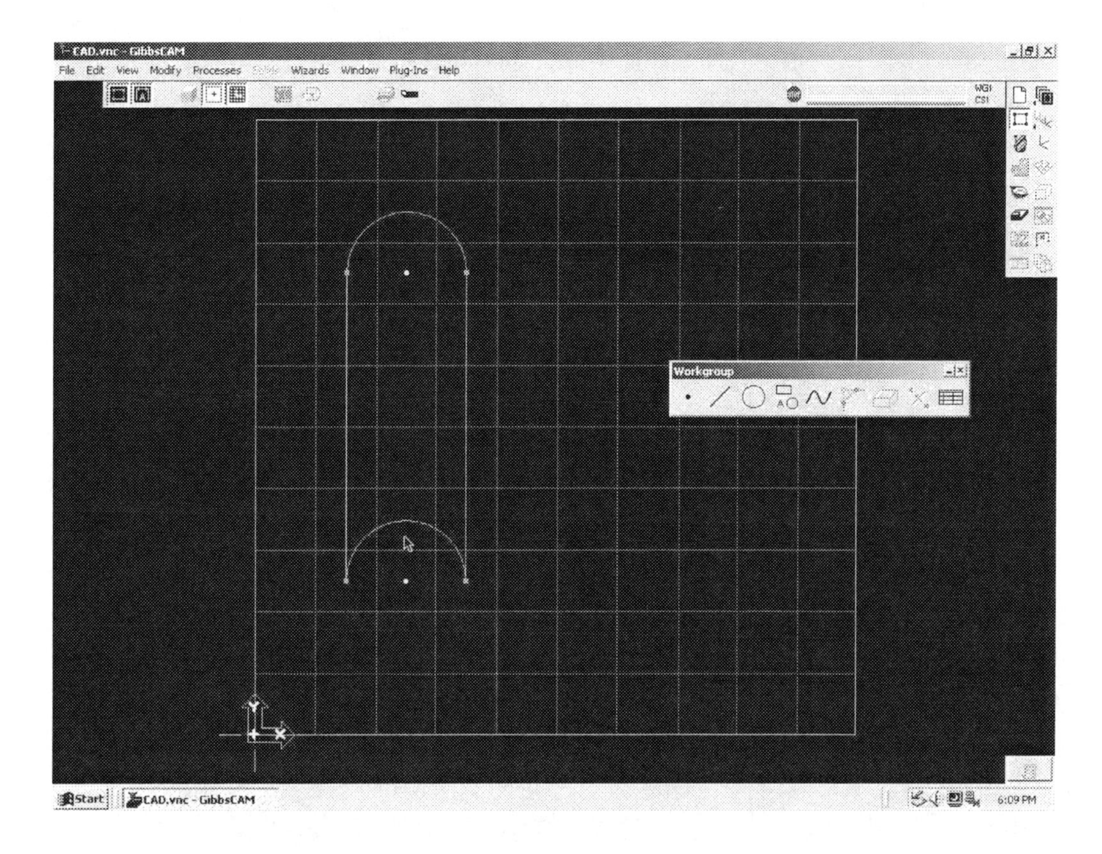


Select the arc that needs to be reversed.

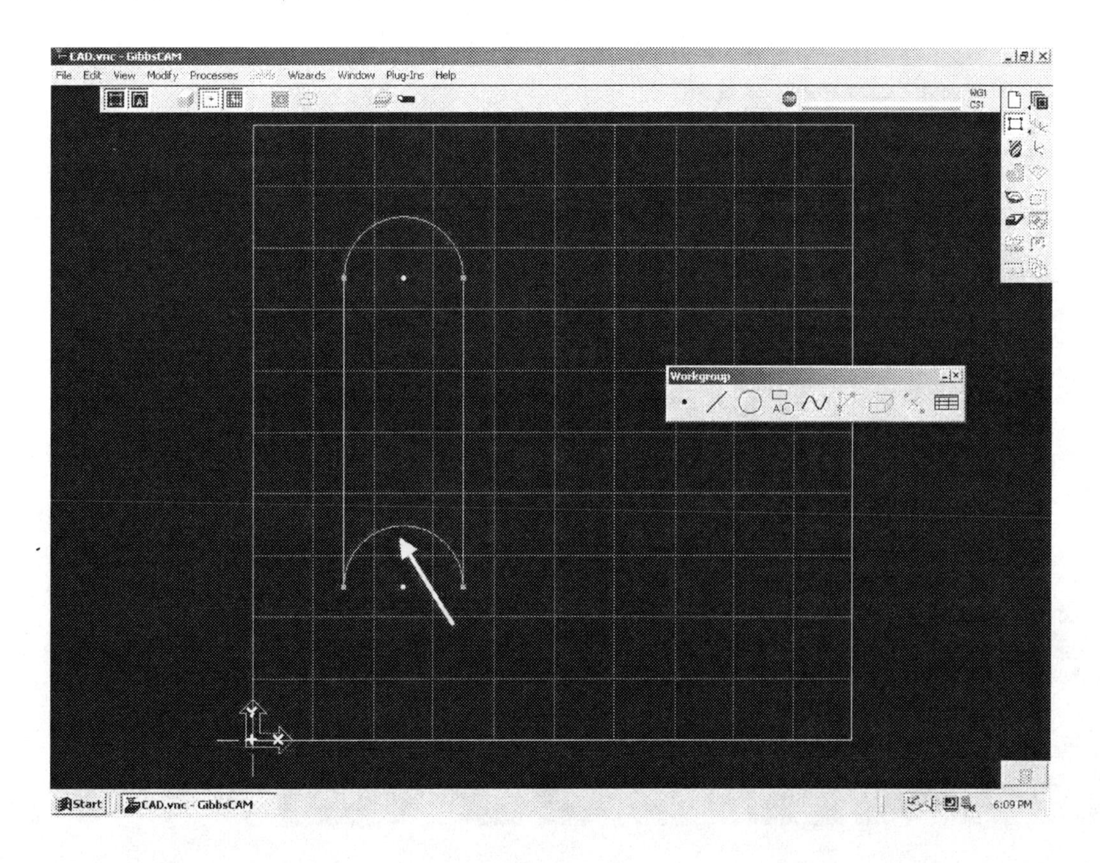

Choose *Modify-Reverse Arc* from the *Main Menu.*

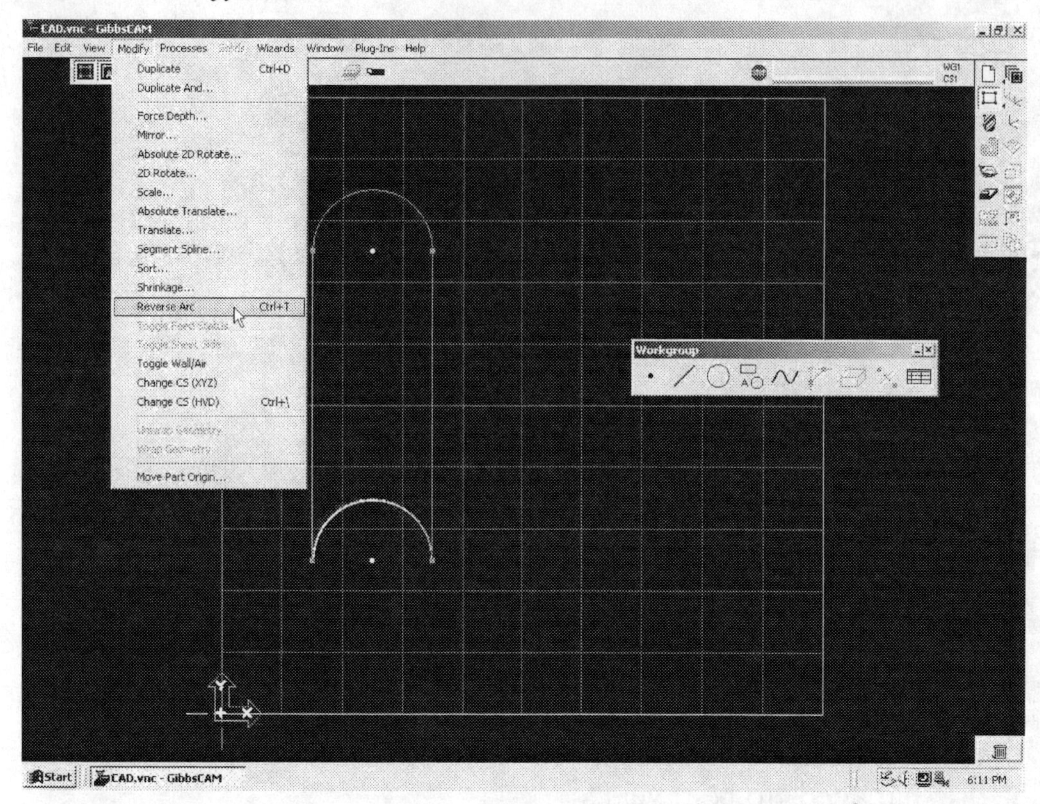


Your geometry should now look like this.

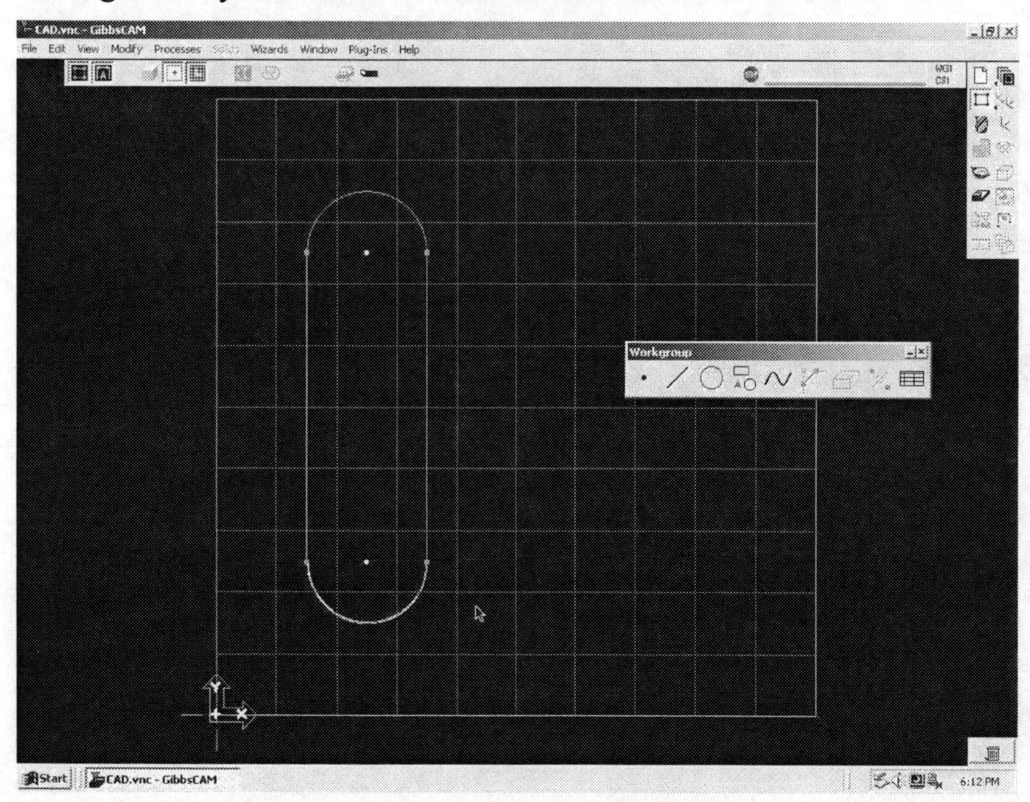

You will now draw a “freehand” T-shaped profile. When drawing the profile don’t worry about exact line sizes and locations. Draw the general shape and then you will fully define the profile in the Geometry Expert.

Select the line icon from the Geometry Palette.

The Line Palette will appear.

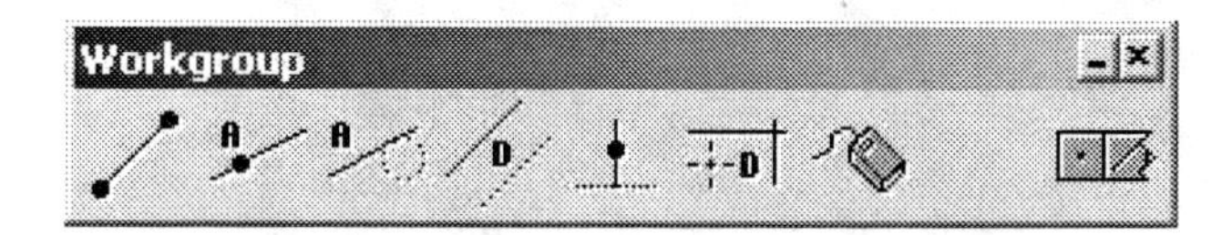


Select the Mouse Line icon.

Start the profile by clicking approximately where you see point 1 in the following image. Move the mouse vertically and click again at the approximate location of point 2. Keep moving around the profile in a clockwise direction and click at the approximate location of each of the remaining points to create the profile. Important: Click on point 9 directly over the first point you created to connect these 2 lines.

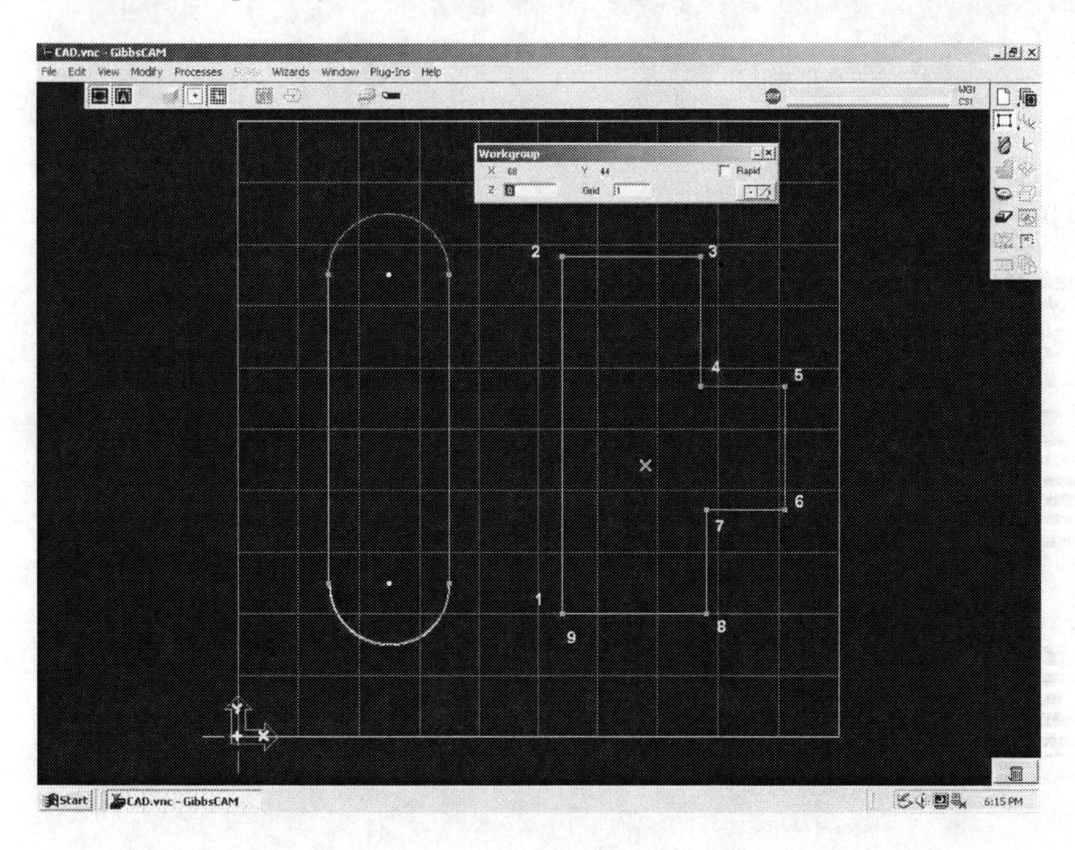


You will now fully define the profile by editing it in the Geometry Expert. Select the Geometry Expert icon in the Geometry Palette.

Double click on the T-Shaped profile to load it into the Geometry Expert.

Edit the values in the Geometry Expert to match those below:

| Line | Angle | LPX | LPY |
|---|---|---|---|
| L4 | | | 75 |
| L5 | 270 | 75 | |
| L6 | 0 | | 60 |
| L7 | 270 | 90 | |
| L8 | 180 | | 40 |
| L9 | 270 | 75 | |
| L10 | 180 | | 25 |
| L3 | 90 | 55 | |

Your geometry should look like the following image.

You will now complete the exercise by adding fillets. Hold down the shift key and drag a selection box around the profile as shown in the next graphic.

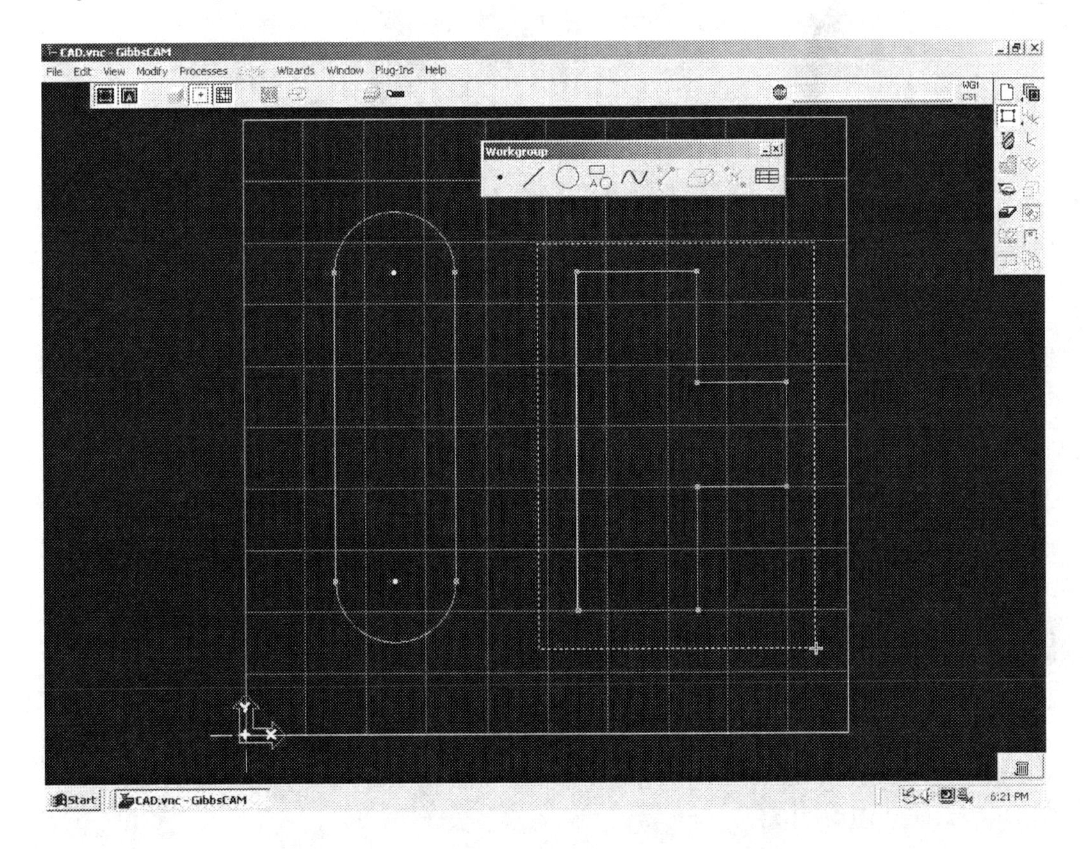

Select the Chamfer icon from the Geometry Palette.

Select the radius icon and enter 5 for the value. Click on the circle icon.

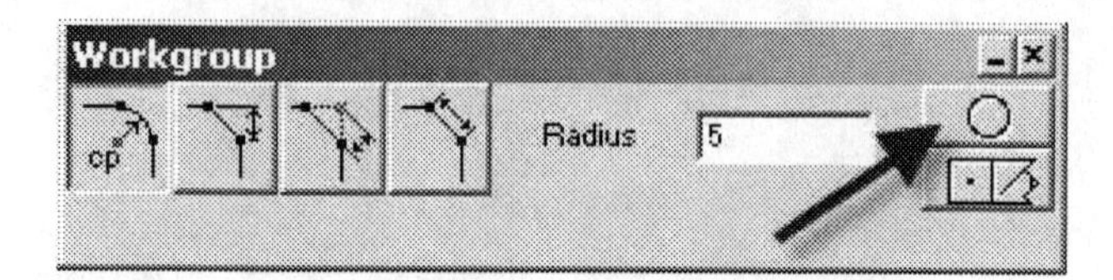


Your geometry should look like the following image.

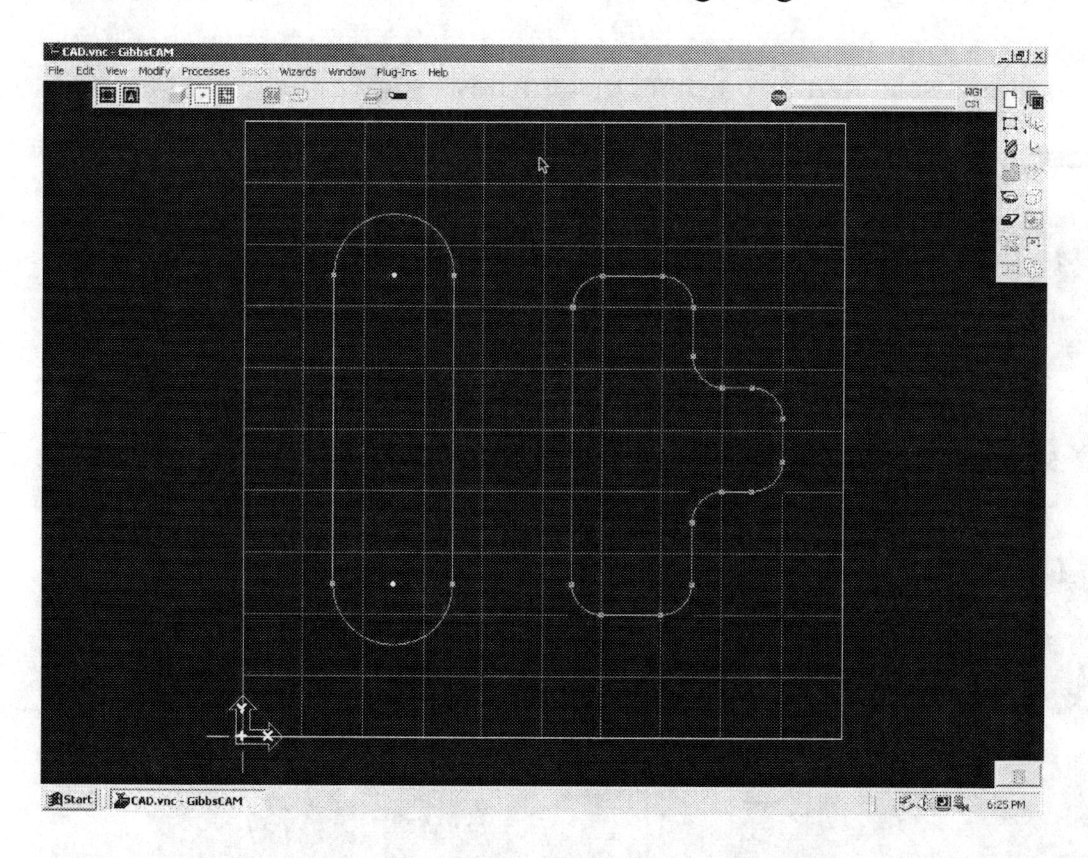

This completes the exercise. Save your file. Choose *File-Save* from the *Main Menu*. Close the file.

## Exercise 7 – Workgroups

In this exercise you will create 2 layers of geometry for the same part file using Workgroups. Using Workgroups can help keep your geometry organized when you are working on a part file with multiple machining operations. Both the CAD tools and the Geometry Expert will be utilized in creating the geometry for this exercise.

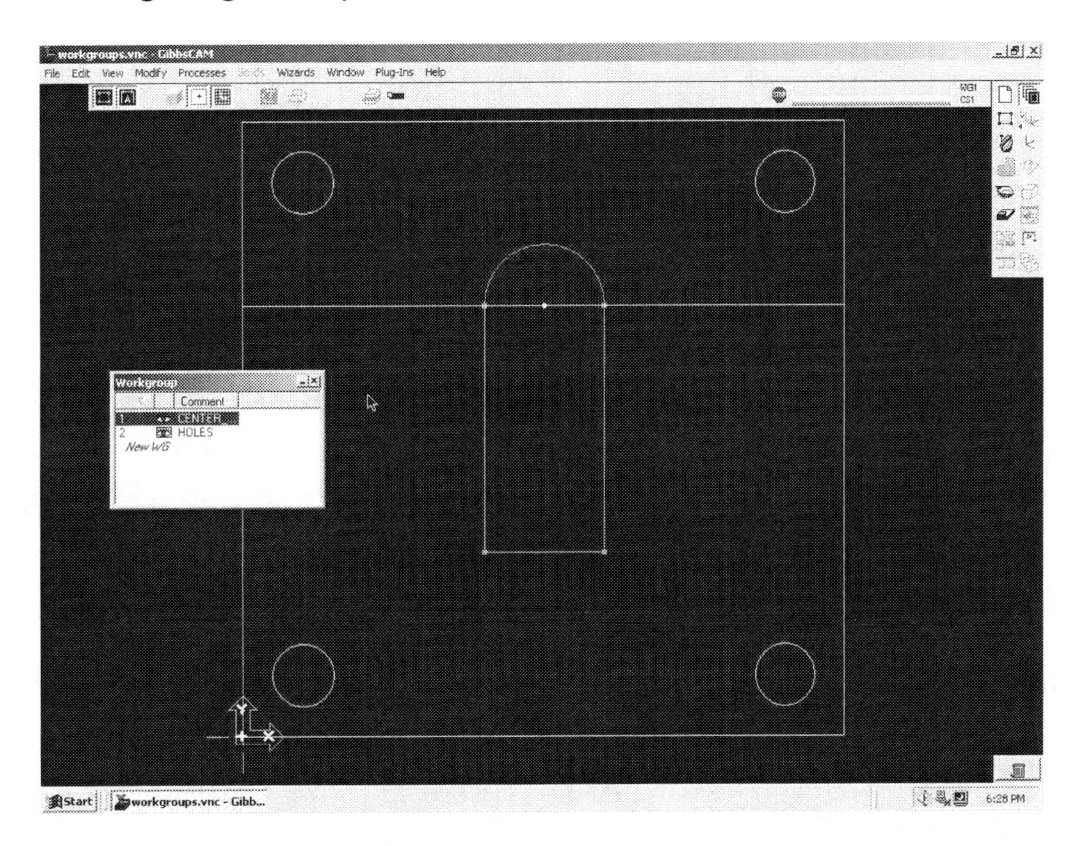


Create a new file named WORKGROUPS. Set the machine type to 3 axis vertical mill and the units to mm. Enter the values as shown in the next graphic for the stock size.

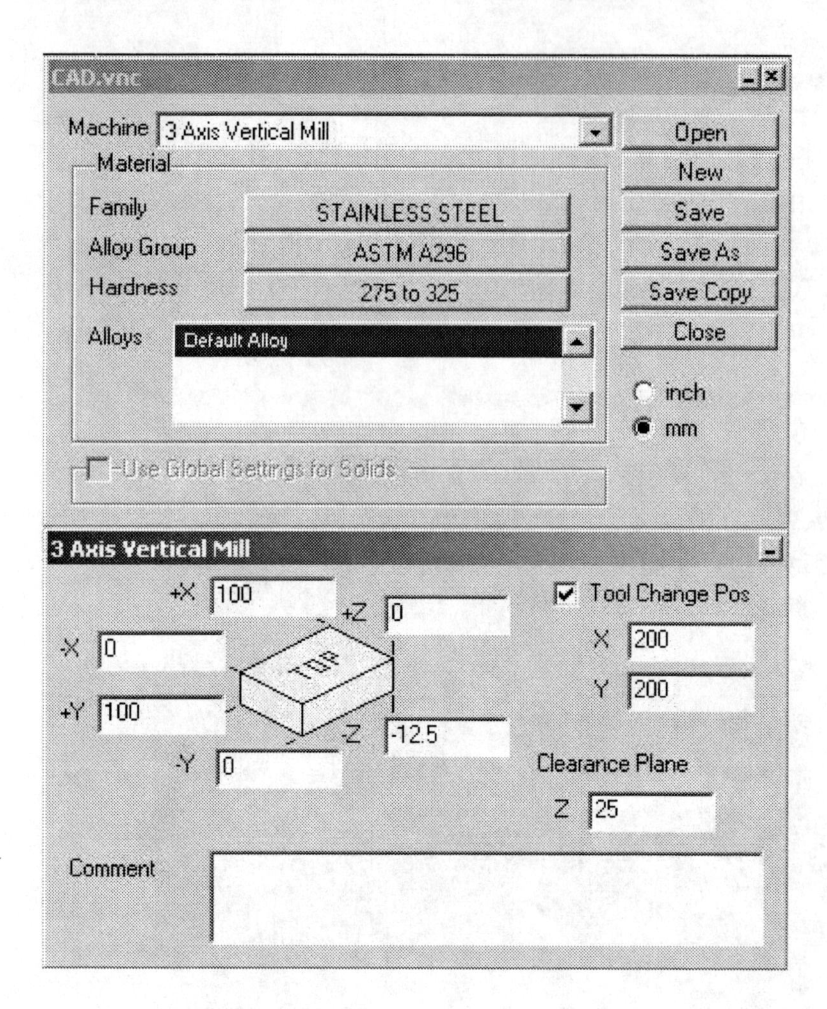


Select the Geometry icon from the Top Level Palette.

The Geometry Palette will appear.

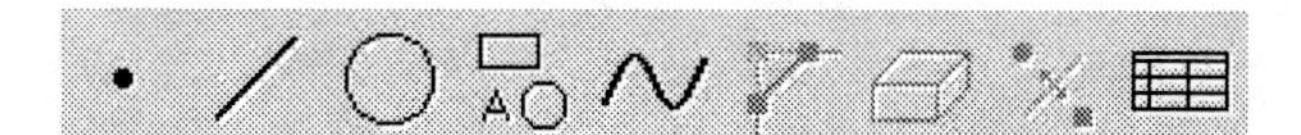

Select the Shapes icon.

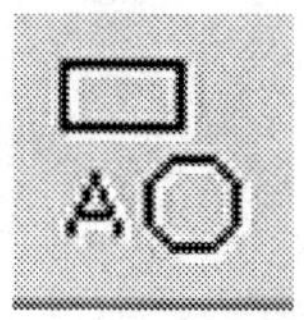

The Shapes Palette will appear.

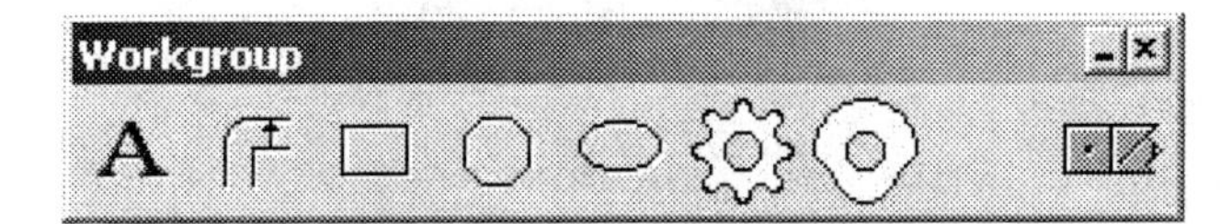


Select the Box icon.

Enter the values below and click on Do It.

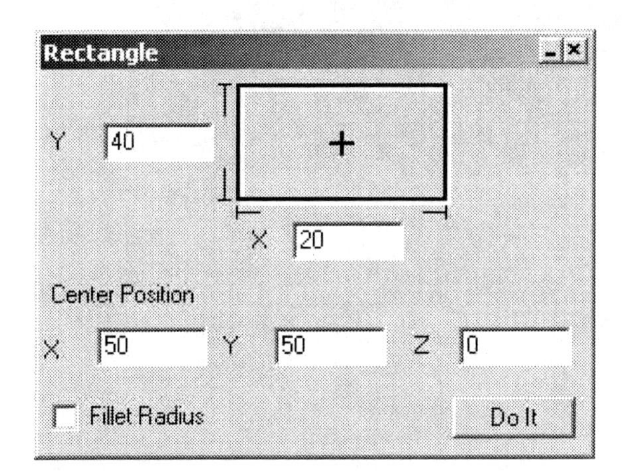


Your screen should look like the following graphic.

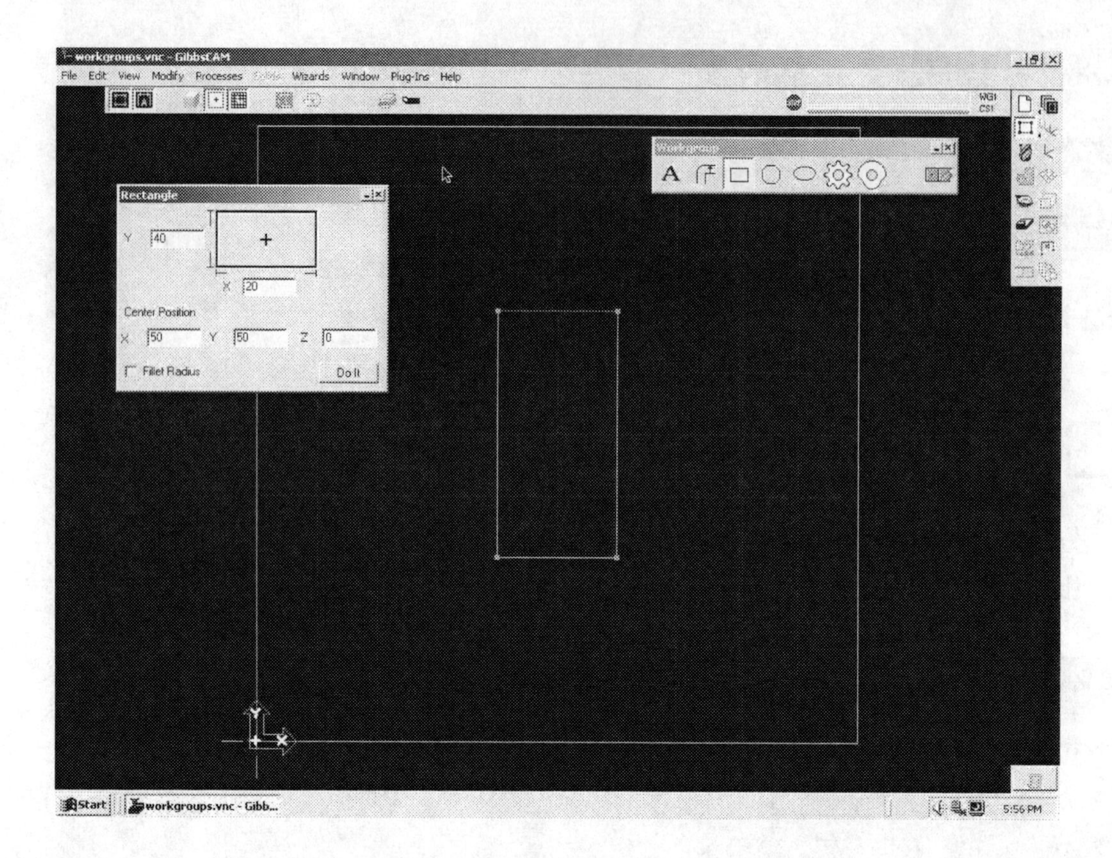


You will now add a midpoint to the top horizontal line. This will be used for the center point of an arc that you will add to this profile. Select the point icon from the Geometry Palette.

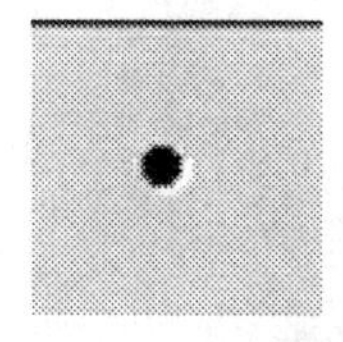

The Points Palette will appear.

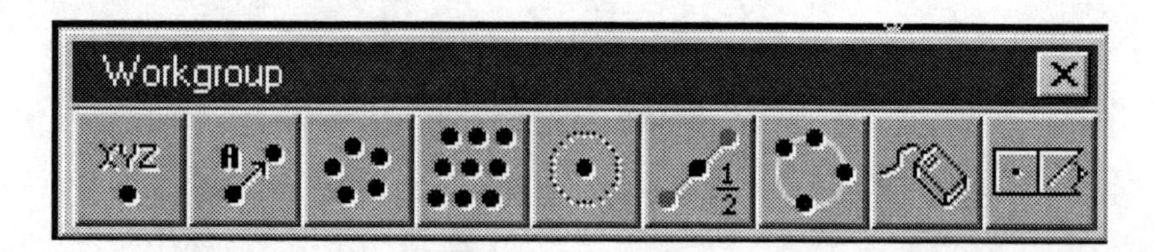


Select the Midpoint icon.

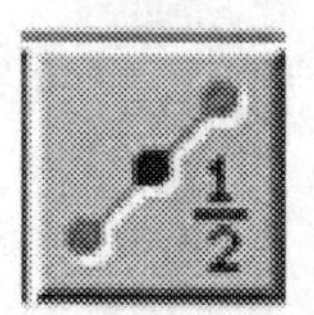

Select the line as shown and click on the Single Point icon to create a midpoint on that line.

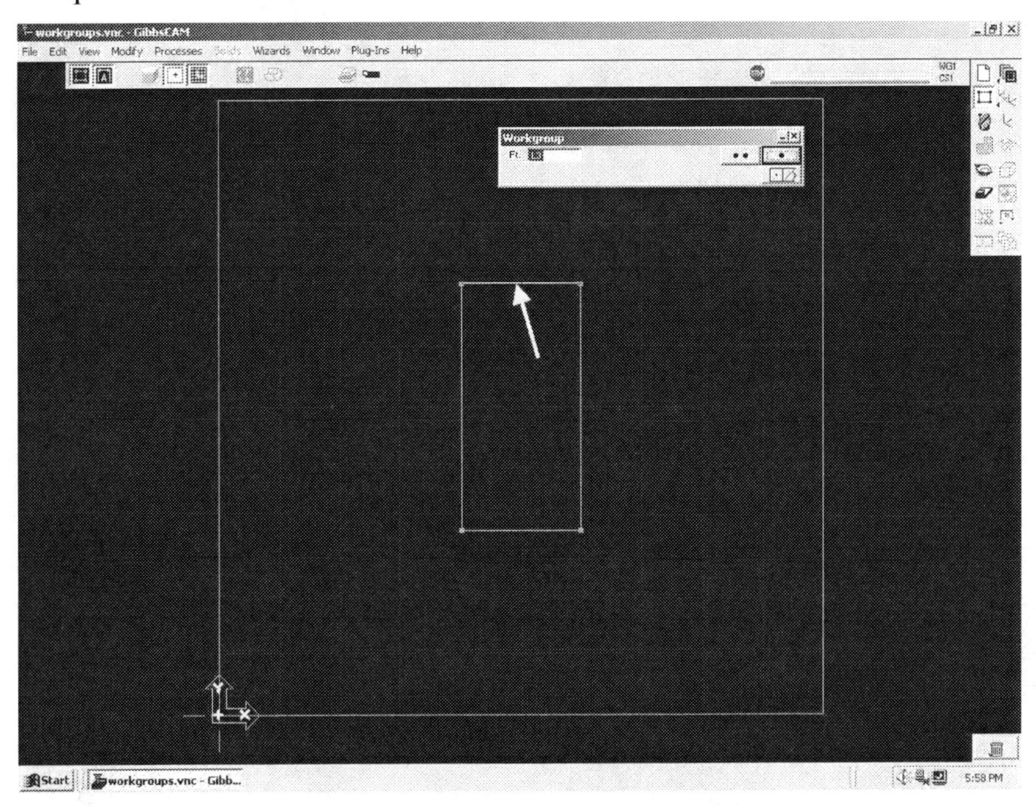


Your screen should look like the following image.

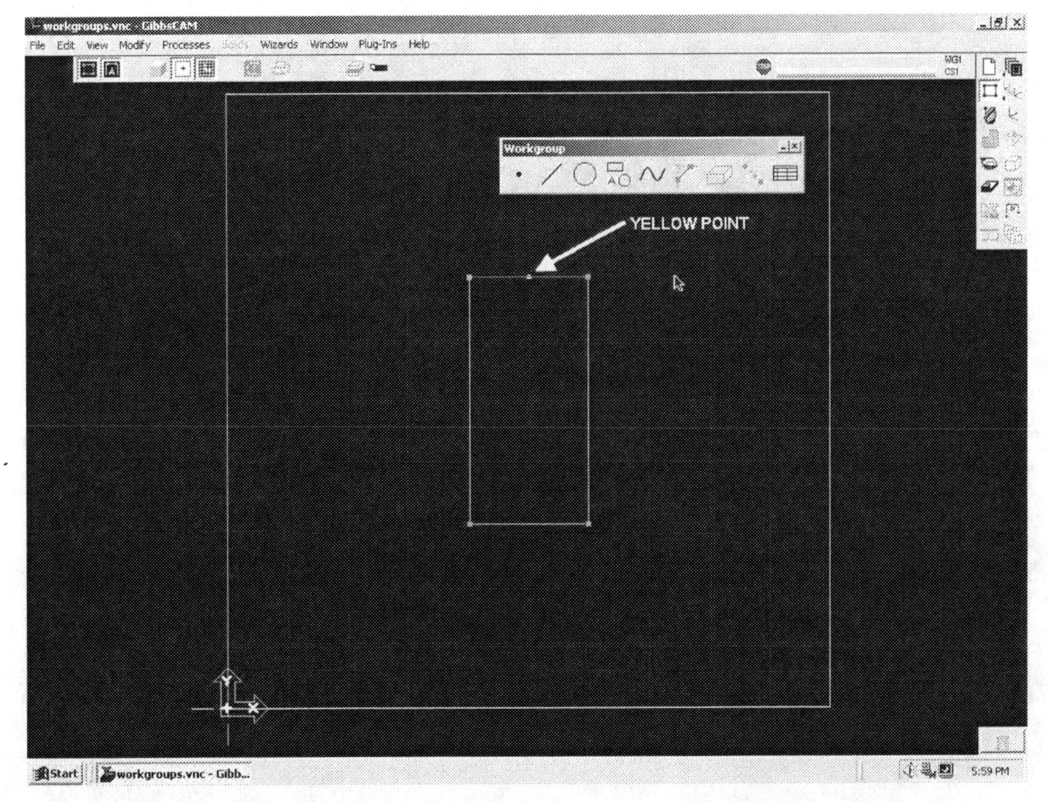

Select the Circle icon from the Geometry Palette.

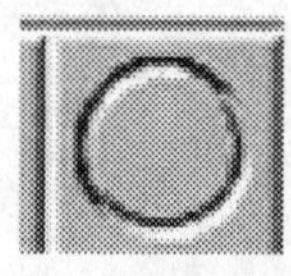

The Circle Palette will appear.

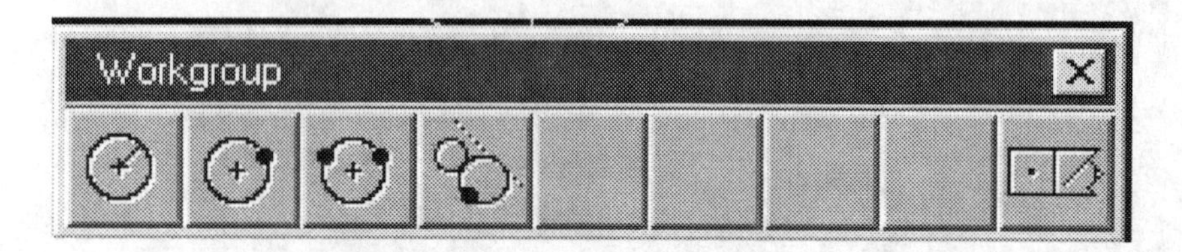


Select the Point & Center point icon.

Select the midpoint you created as the CP.

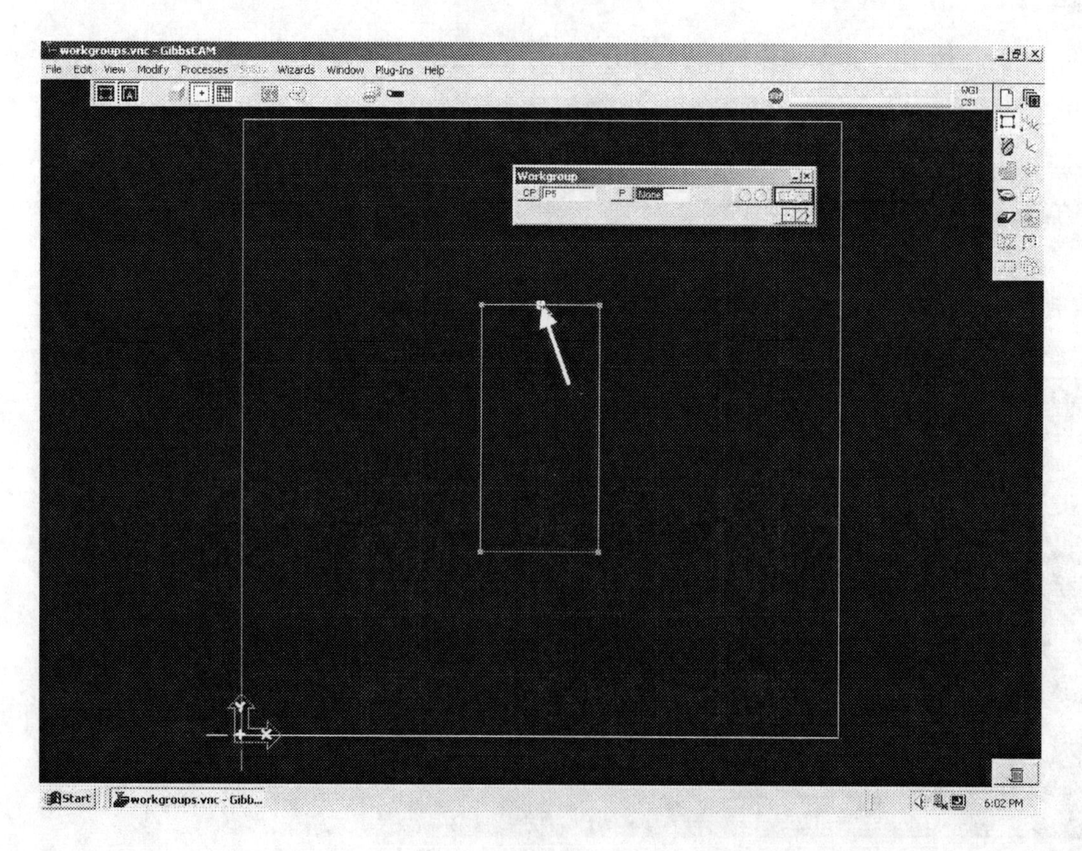

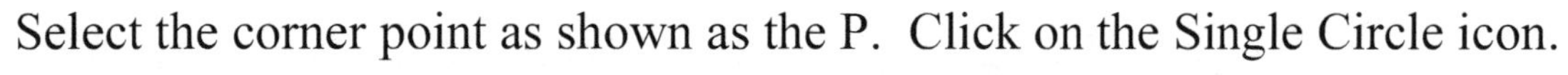

Select the corner point as shown as the P. Click on the Single Circle icon.

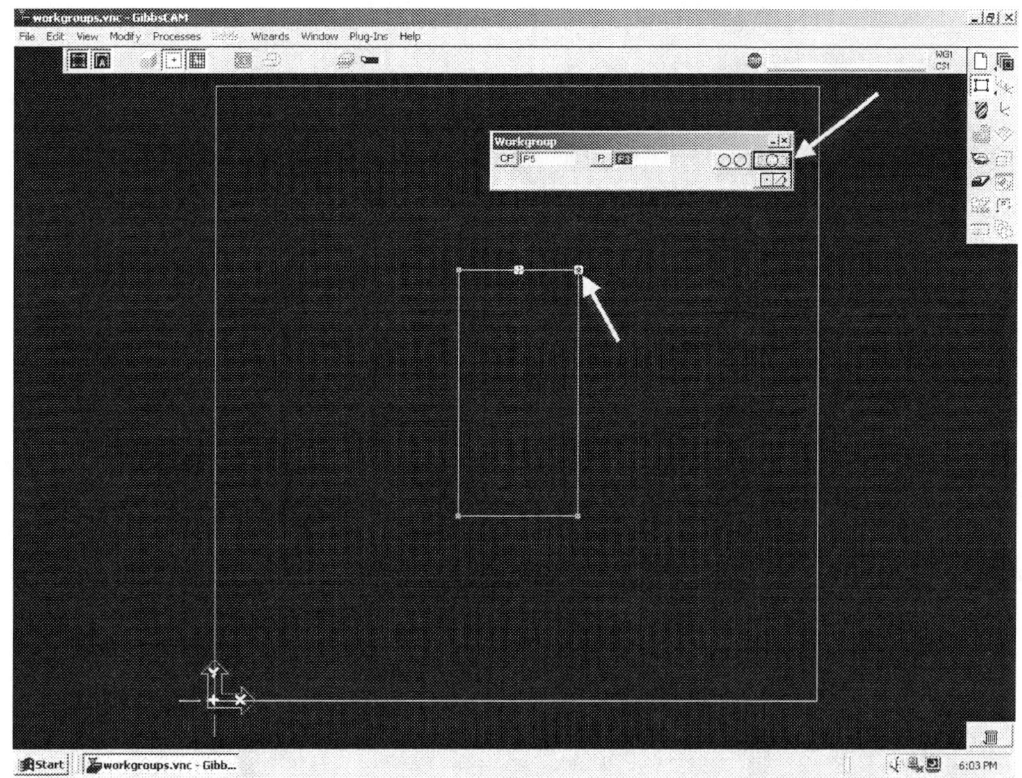


Your screen should look like the following image.

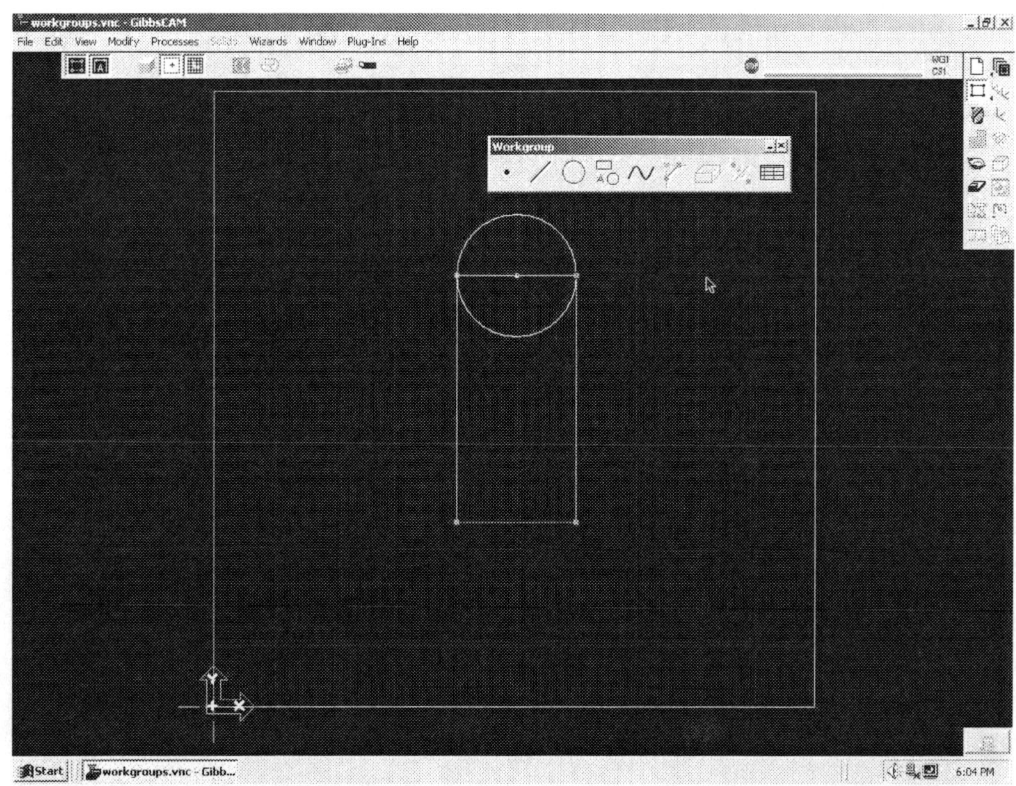

You will now connect/trim the arc to the rectangle profile. Hold down the control key and select the 2 blue corner points as shown.

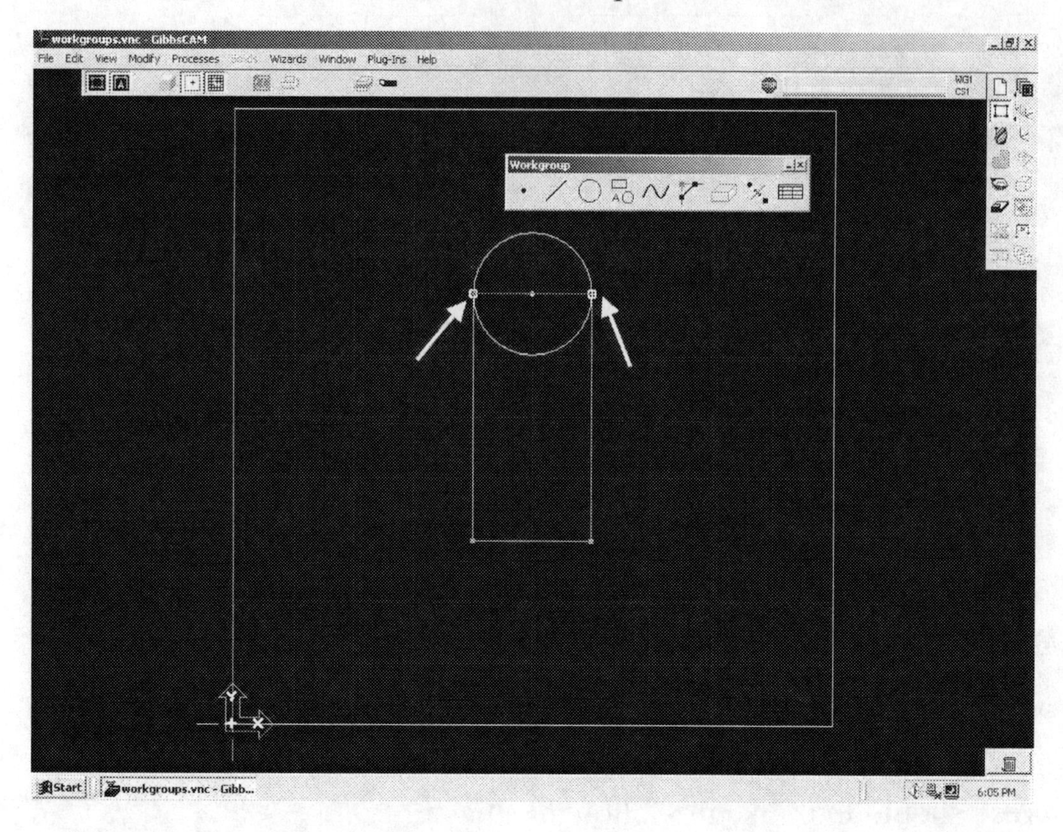


Select the connect/disconnect icon to disconnect the connector points from the profile.

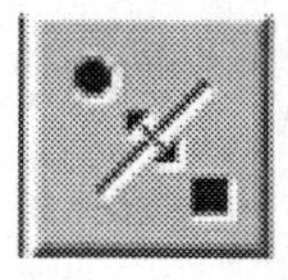

Your screen should look like the following image.

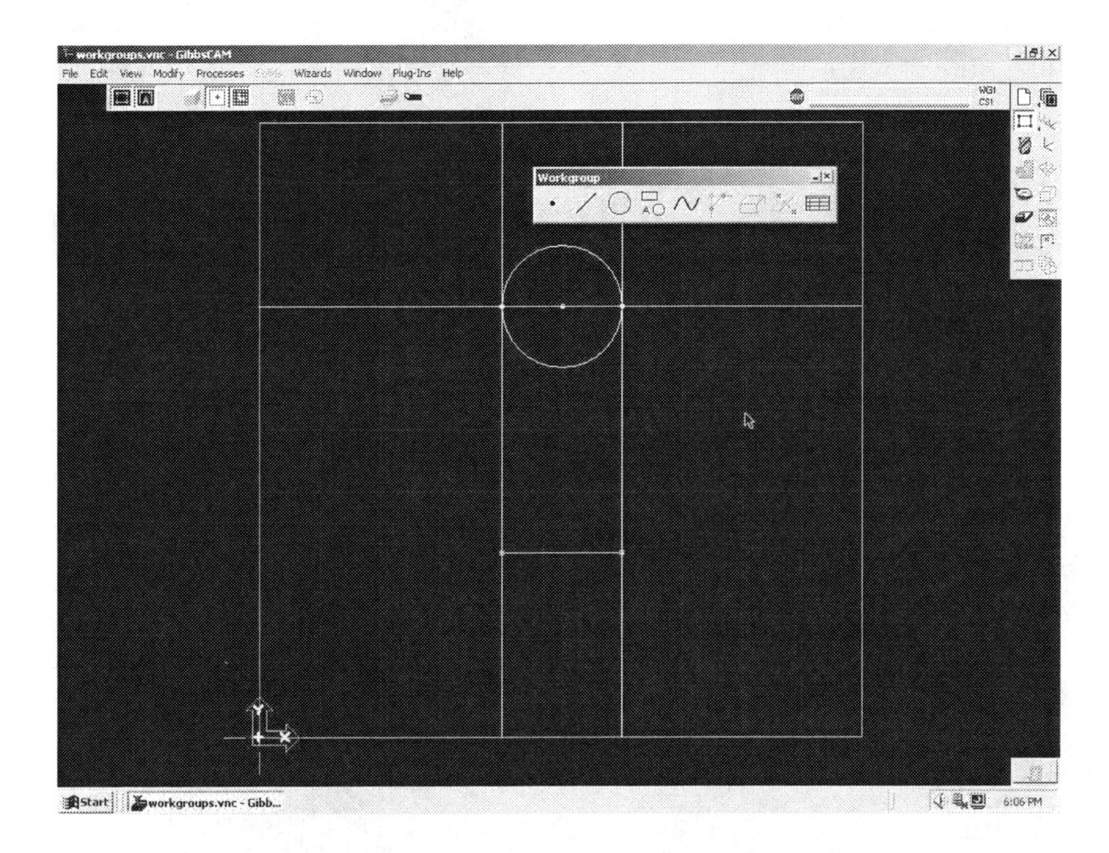


You will now connect the arc to the profile. Hold down the control key and select the circle and the yellow vertical line on the left as shown.

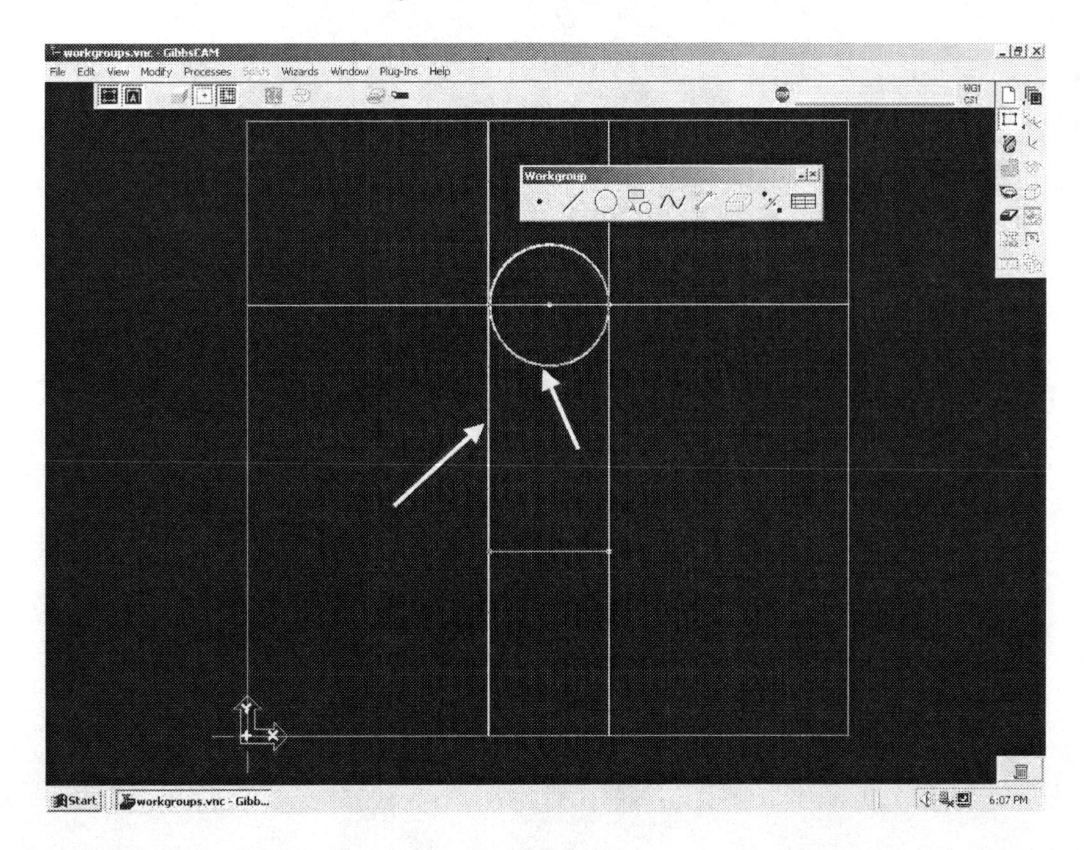

Select the Connect icon.

A blue connector point will be placed at the intersection of the line and the circle. Your screen should look like the following image.

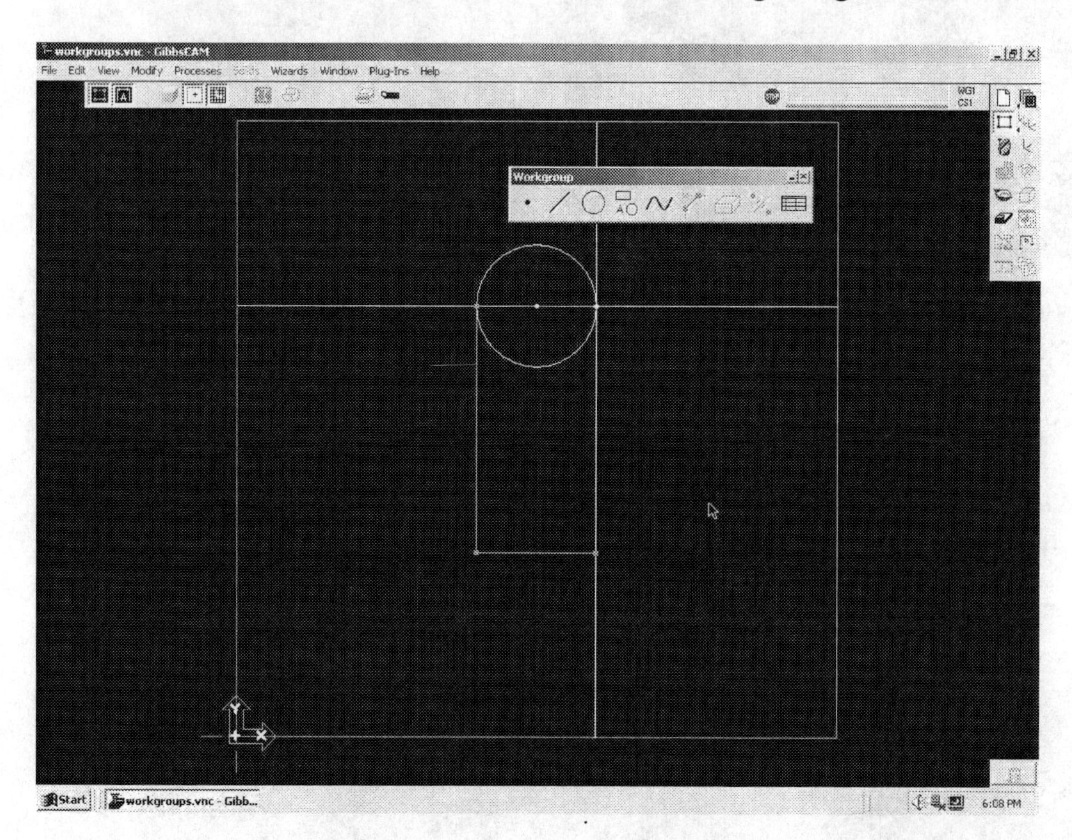

Hold down the control key and select the circle and the other vertical yellow line as shown.

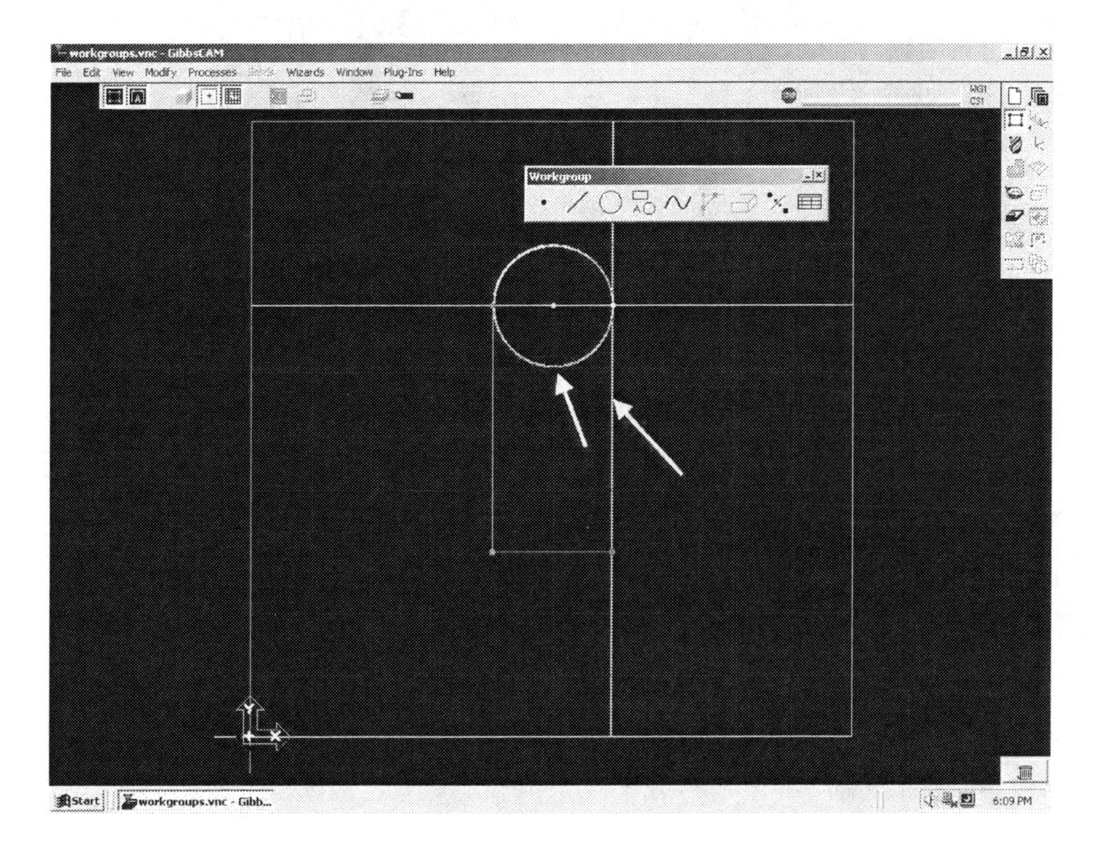

Select the Connect icon.

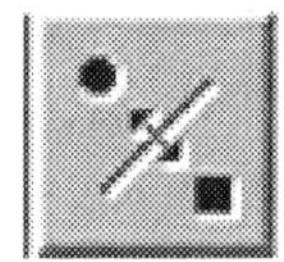

Your screen should look like the following graphic.

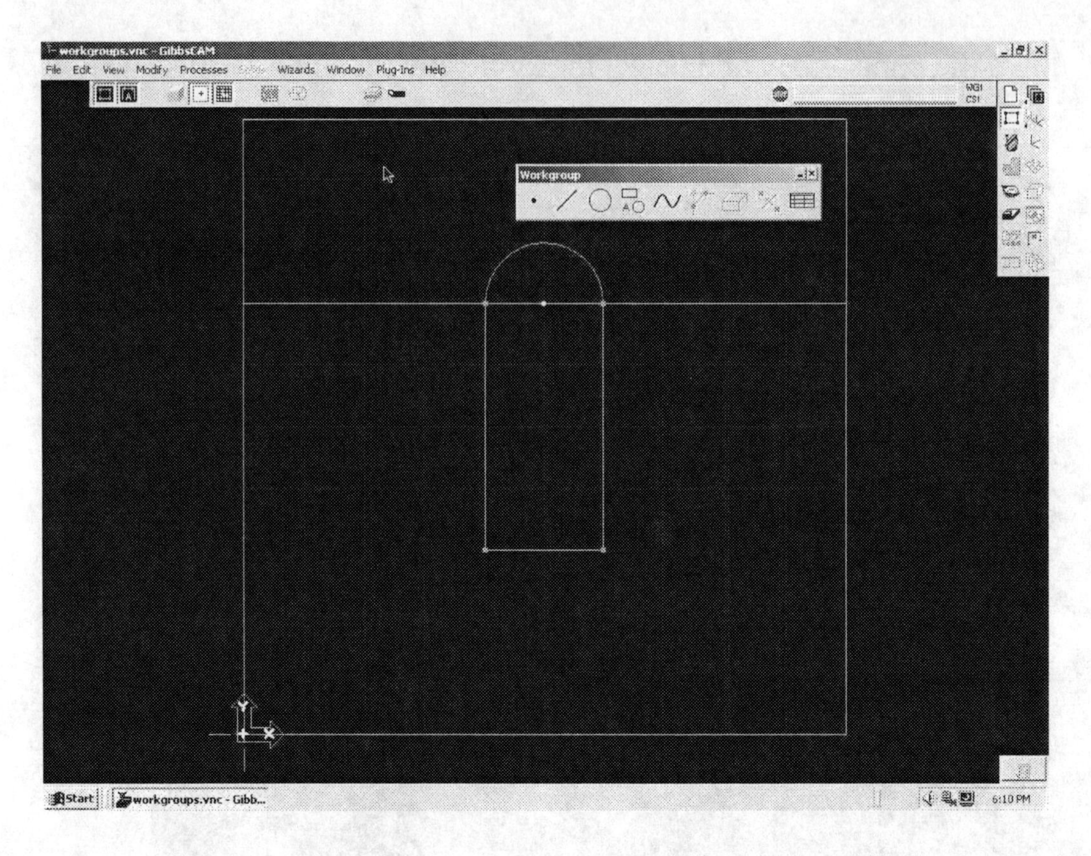

Note: If your arc is reversed from the image above, select it and choose *Modify-Reverse Arc* from the *Main Menu*.

This completes the profile on the first workgroup. You will create a second workgroup that will contain geometry for 4 holes.

Single-click on the Workgroups icon from the Top Level Palette.

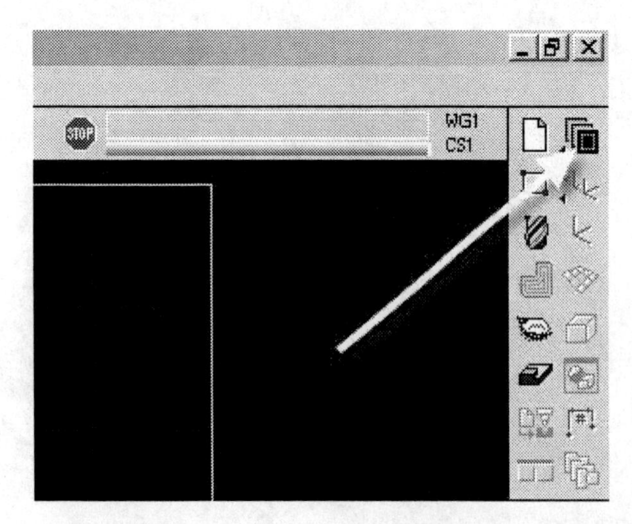

The Workgroup dialog box will appear. Double-click on *Workgroup* and change the name to CENTER_POCKET as shown.

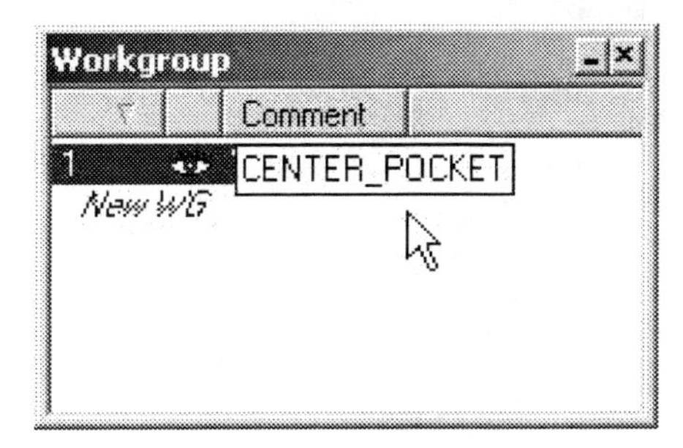


Select *New Workgroup* to create a new workgroup.

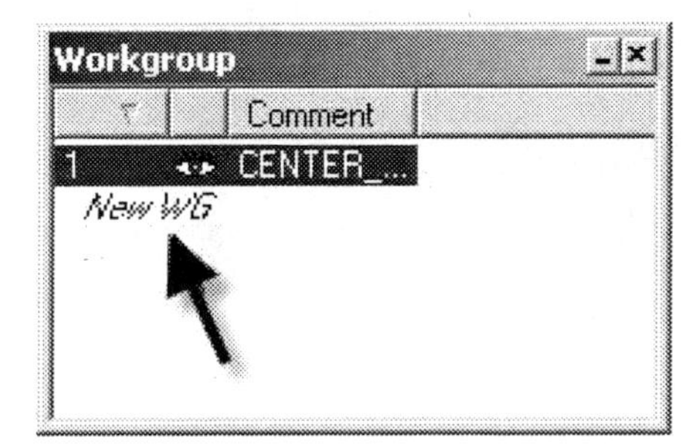


The new workgroup is highlighted and now the active layer. The geometry you created should no longer be on the screen. Change the name of the new layer to HOLES.

You will create 4 holes for this workgroup using the Geometry Expert.

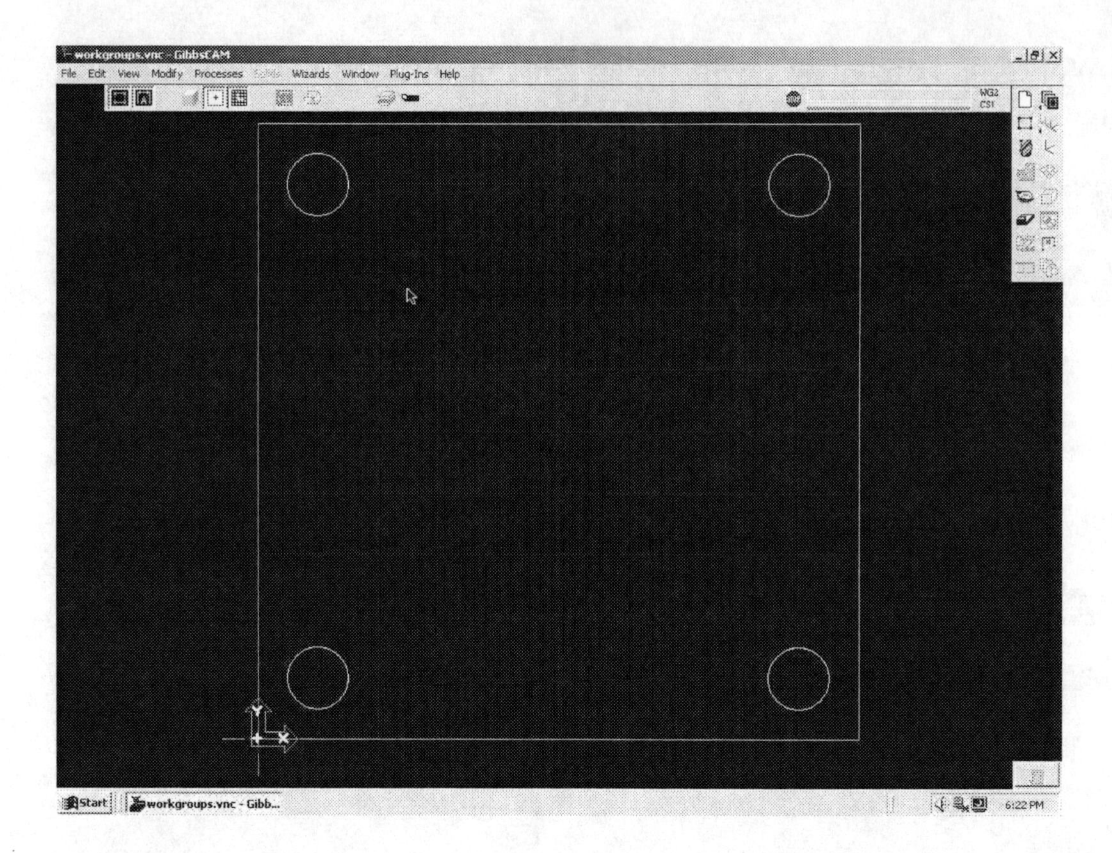


Select the Geometry Expert icon from the Geometry Palette

Enter the following rows into the Geometry Expert spreadsheet.

| Feature Type | Ref. | Radius | CP X | CP Y |
|---|---|---|---|---|
| Clockwise Arc | C1 | 5 | 10 | 10 |
| Clockwise Arc | C2 | 5 | 10 | 90 |
| Clockwise Arc | C3 | 5 | 90 | 90 |
| Clockwise Arc | C4 | 5 | 90 | 10 |

The Geometry Expert should look like this.

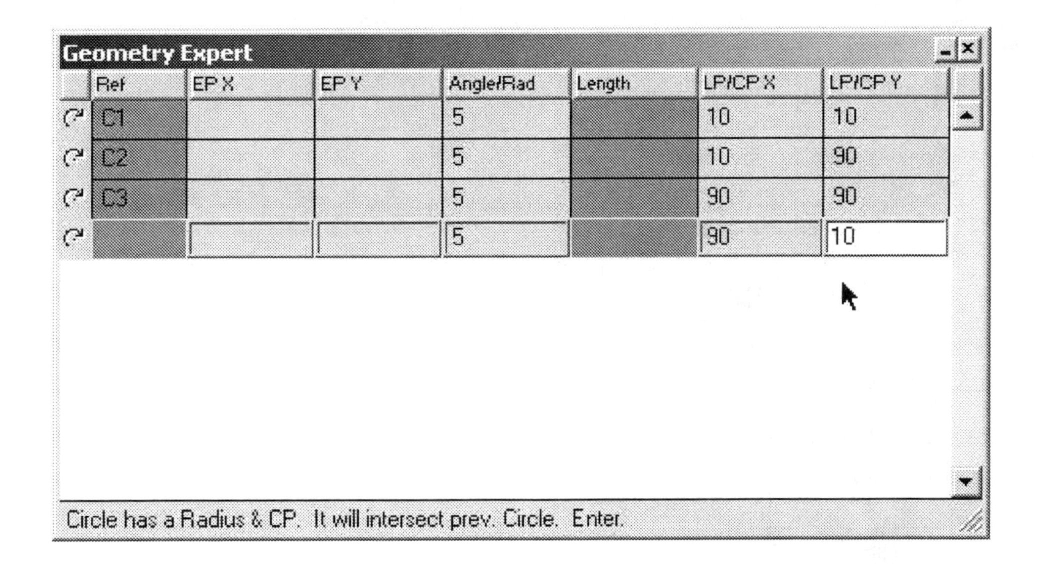


Hit enter. Your screen should look like the next graphic.

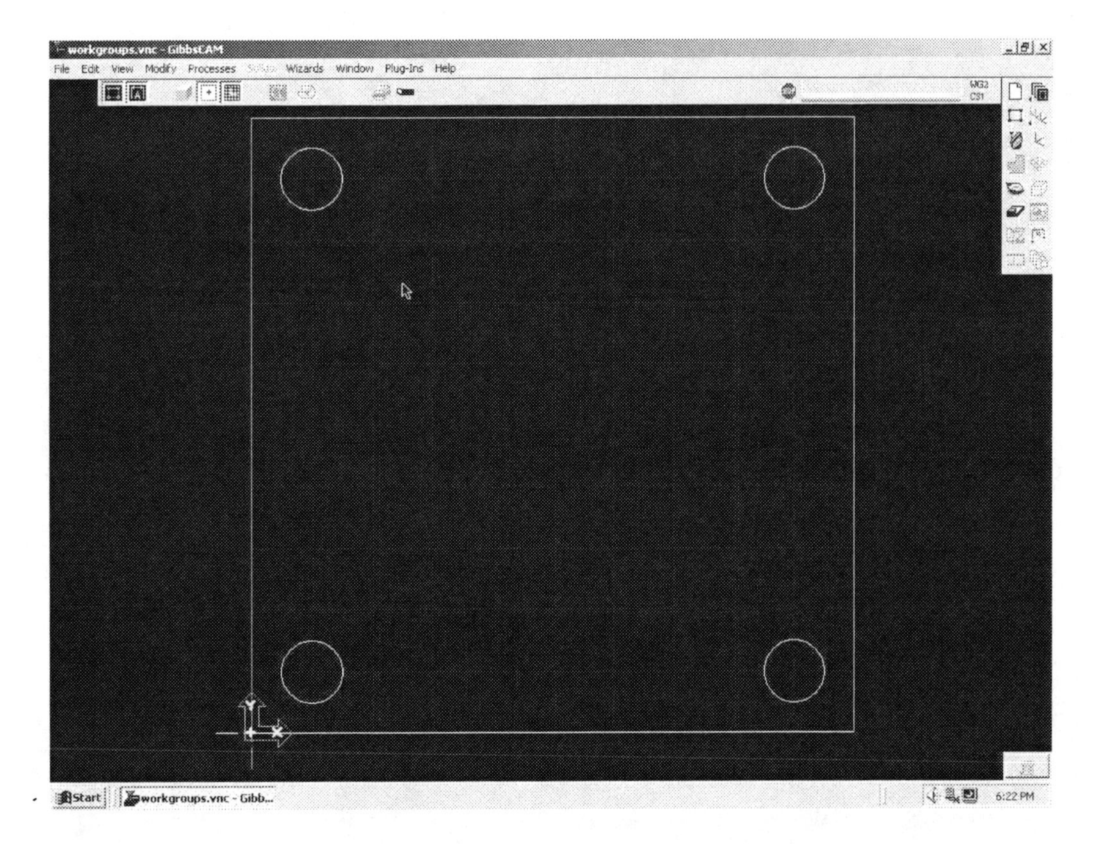

To view the first layer of geometry you created, select the Workgroup icon from the Top Level Palette.

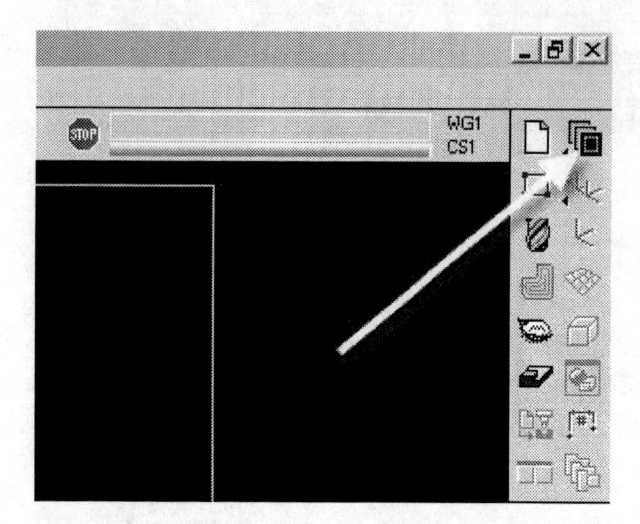


Select the CENTER_POCKET workgroup.

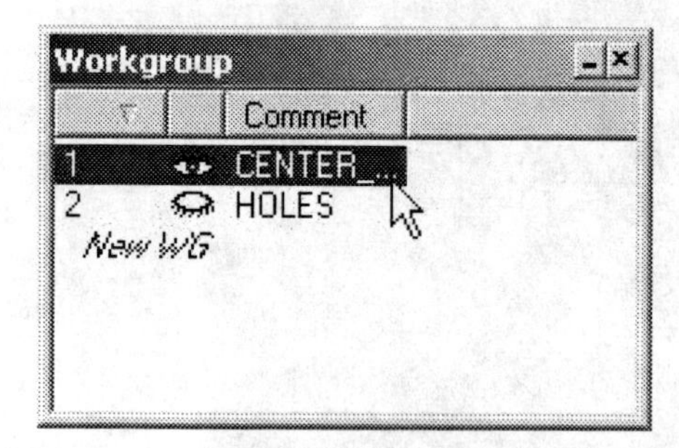


You should now see the first layer of geometry you created.

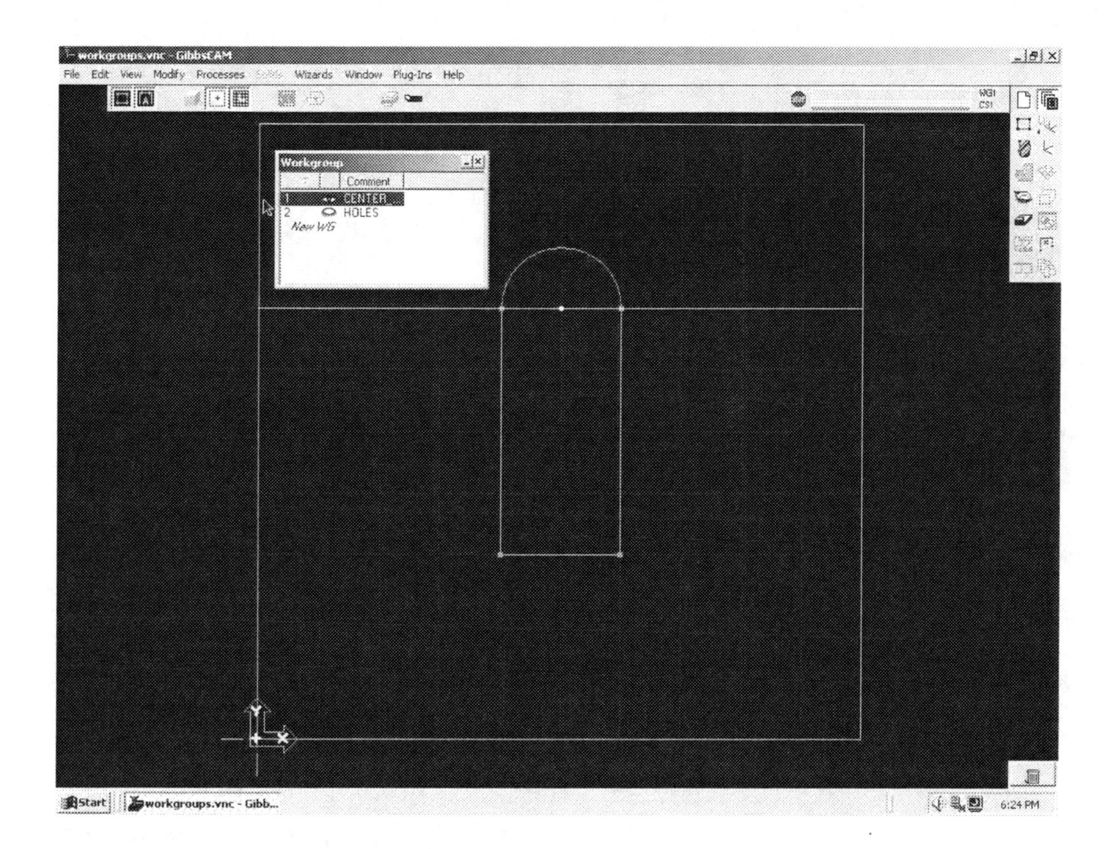

You can also view different workgroups at the same time. Double-click on the eyes next to the HOLES workgroup to view them.

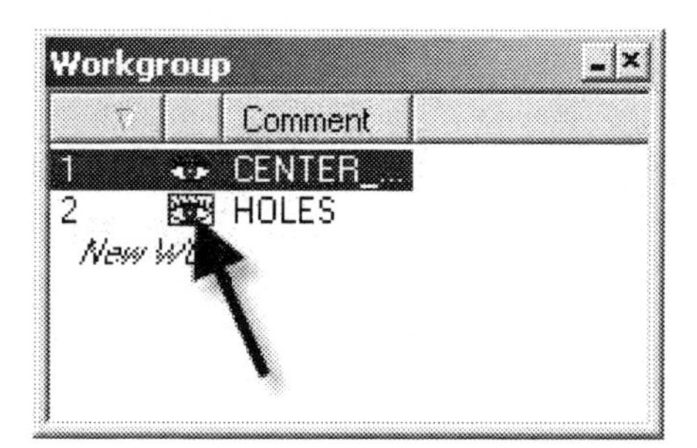


This completes the exercise. Save the file. Choose *File-Save* from the *Main Menu.*

## Exercise 8 – Text

In this exercise you will create text for engraving.

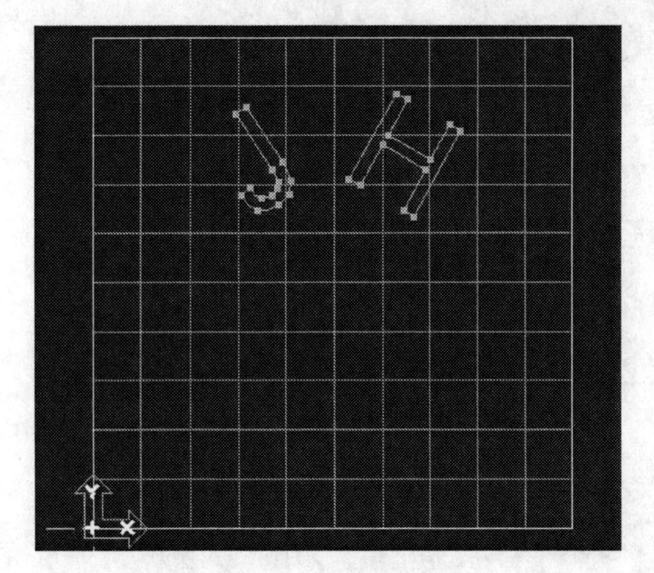

Create a new file named TEXT.vnc. Set the machine type to 3 axis vertical mill and the units to mm. Enter the values as shown in the next graphic for the stock size.

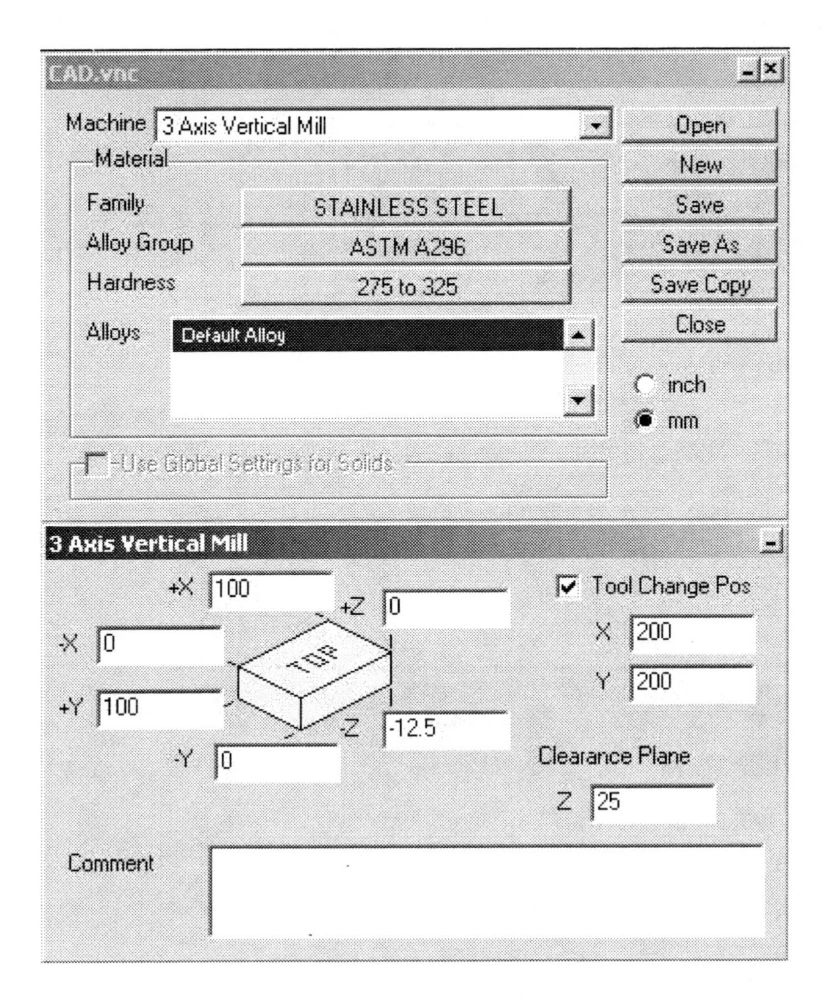


Select the Geometry icon from the Top Level Palette.

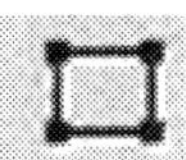

The Geometry Palette will appear.

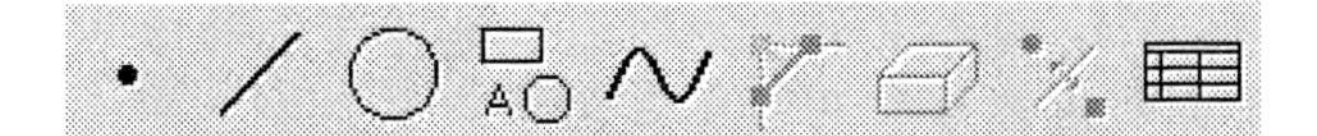

Select the Shapes icon.

The Shapes Palette will appear.

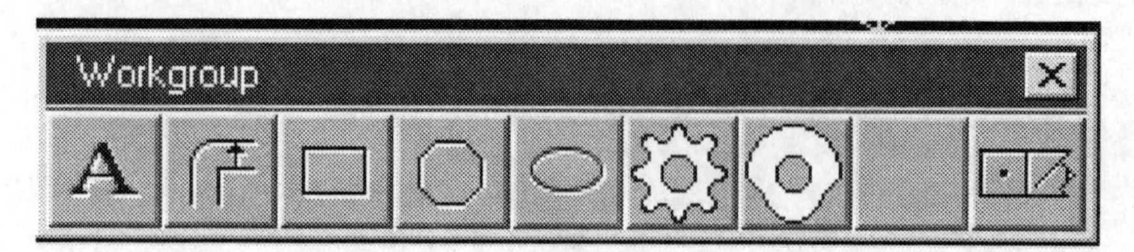


Select the Engraving icon.

The Text Creation dialog box will appear.

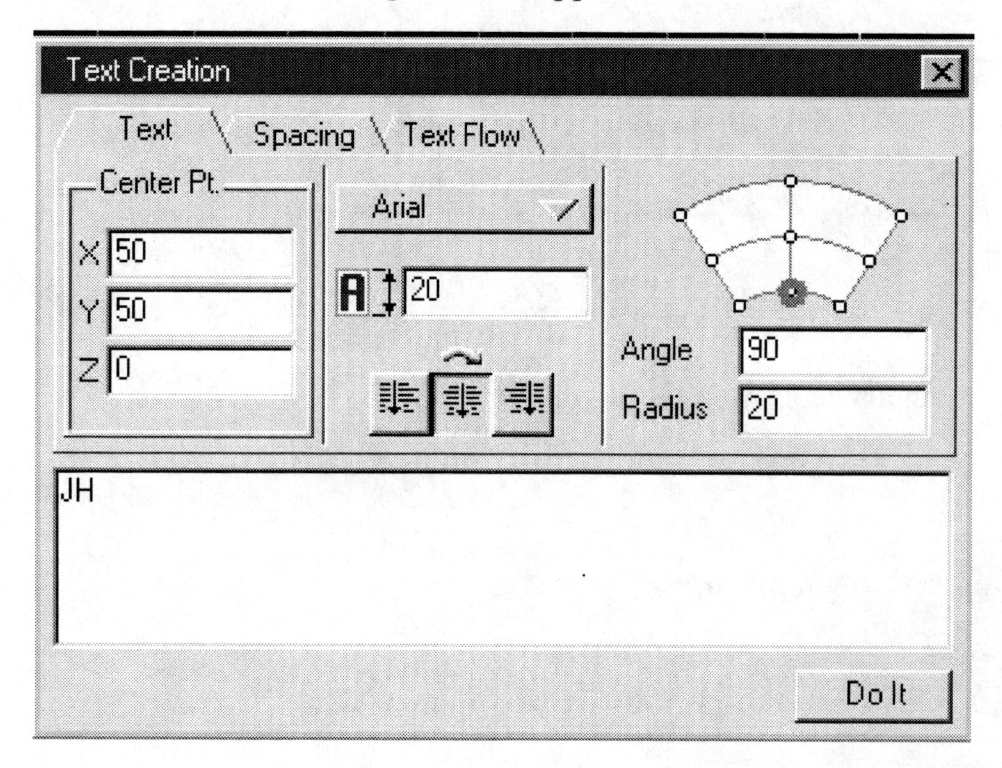


The Text Creation dialog box has 3 tabs that contain windows that allow the user to specify text, spacing, and text flow. They are the *Text* tab, the *Spacing* tab, and the *Text Flow* tab. Select the *Text* tab if it isn't already active.

The Center Pt. Field determines where the text will be located. Enter the values as shown.

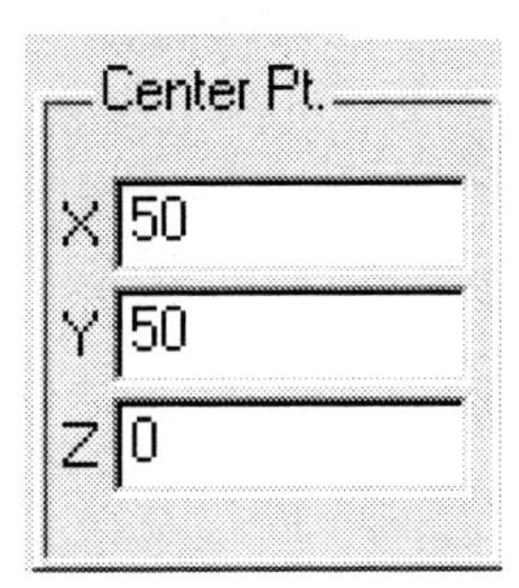


The type of font can be selected from a pull down menu. The text size is determined by the value in the Text Size Field. Select Arial as the font and 20 for the text size as shown.

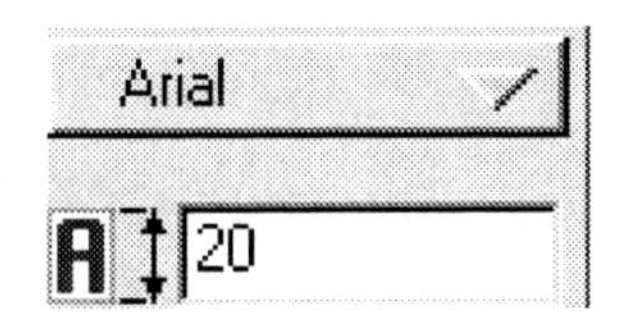


The Justification icons are used to specify left justified, center justified, or right justified. Select Center-Justified as shown.

The look of the text is determined by what is selected in the Text Flow tab. Select the *Text Flow* tab and pick the Clockwise Arc icon in the Shapes field.

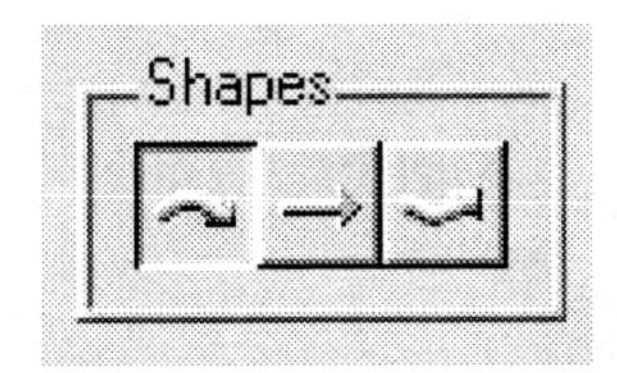

Select the *Text* tab. Enter the values below in the Text Alignment field.

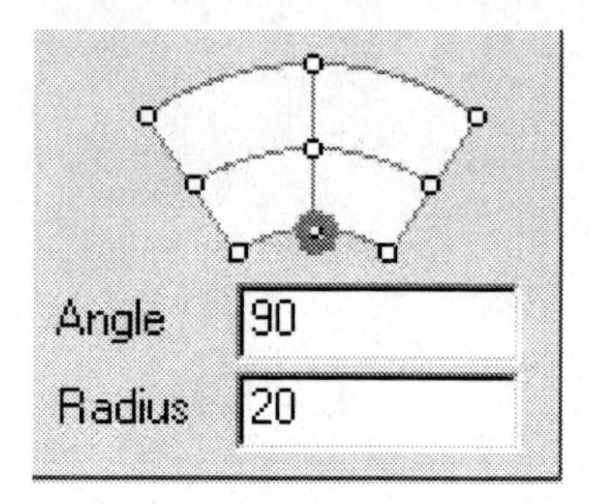


The last field in the Text tab is the Text Entry Box. This is where you will enter your text. Enter your initials in the Text Entry Box.

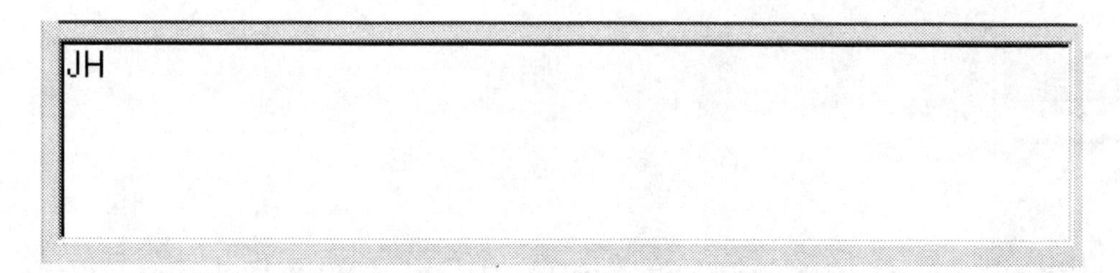


Click on the Spacing Tab. This tab controls spacing distance between letters, words, and lines of text. Enter the values as shown.

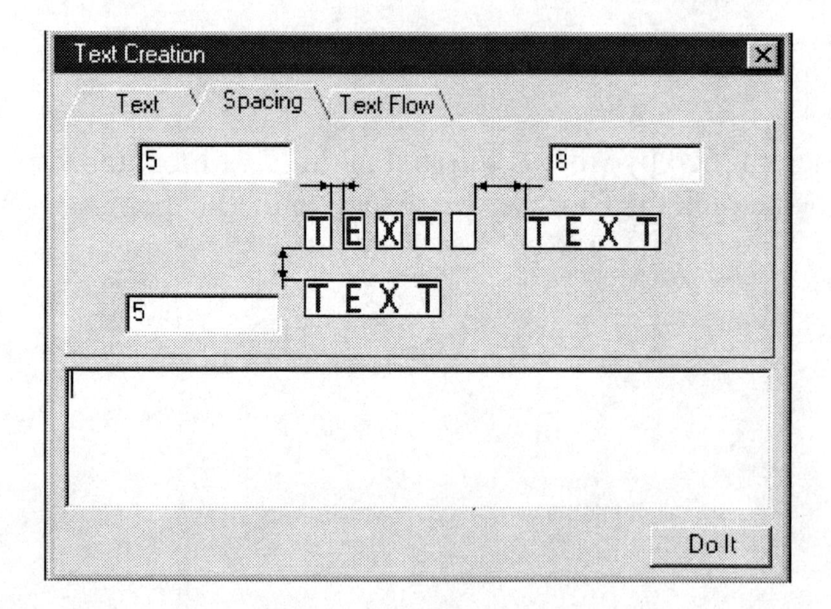


Click on the Text Flow tab.

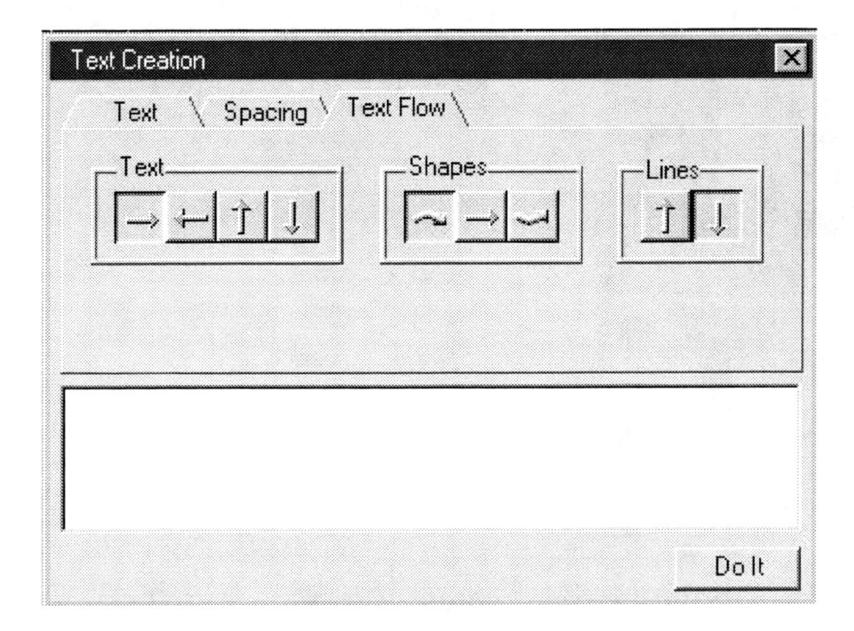


The icons in the Text field control text direction. Choose left to right as shown.

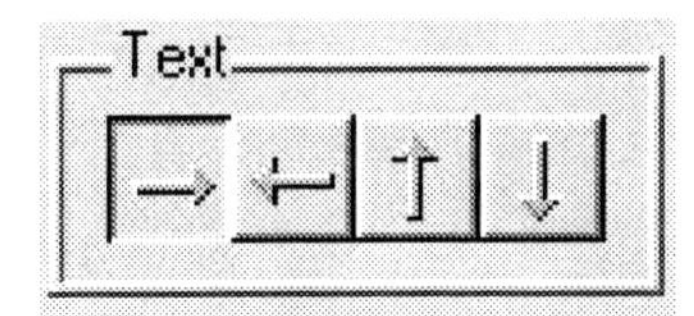


The icons in the Shapes field can be set to clockwise arc, straight line, or counter-clockwise. You picked clockwise arc earlier. Leave this selection as it is.

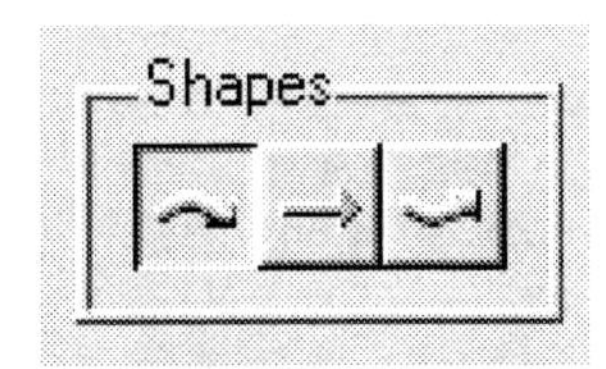


The icons in the Lines field determine whether the lines of text go from top to bottom or bottom to top. Select top to bottom as shown.

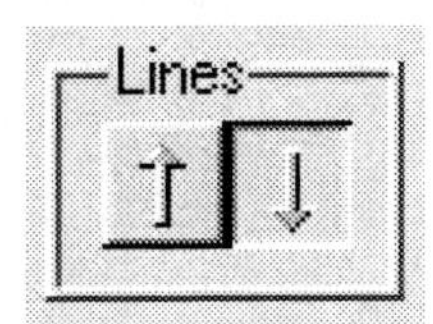


Click on Do It. Close the Text Creation dialog box.

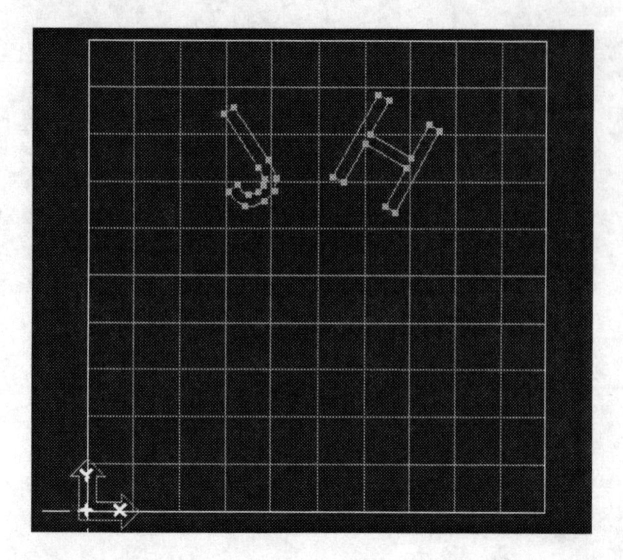


This completes the exercise. Save the file. Choose *File-Save* from the *Main Menu*.

## Exercise 9 – Dimensions

In this exercise you will open a part file and add dimensions to the part geometry.

Select *File-Open* from the *Main Menu*. Choose the file CAD.vnc and click on Open.

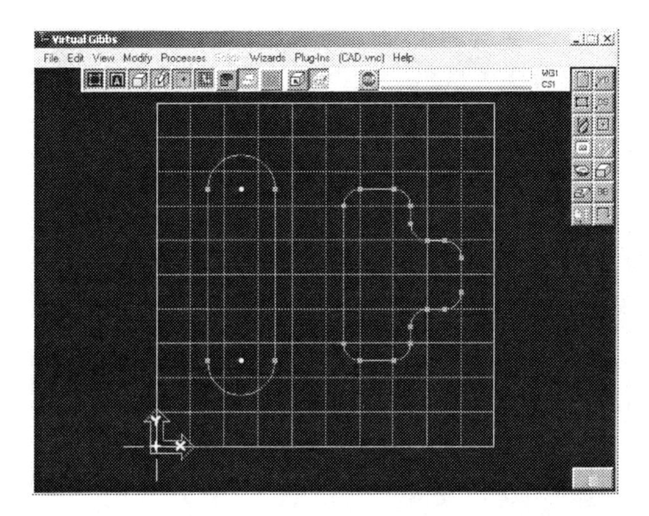

Click on the Dimensions icon on the Top Level Palette.

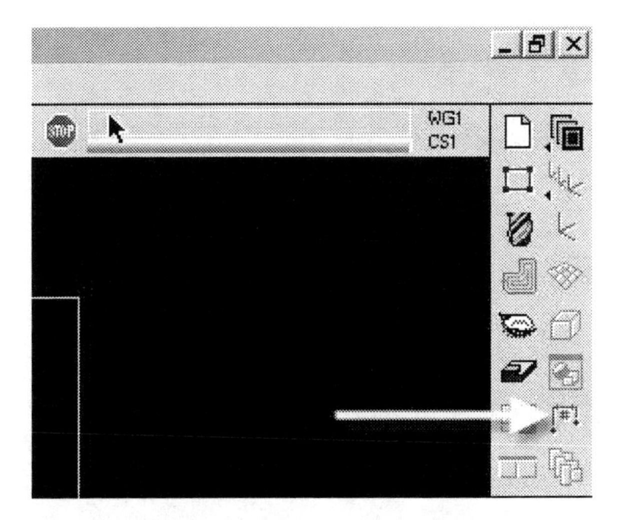


The Annotation Palette will appear.

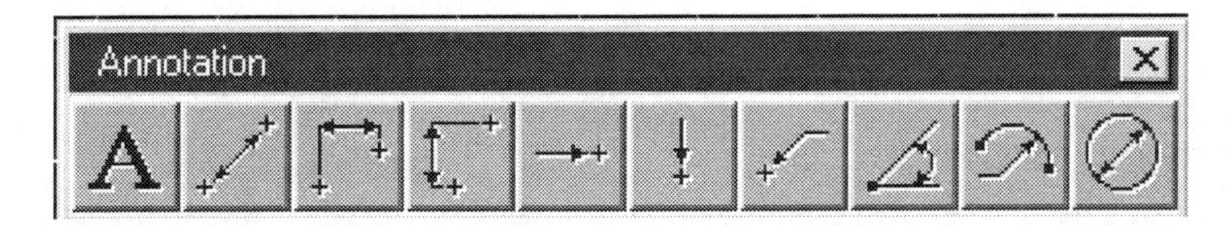

Select the Horizontal Dimension icon.

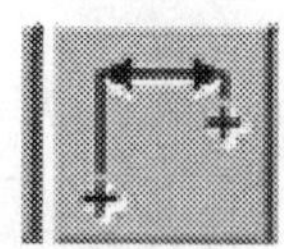

Select the 2 points as shown to place a horizontal dimension.

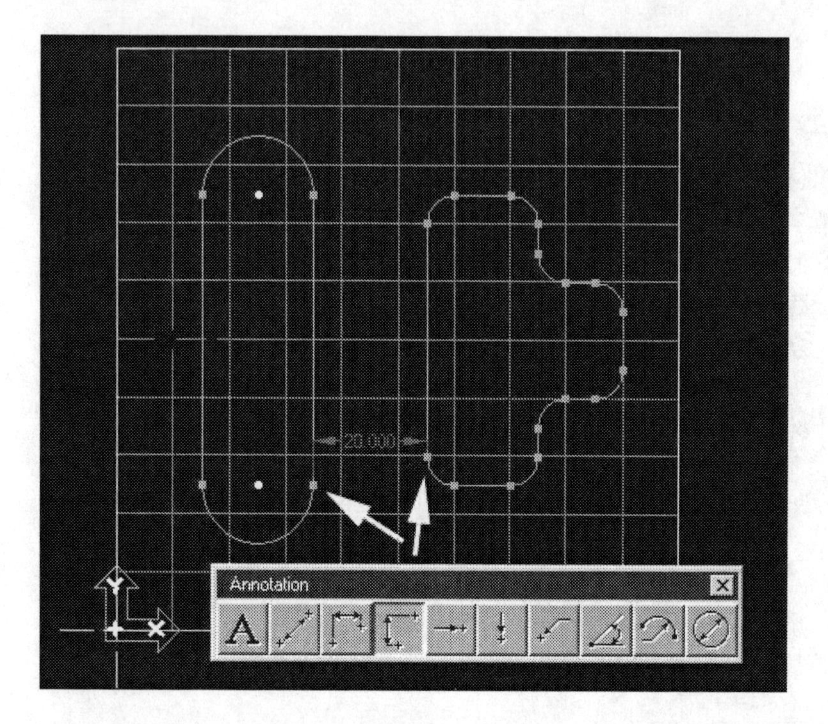


Select the Vertical Dimension icon.

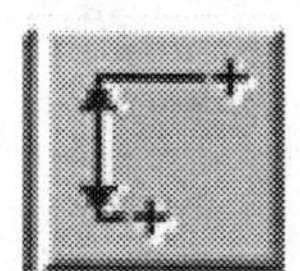

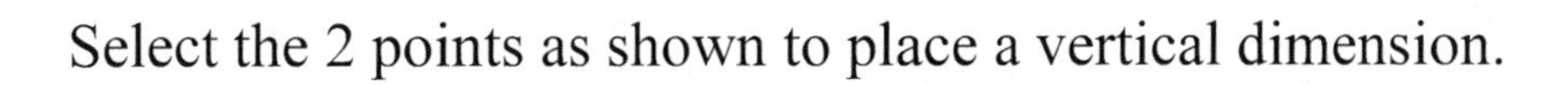

Select the 2 points as shown to place a vertical dimension.

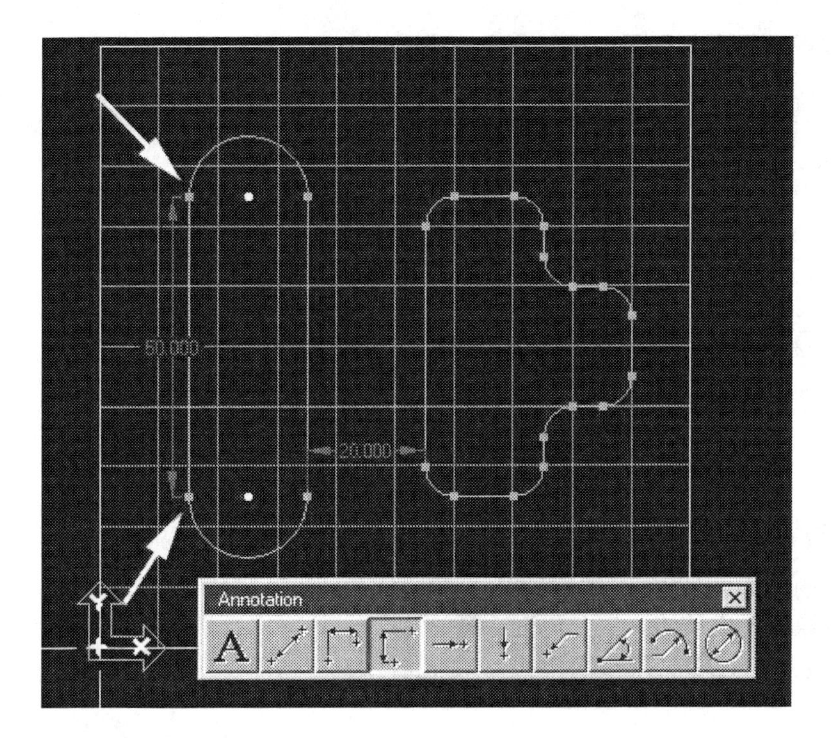


Select the Radius icon.

Select the arc as shown.

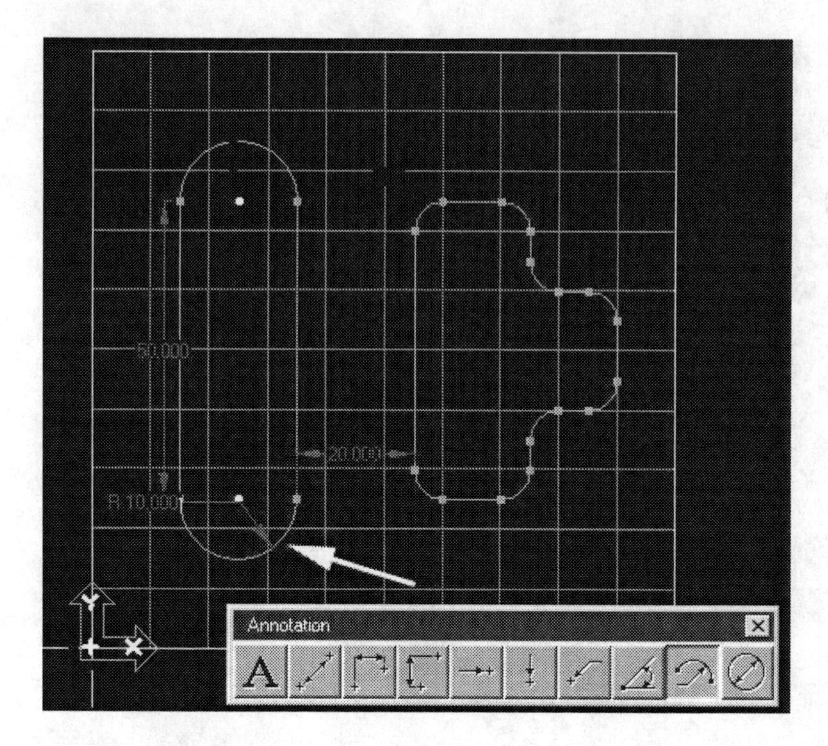


Select the Horizontal From Origin icon.

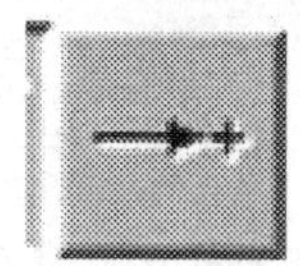

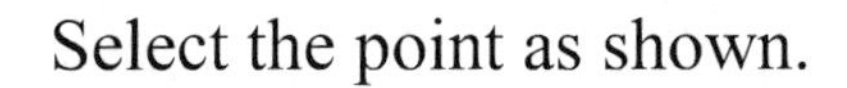

Select the point as shown.

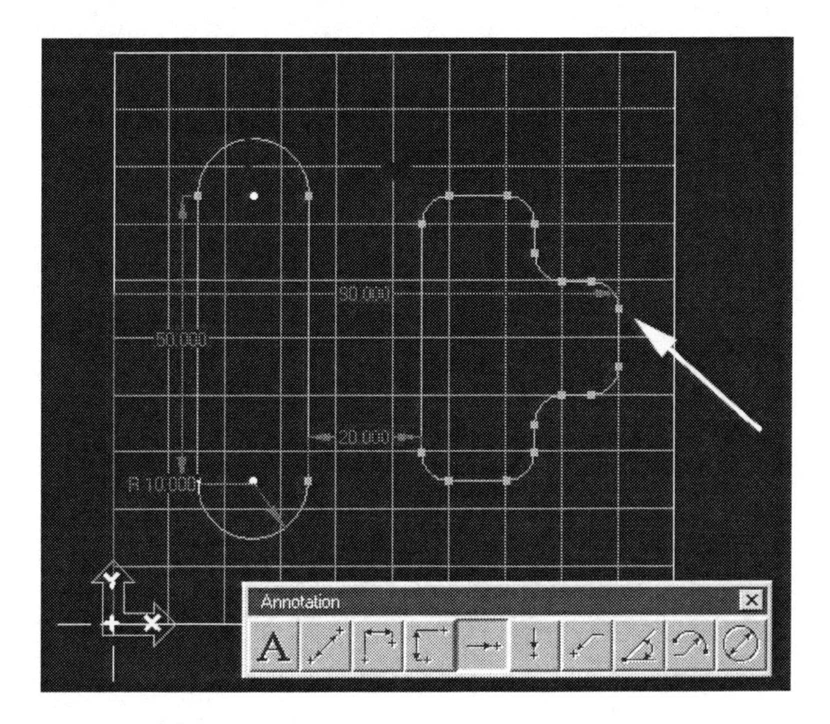


You can click and drag on any dimension line to change the location of a dimension.

This completes the exercise. Do not save the file. Choose *File-Close* from the *Main Menu*.

# NOTES:

Chapter 3

# Importing CAD Geometry

## Introduction

Part geometry can be created within GibbsCAM or imported through another CAD program. GibbsCAM software can open AutoCAD .dwg files and import and export .dxf files. The SolidSurfacer option allows users to open 3D solid model files directly from Autodesk Inventor, SolidEdge, SolidWorks, and other CAD programs. These options allow GibbsCAM users to use their native CAD data for machining.

## Exercise 1 – Opening AutoCAD .dwg Files

In this exercise you will open an AutoCAD .dwg file.
*This file can be downloaded at www.schroff1.com.*

From the *Main Menu*, select *File-Open.*

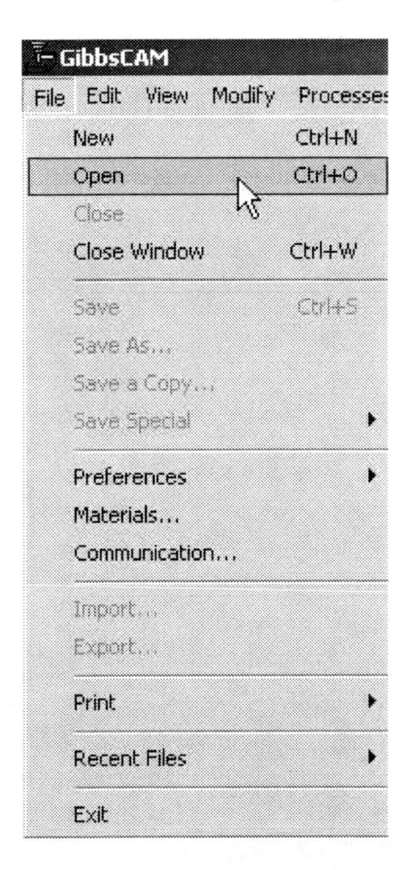

Select the File AUTOCAD_FILE.dwg. Under *Files of Type*, select DWG as shown.

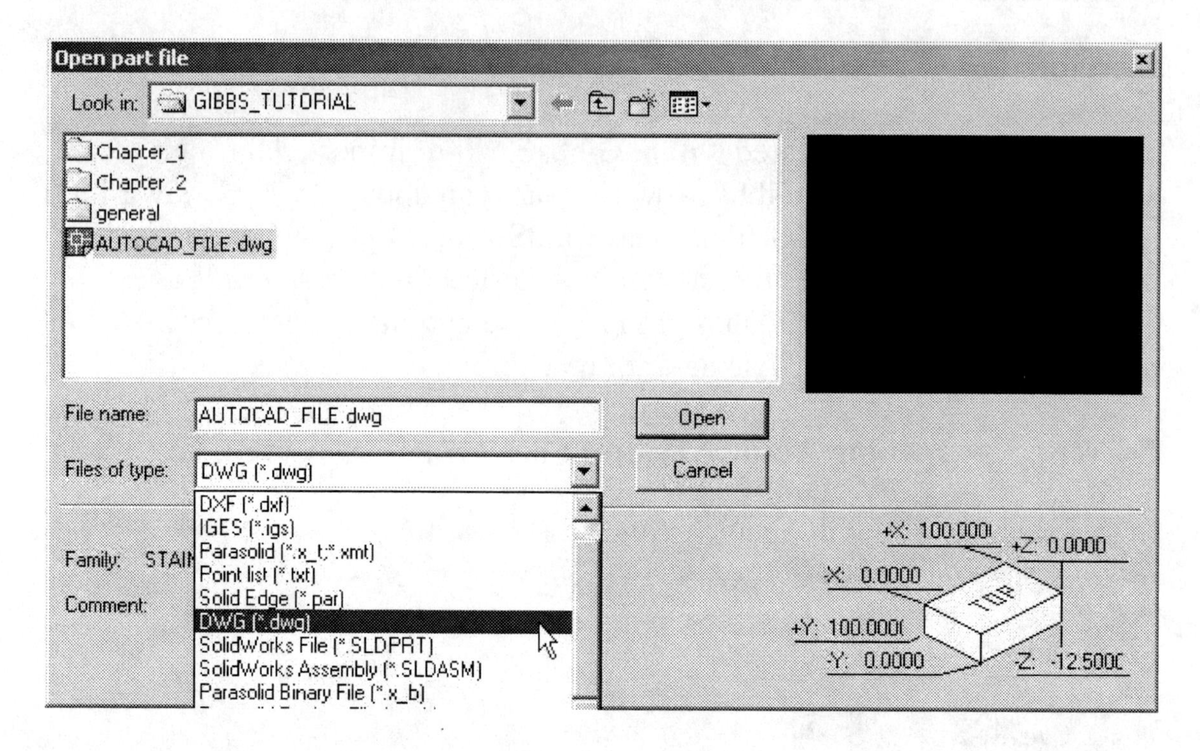


The *Solid File Open Options* dialog box will appear. Select the options shown in the following image and click on OK.

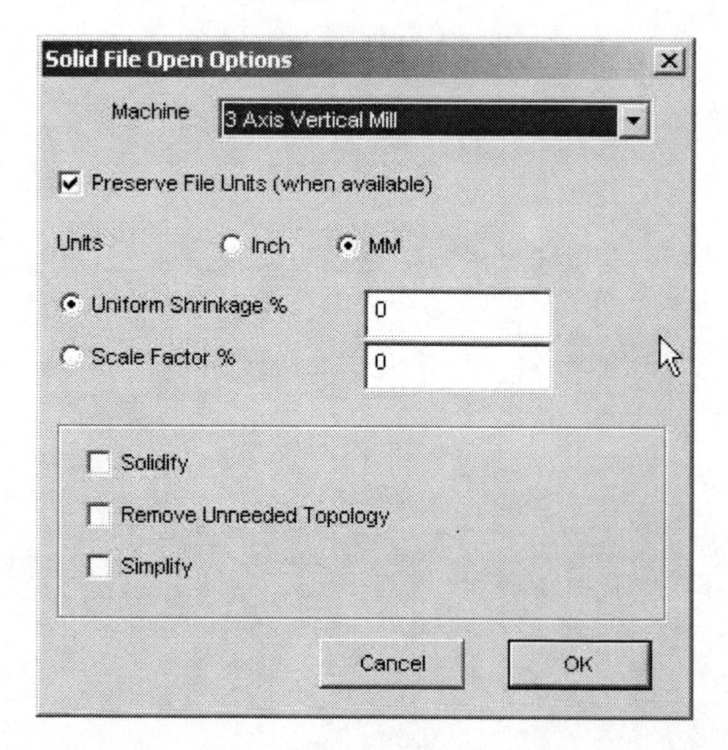

The DWG Direct dialog box will appear. Select the options as shown and click on Process.

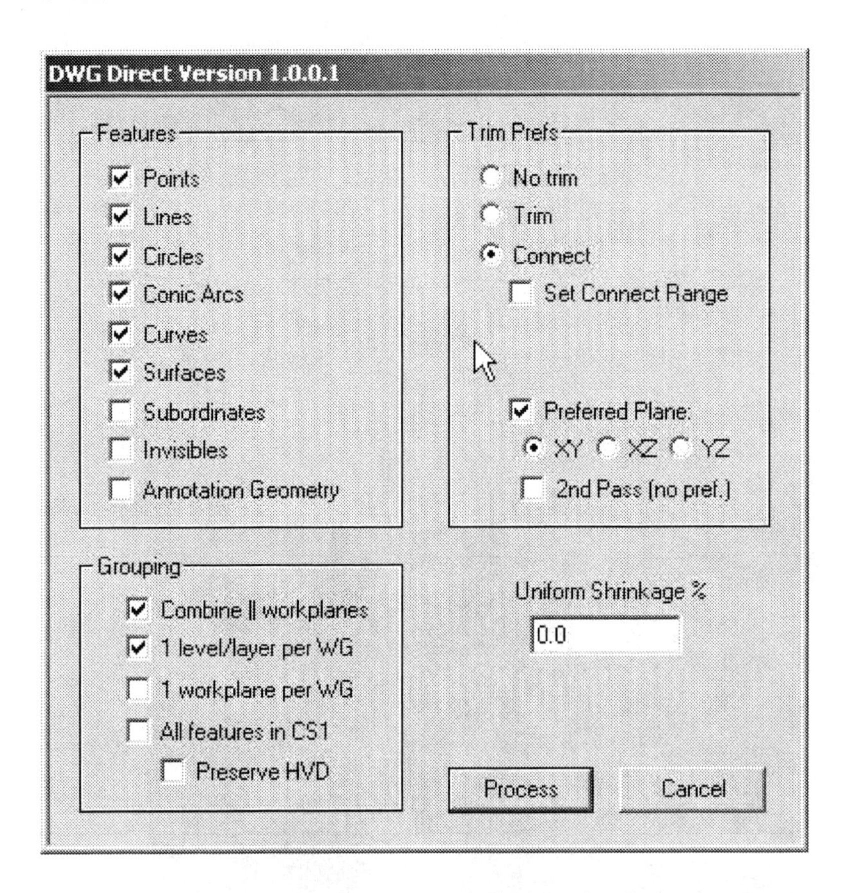


The imported geometry will appear in a new workgroup. Select the Workgroups icon from the Top Level Palette.

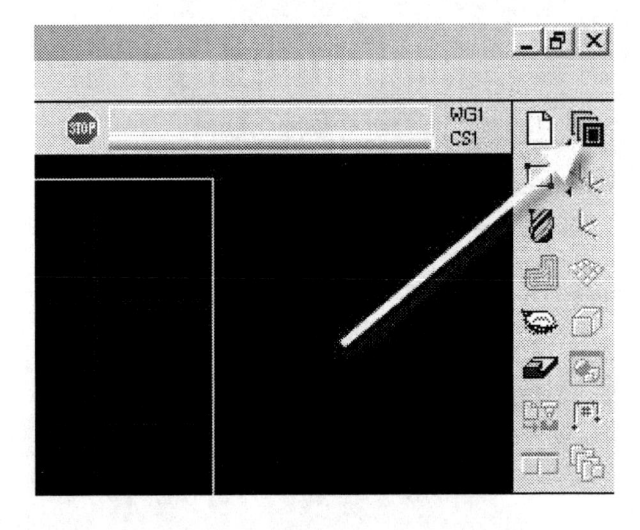

The Workgroups dialog box will appear. Click on the eyeball next to the DXF layer to open it. The imported geometry layer is now active in GibbsCAM and will turn blue.

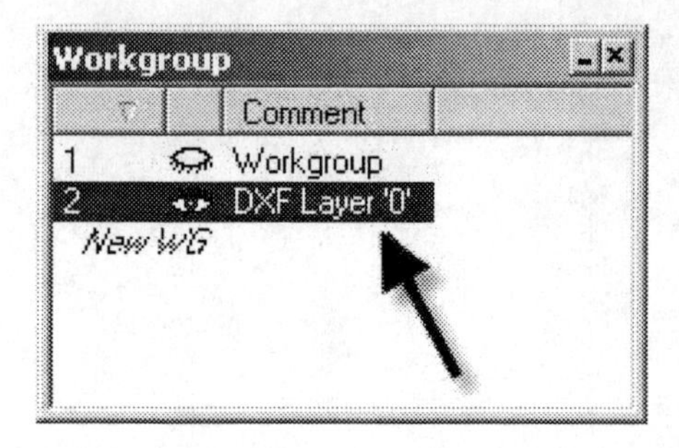


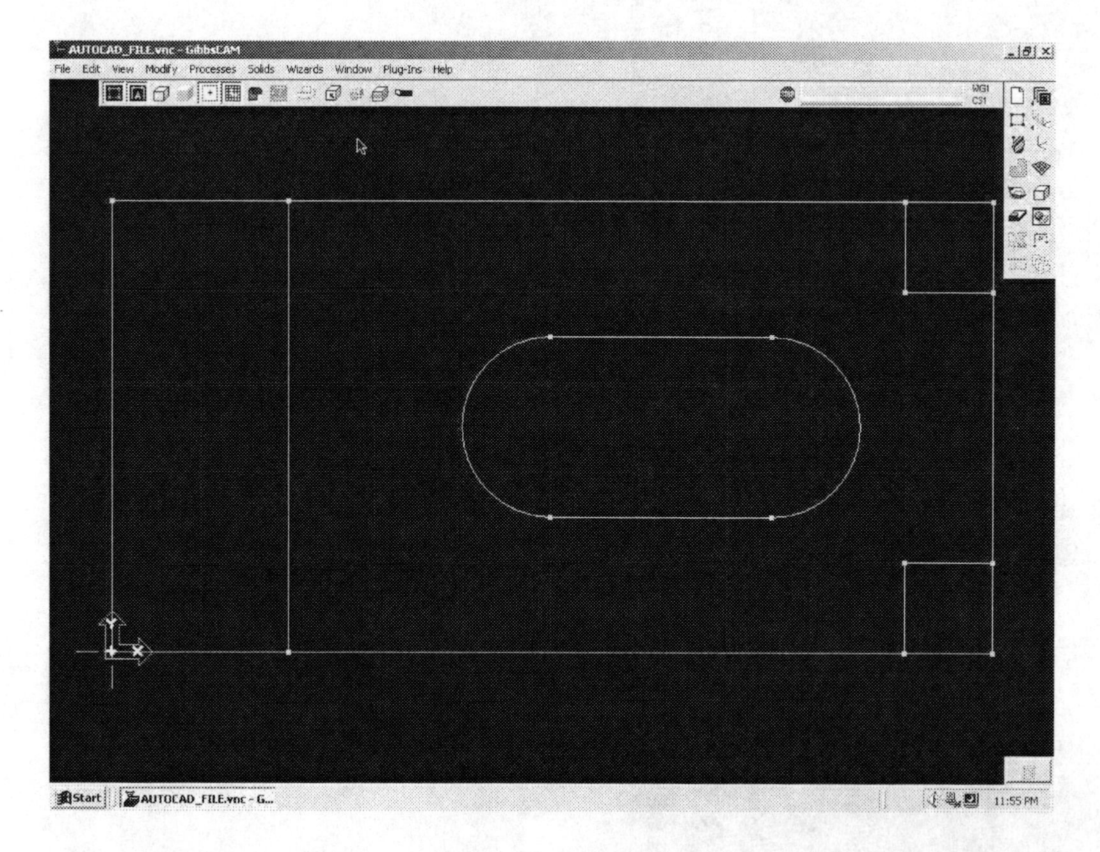


This completes the exercise. Save your file. Choose *File-Save* from the *Main Menu*.

## Exercise 2 – Extracting Geometry from a 3D CAD Solid Model

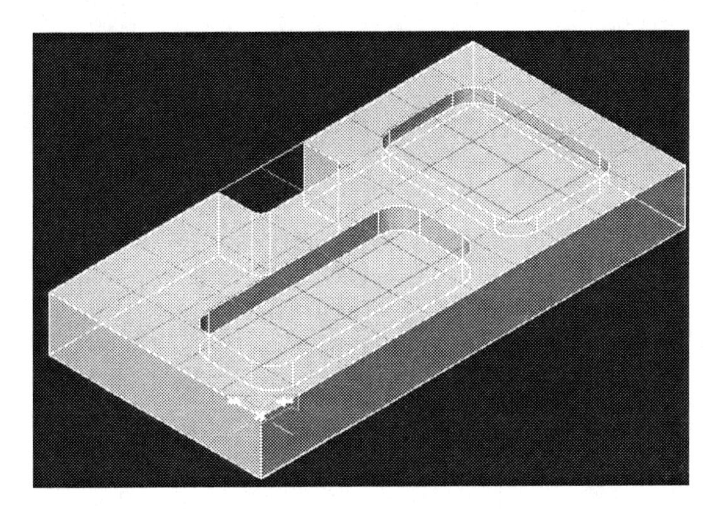

Note: You must have the SolidSurfacer option in GibbsCAM in order to complete this exercise.

In this exercise you will use two methods to extract geometry from a 3D CAD solid model. The first method involves using the *Profiler*, which acts as a moveable cross section. The second method is to extract geometry from the edges of the part itself.

You will not see the geometry from a solid model appear in a new workgroup the way that .dwg or .dxf files do.

### Method 1 – The Profiler

The solid model used in this exercise is a SolidWorks .sldprt file. GibbsCAM can open AutoDesk Inventor, MDT, IGES, Parasolid, SolidEdge, and SolidWorks files.

To open the solid model directly in GibbsCAM choose *File-Open* from the *Main Menu*. Select the pulldown arrow in the *Files of Type* dialog box and select SolidWorks files (.sldprt). Select the file SOLIDMODEL.sldprt and click on Open.

*This file can be downloaded at www.schroff1.com.*

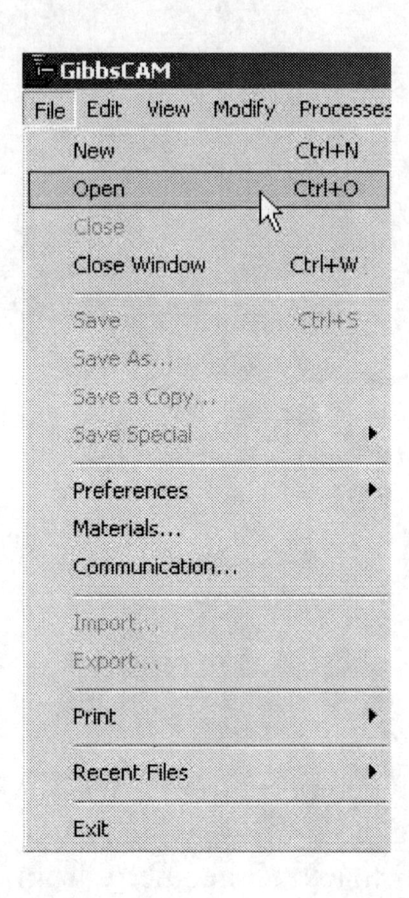


Click on the pull-down arrow in the Files of type dialog box and select SolidWorks File (*.SLDPRT). Select the file SOLIDMODEL.SLDPRT and click on Open.

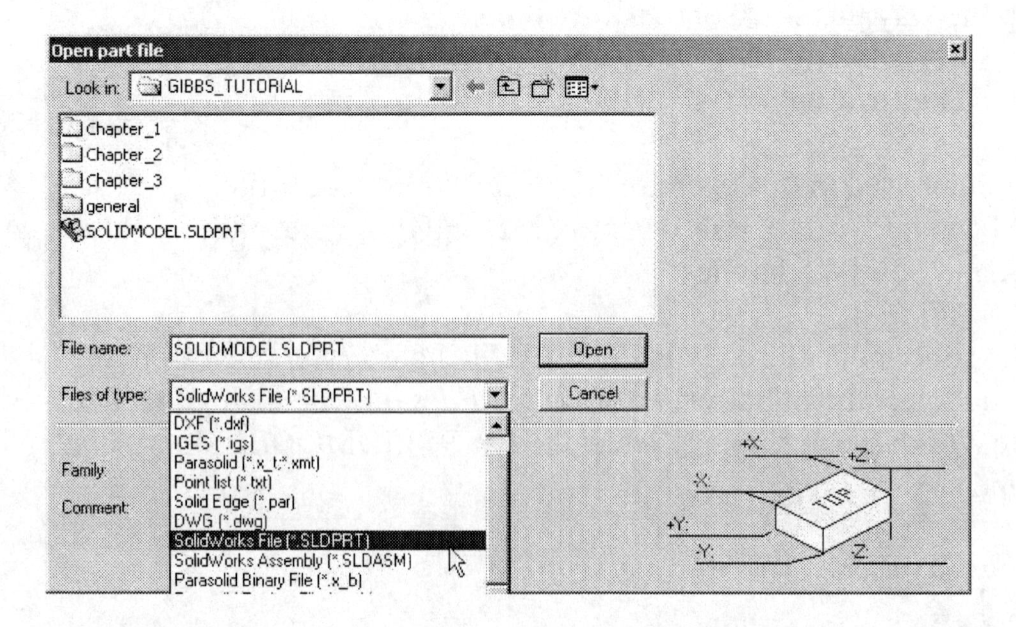

The Solid File Open dialog box will appear. Enter the values as shown and click on OK. Be sure that mm is selected as the units for this part.

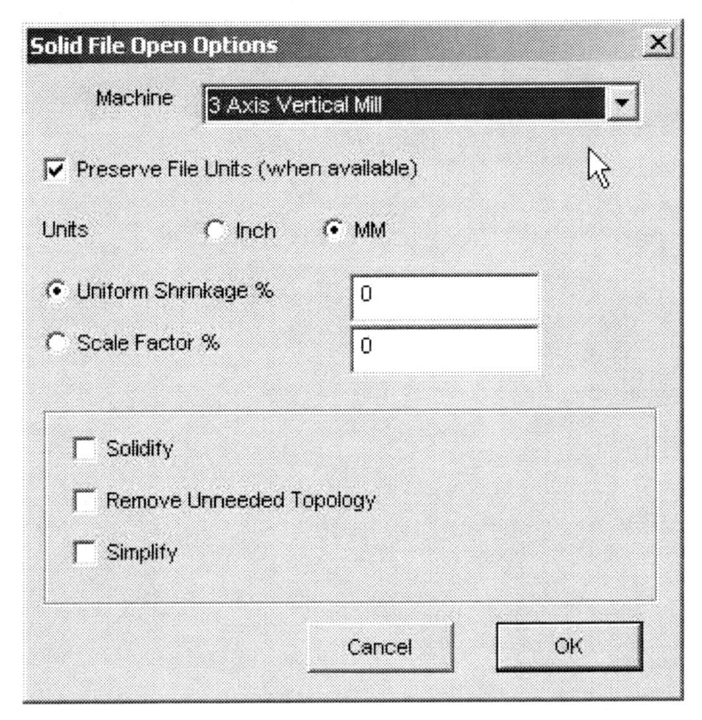


The Select Configuration dialog box will appear. This will list any configurations of the part. Select the configuration as shown in the next graphic and click on OK.

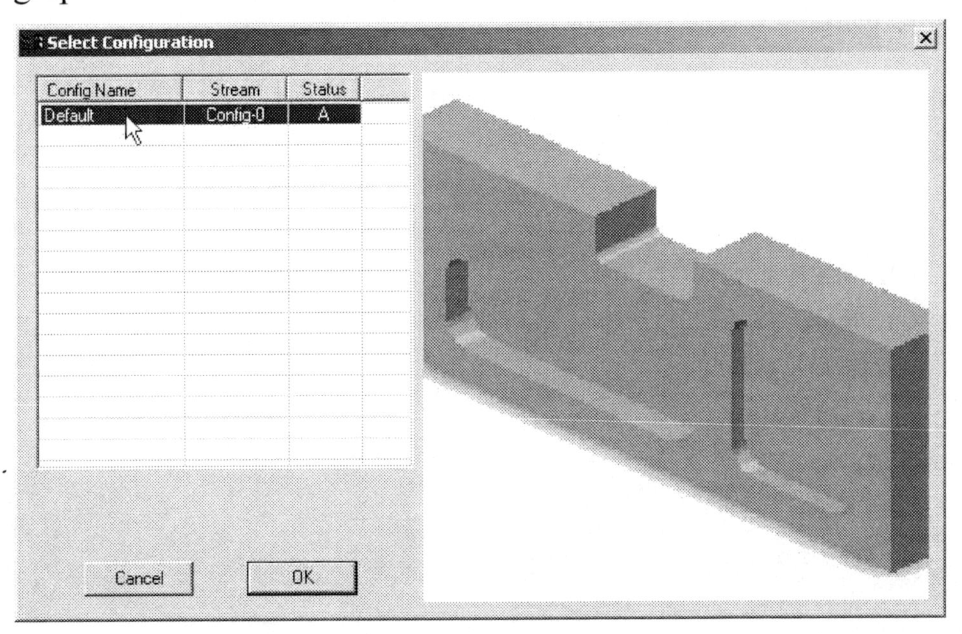


GibbsCAM will take the information from the solid model and generate the values for the stock size.

Make sure that the Show Solids icon (located under the *Main Menu*) is depressed to view the solid model.

To see an isometric view of the part, select the View Icon from the Top Level Palette.

The View Control Palette will appear.

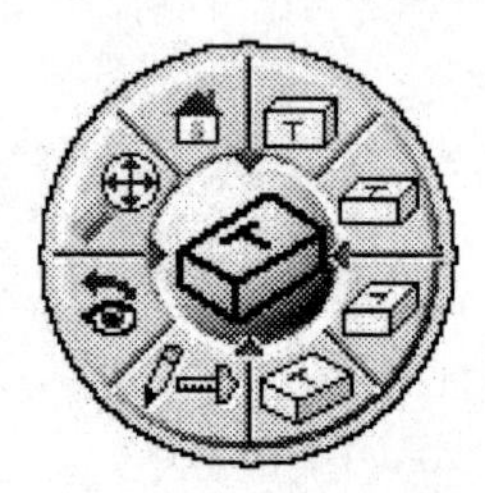

Select the Isometric View Icon.

Select the Profiler icon as shown in the next graphic.

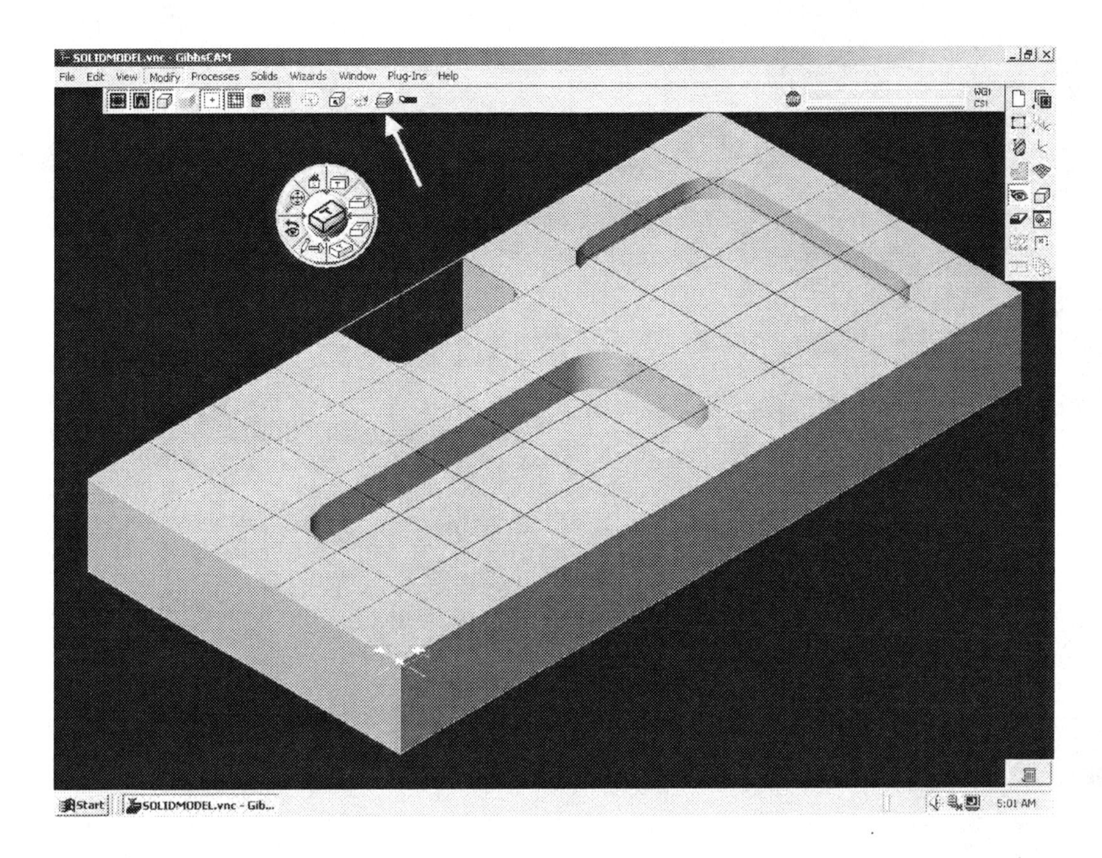


The Profiler grid will appear on the part. The Profiler acts as a cross section tool. Any geometry that is in contact with the Profiler can be selected for machining operations (except holes as of Version 7.31).

Click and drag on the green Profiler to move it up and down. Click with the right mouse button on the profile and select *Profile Depth* to move it to an exact depth as shown in the next graphic.

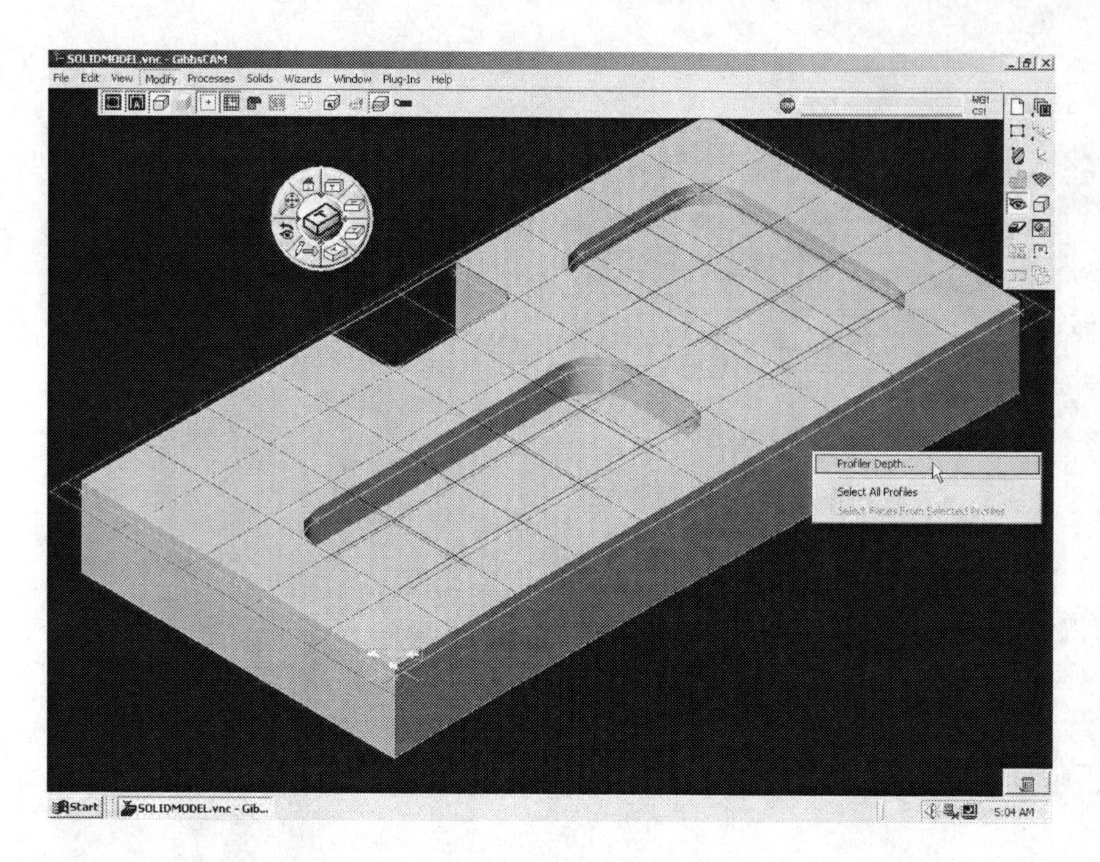


Enter –3 in the Profiler Depth dialog box and click on *Apply*.

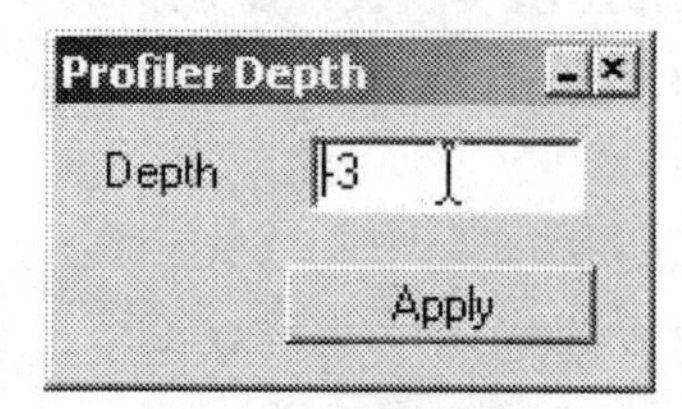


Close the Profiler Depth dialog box.

Click on the profile for the pocket as shown in the next graphic and notice that it will turn blue. This means that it can be selected for machining operations.

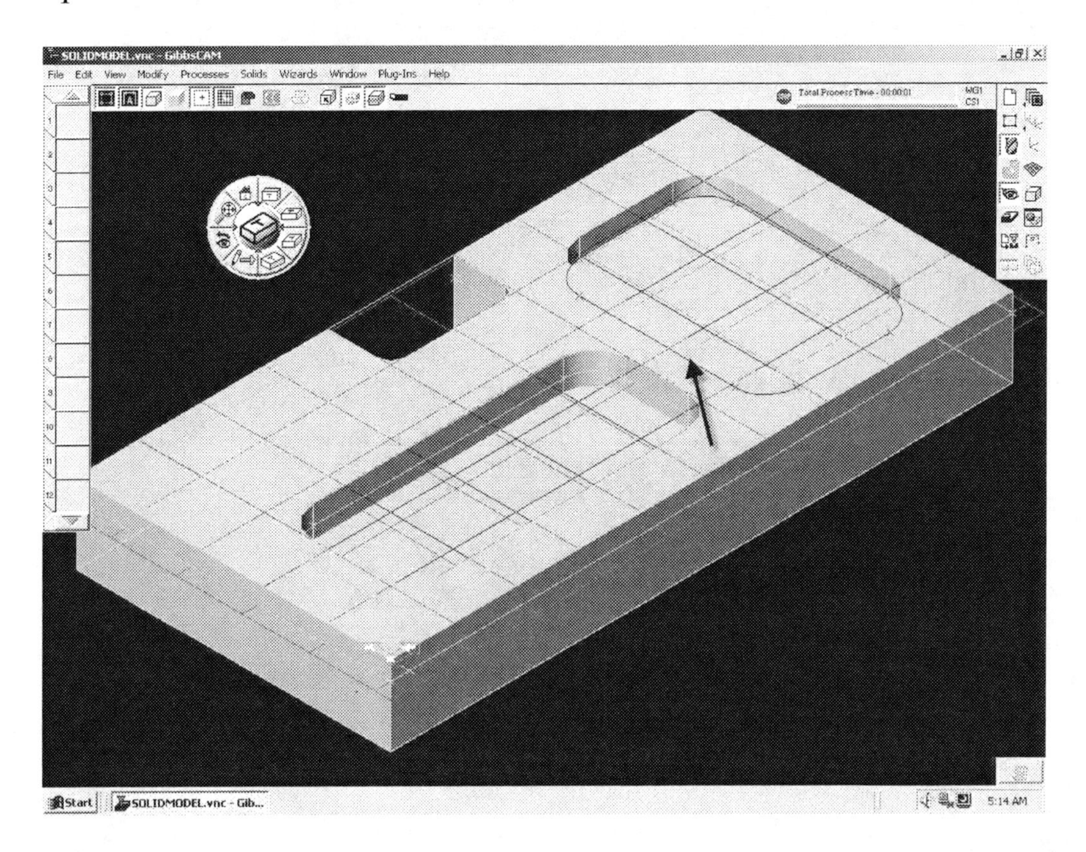


**Method 2 – Extraction**

Turn off the Profiler by clicking on the icon so that it is not depressed.

Select the Edge Selection icon (located under the *Main Menu*) to allow part edges to be selected. This will be active when this icon is depressed.

You will now extract the geometry from the solid model. Click on the Geometry icon from the Top Level Palette.

The Geometry Palette will appear.

Select the Geometry From Solids Icon.

The Solids Palette will appear.

Select the Geometry Extraction icon.

You will first extract the profile for the cutout. Click and drag a box around the cutout to zoom in on that area. Hold down the control key and select all of the segments of the outline of the cutout as shown. Click on **Do It**. This will extract the profile as a connected entity. The profile should turn blue.

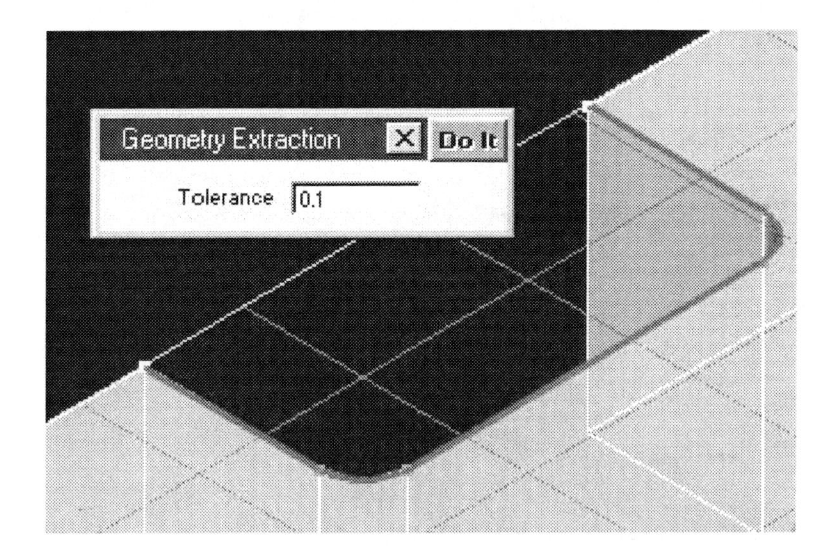


Leave the Geometry Extraction dialog box open. Return to the previous view. Select the Previous View icon from the View Palette or *View – Unzoom* from the *Main Menu*.

Now you will extract the profile for a pocket. Zoom in on the pocket shown. Hold down the control key and select all of the segments of the profile. Click on **Do it**. This will extract the profile as a connected entity. The profile should turn blue.

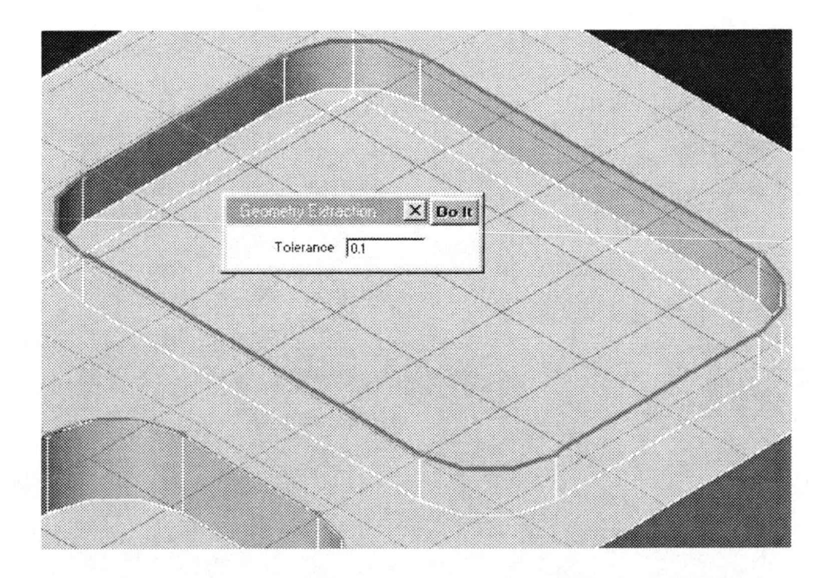

Return to the previous view.

Zoom in on the other pocket as shown. Hold down the control key and select all of the segments of the profile. Click on **Do it**. The profile will be extracted as a connected entity and turn blue.

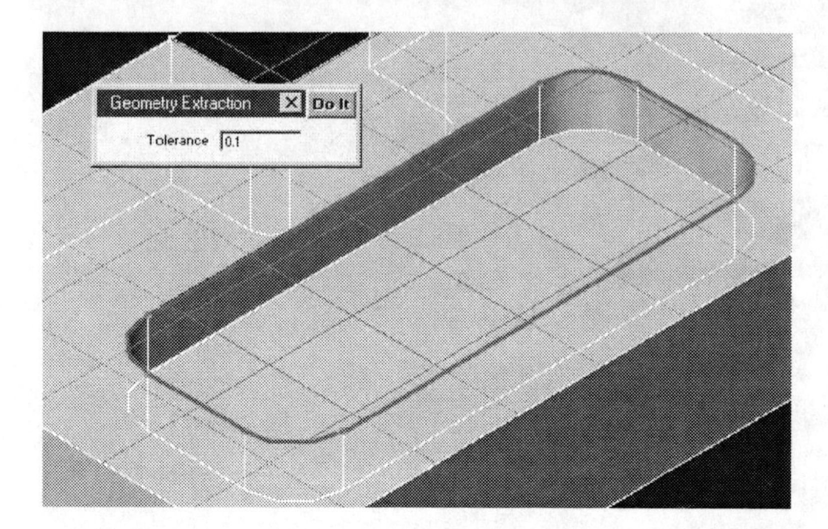


The profiles needed to machine this part have been extracted. This completes the exercise. Save the file. Choose *File-Save* from the *Main Menu*. This will save the file as a GibbsCAM .vnc file.

Chapter 4
# Mill

## Introduction

For each exercise in this chapter you will open a part file that contains CAD geometry. You will then set up the tools needed to machine the part, create machining processes, simulate them, and post process the machining operations to generate CNC code.
*Part files for the exercises in this chapter can be downloaded at www.schroff1.com.*

**Note: For the exercises in this chapter, activate balloons (found under *Help* in the *Main Menu).* This will allow you to view a full explanation of all of the data entry fields you will see in the tool creation and machining dialog boxes as you complete the exercises.**

## Exercise 1 – Drilling

In this exercise you will drill holes for the points you created in Chapter 2 – Exercise 2.

Choose *File-Open* from the *Main Menu.*

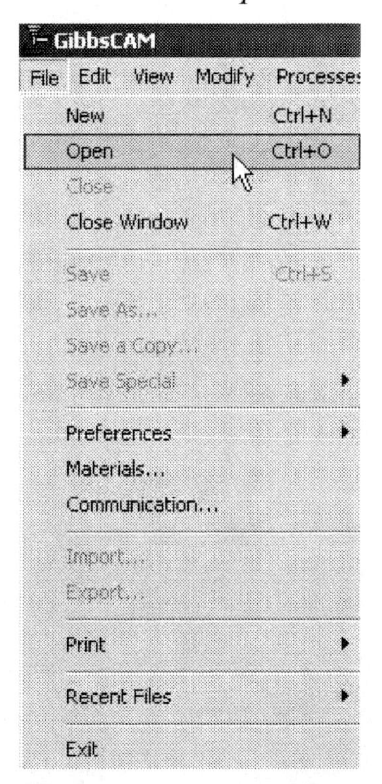

Select the file POINTS.vnc and click on Open. The CAD geometry for this part consists of points that will be used for drilling operations.

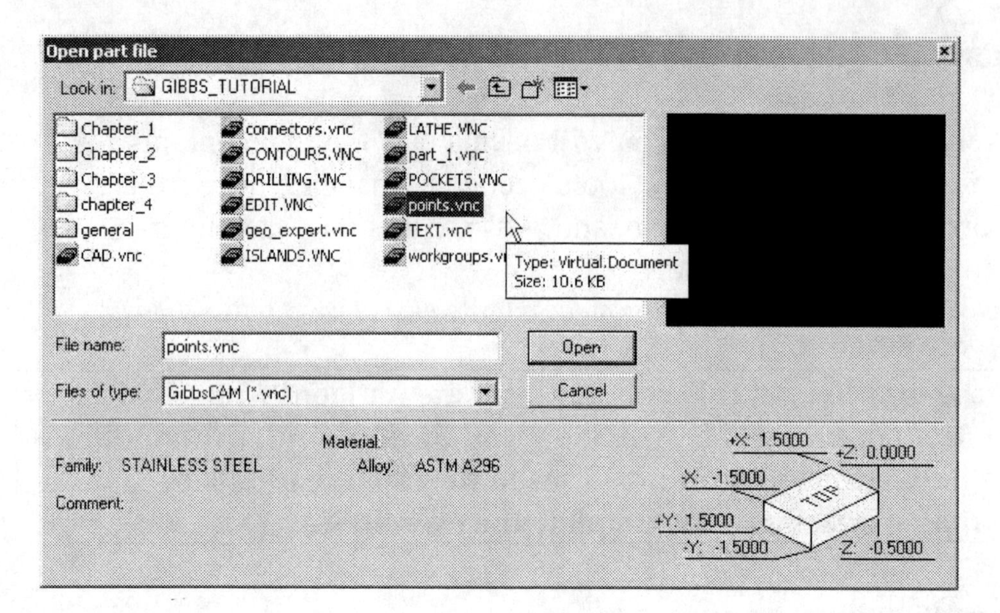


Select the Tools icon in the Top Level Palette to open up the Tool List.

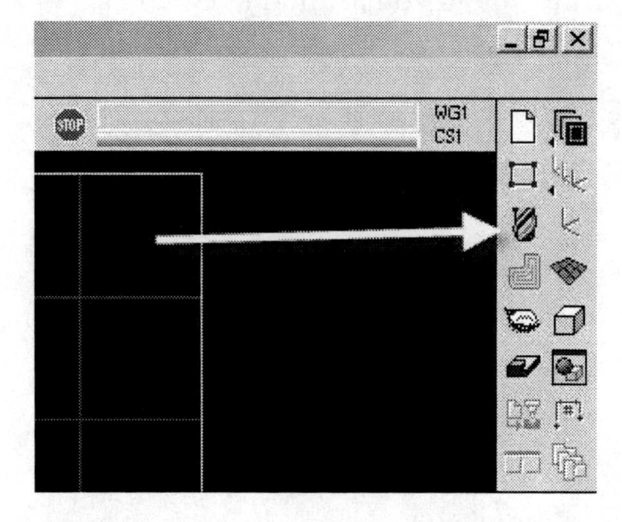


Double click on Tile 1 in the Tool List to create the first tool.

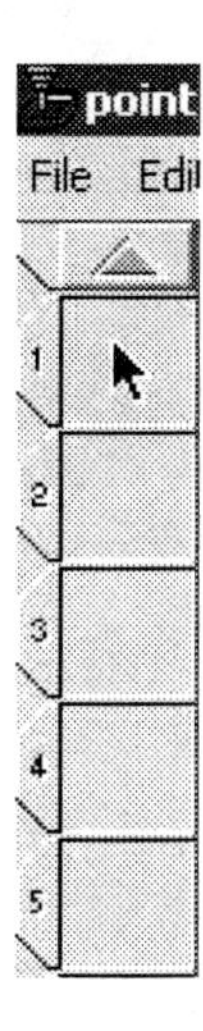


The Tool Creation dialog box will appear. Select the Spot D (spot drill) as the tool type. Enter the values in the diagram to define the size of the tool.

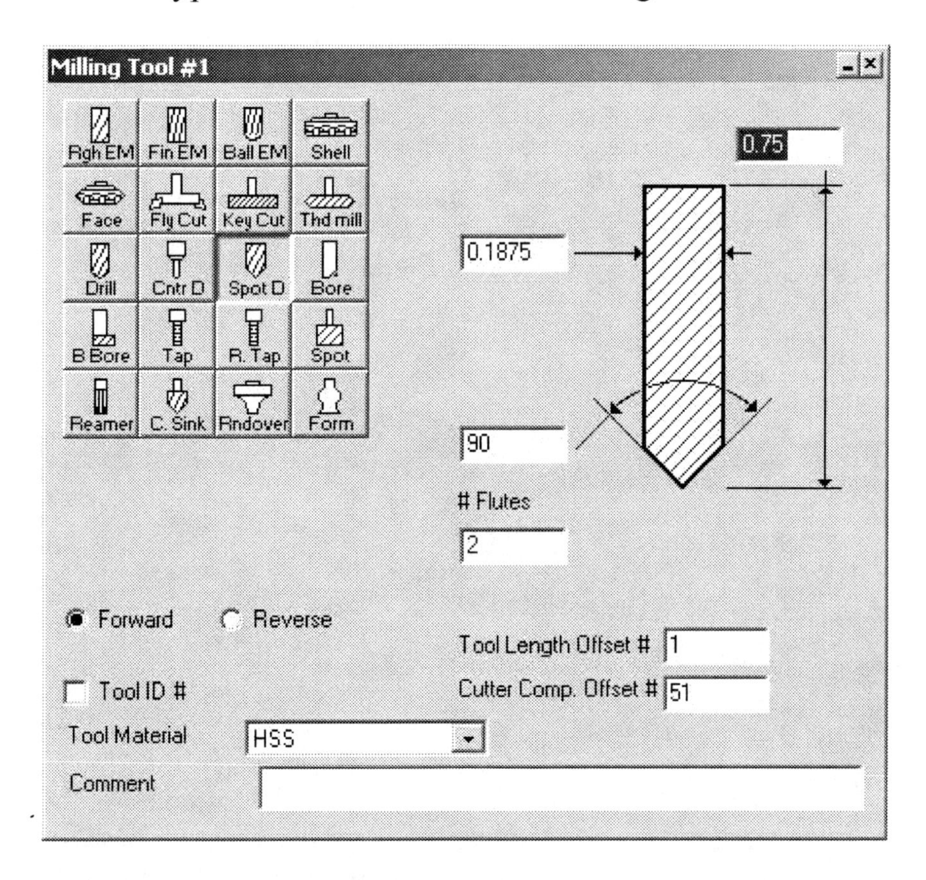


Close the Tool 1 dialog box.
The tool will appear in the Tool List.

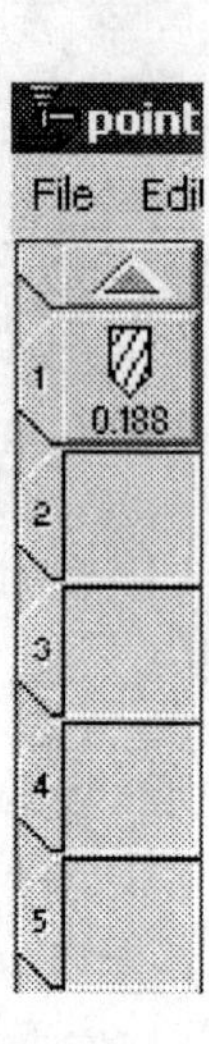


Double click on Tile 2 in the Tool List to create the second tool.

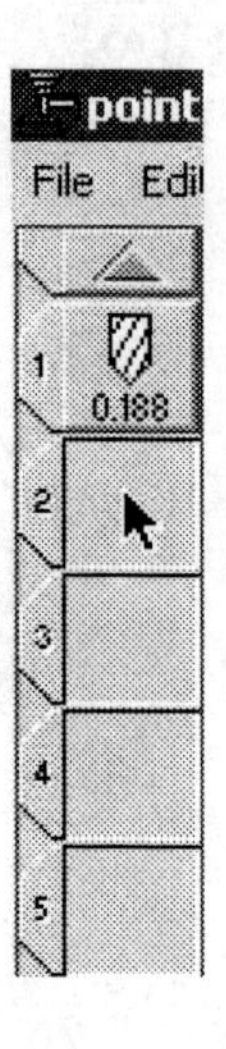

The Tool Creation dialog box will appear. Select Drill as the tool type. Enter the values shown on the next graphic to define the size of the tool.

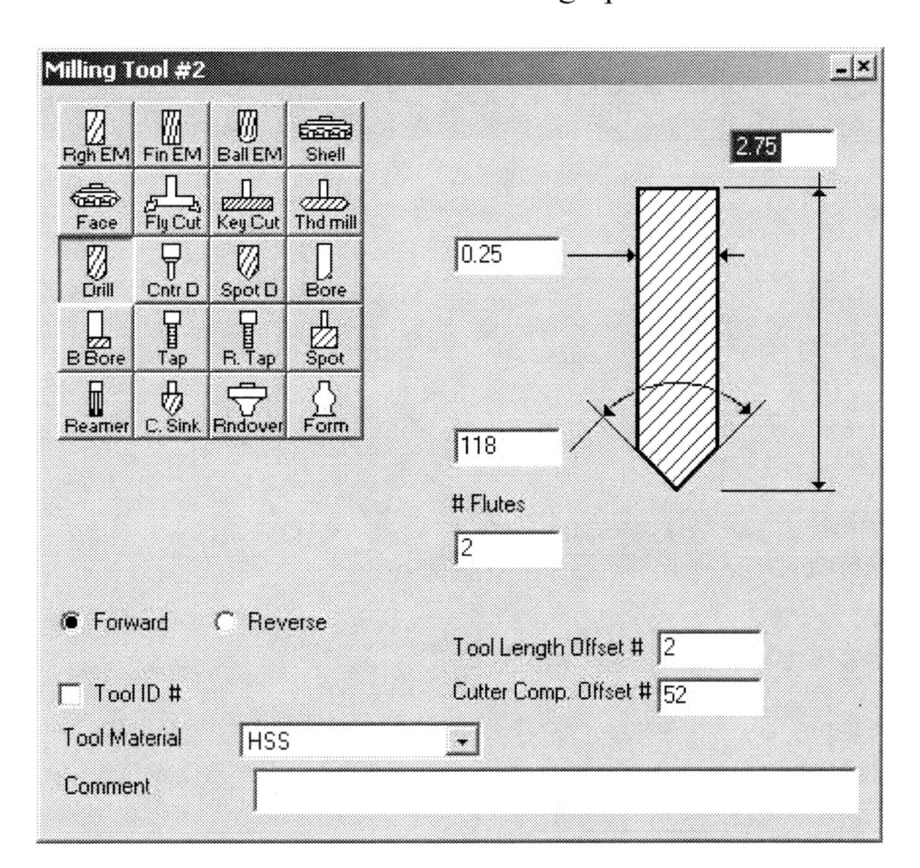


Close the Tool 2 dialog box. The tool will appear in the Tool List.

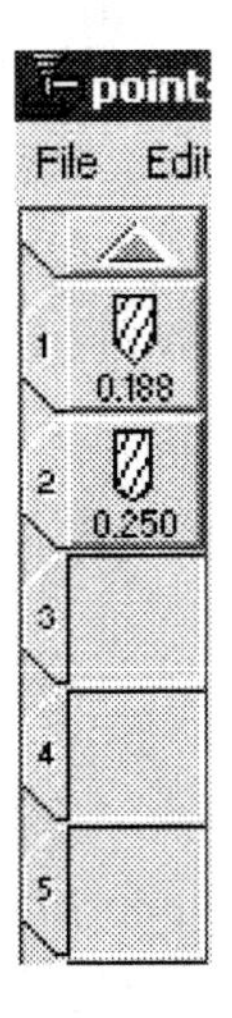

Double-click on Tile 3 in the Tool List.

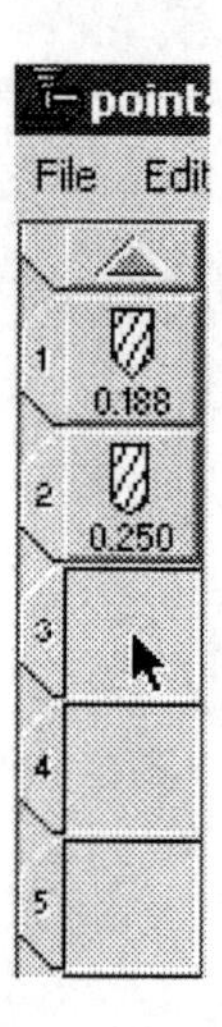


The Tool Creation dialog box wil appear. Select C. Sink (counter sink) and enter the values as shown to define the size of the tool.

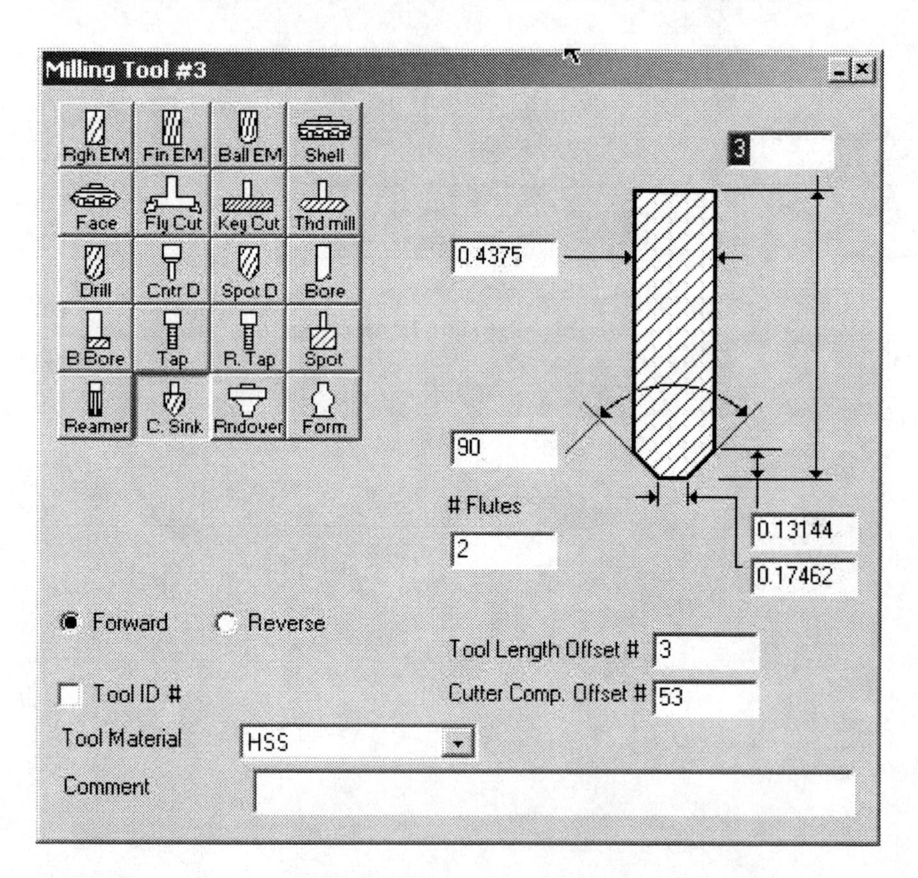


Close the dialog box. Tool 3 will now appear in the Tool List.

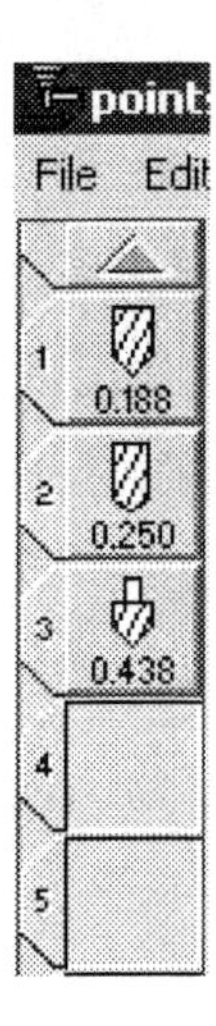


You are now ready to begin the machining operations. Select the CAM icon on the Top Level Palette to open the Machining Palette, Process List, and Operations List.

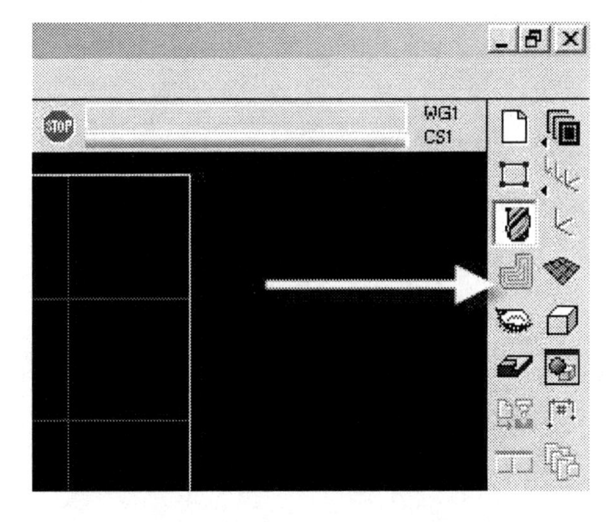


Machining operations are created by dragging a tool from the Tool List and a machining function from the Machining Palette on to the Process List.

Drag Tool 1 from the Tool List to Tile 1 of the process list as shown in the next graphic.

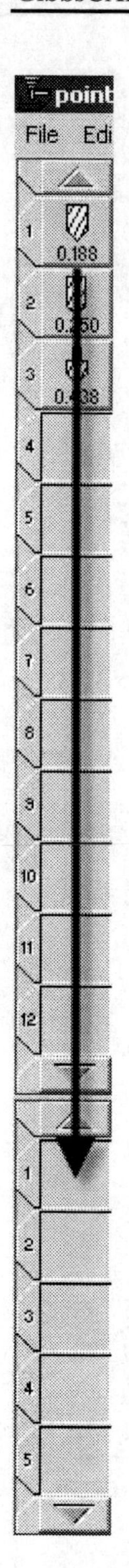


Tool 1 will now appear in the Process List. Drag the Holes icon from the Machining Palette to Tile 1 of the process list as shown in the next graphic.

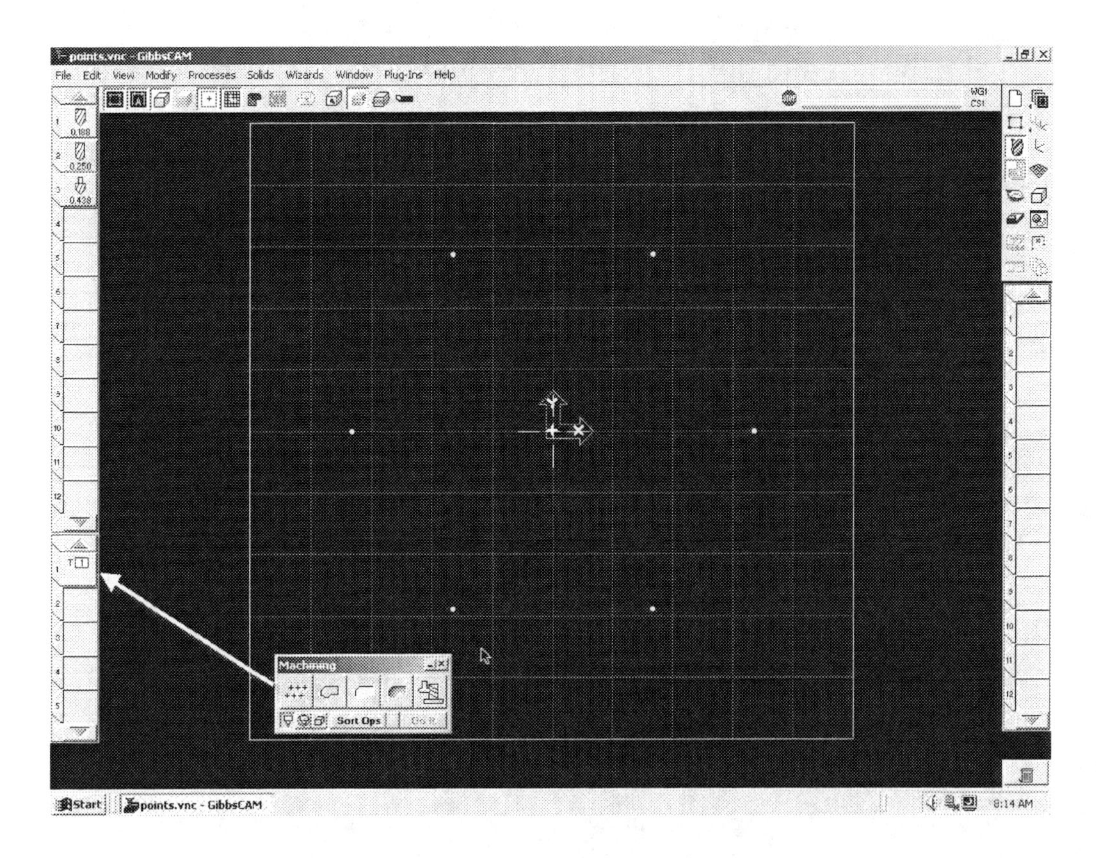


The Process Holes dialog box will appear. Enter the values as shown.

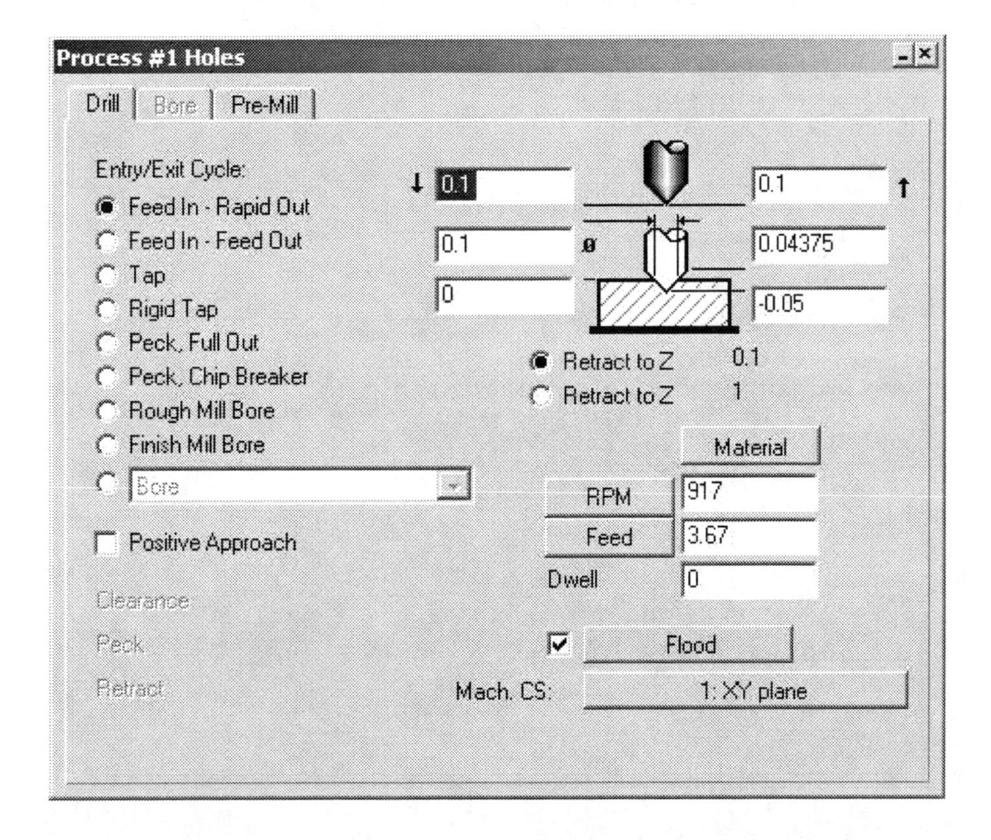

Close the dialog box. Hold down the shift key and drag a box around the points as shown.

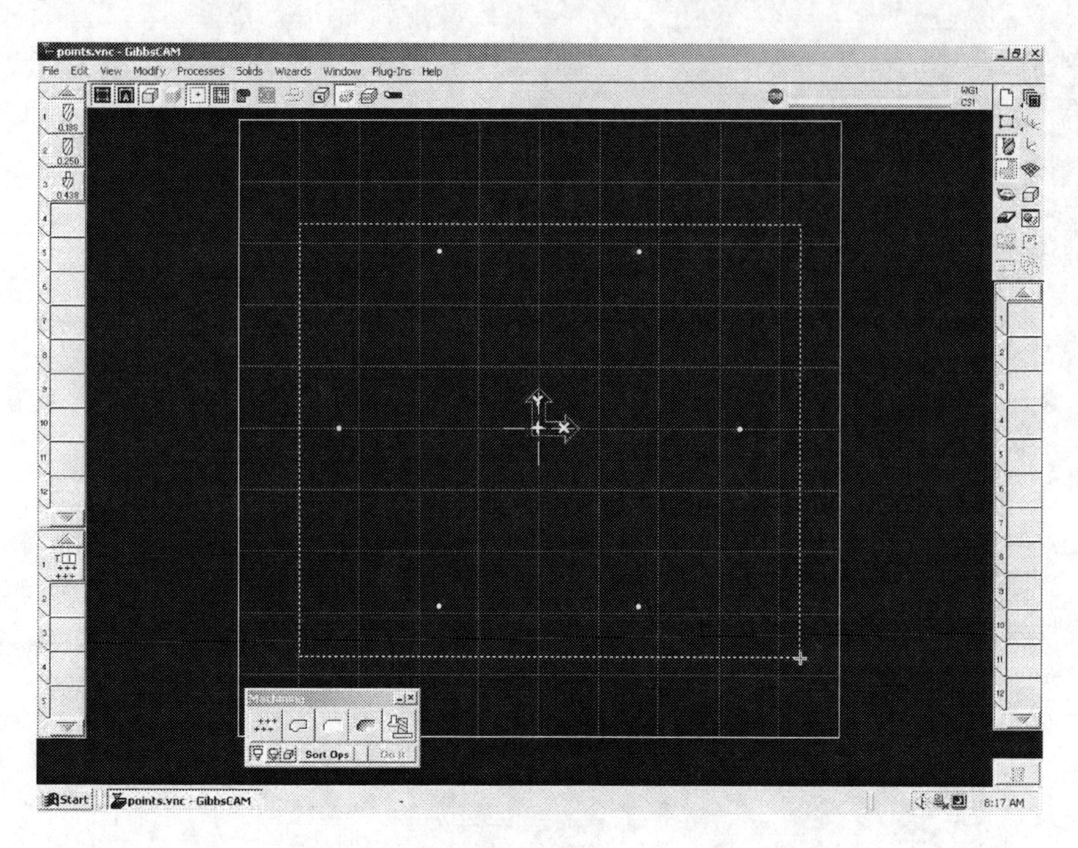


Click on Do It on the Machining Palette.

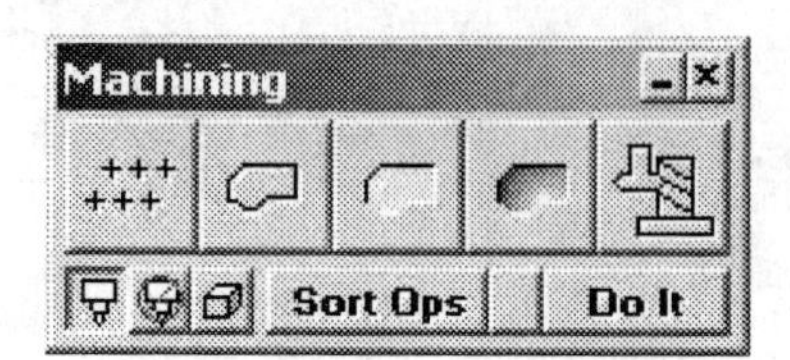


Change your view to an isometric view to see the toolpath.

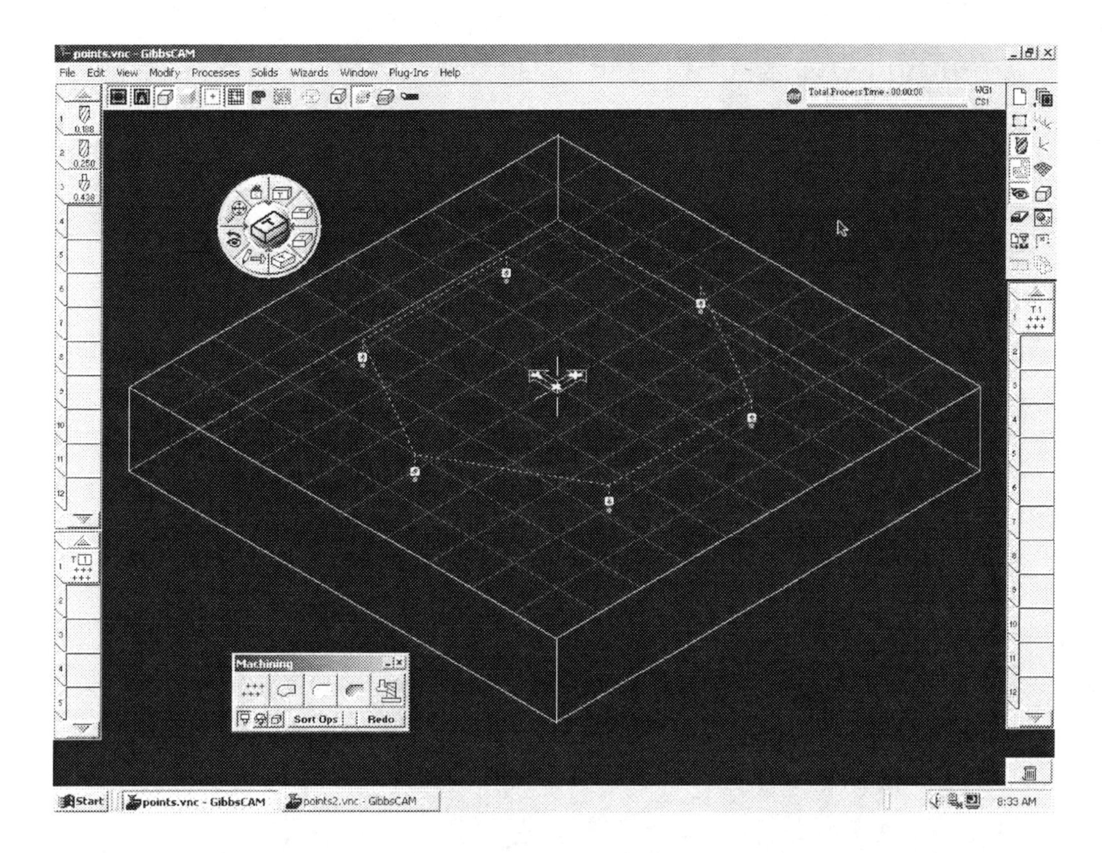


To begin the next operation, select Tile 2 in the Operations List.

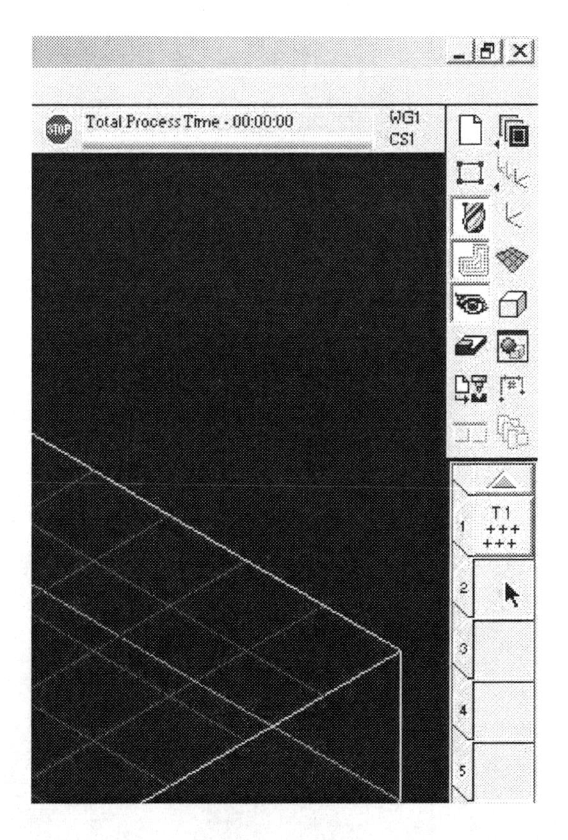

Drag Tile 1 from the Process List to the Trash Can.

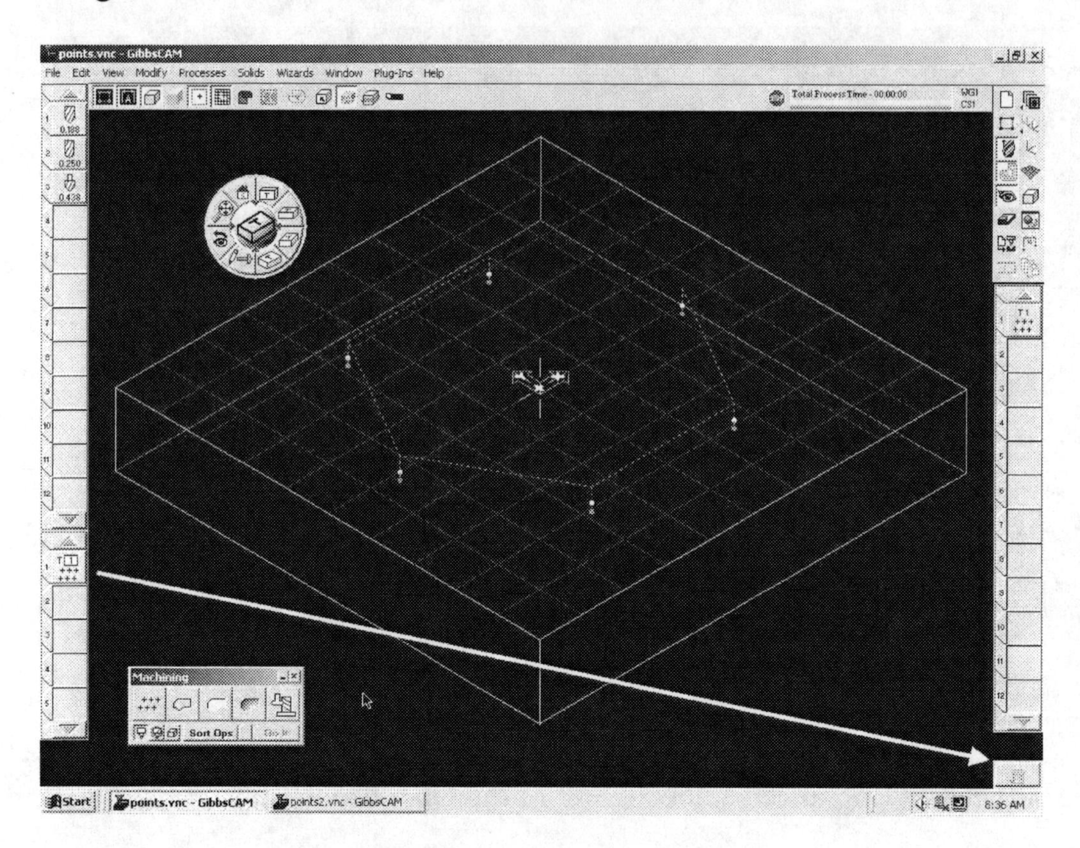


Click on the Render Icon on the Top Level Palette.

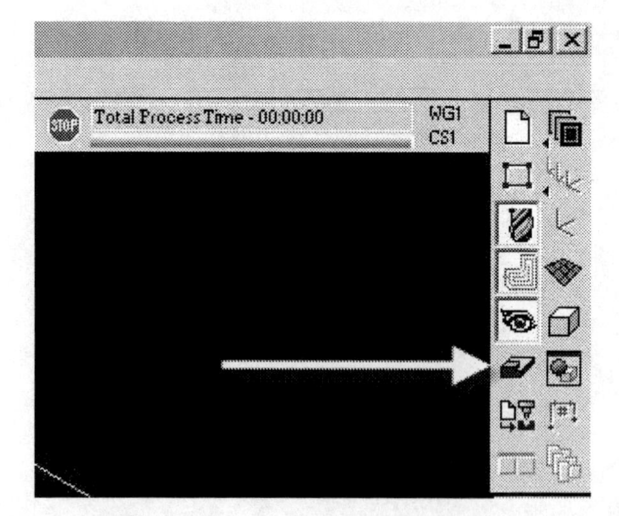

The Render Control dialog box will appear. Click on the Visible Tool icon as shown. Hit the Stop button and then the Rewind button to reset the simulation. Hit the Play button to run the simulation.

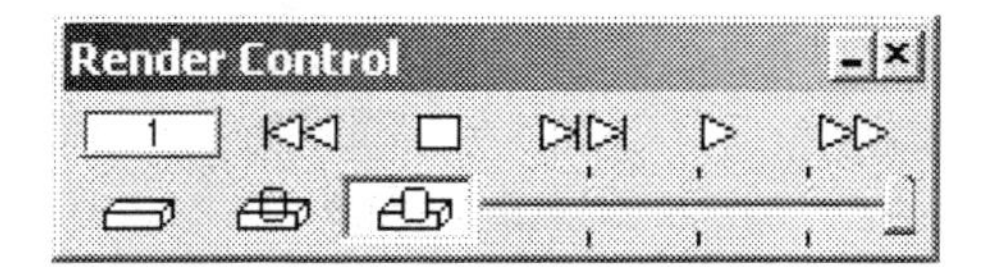


Note: You can drag the slider + or – to speed up or slow down the simulation.

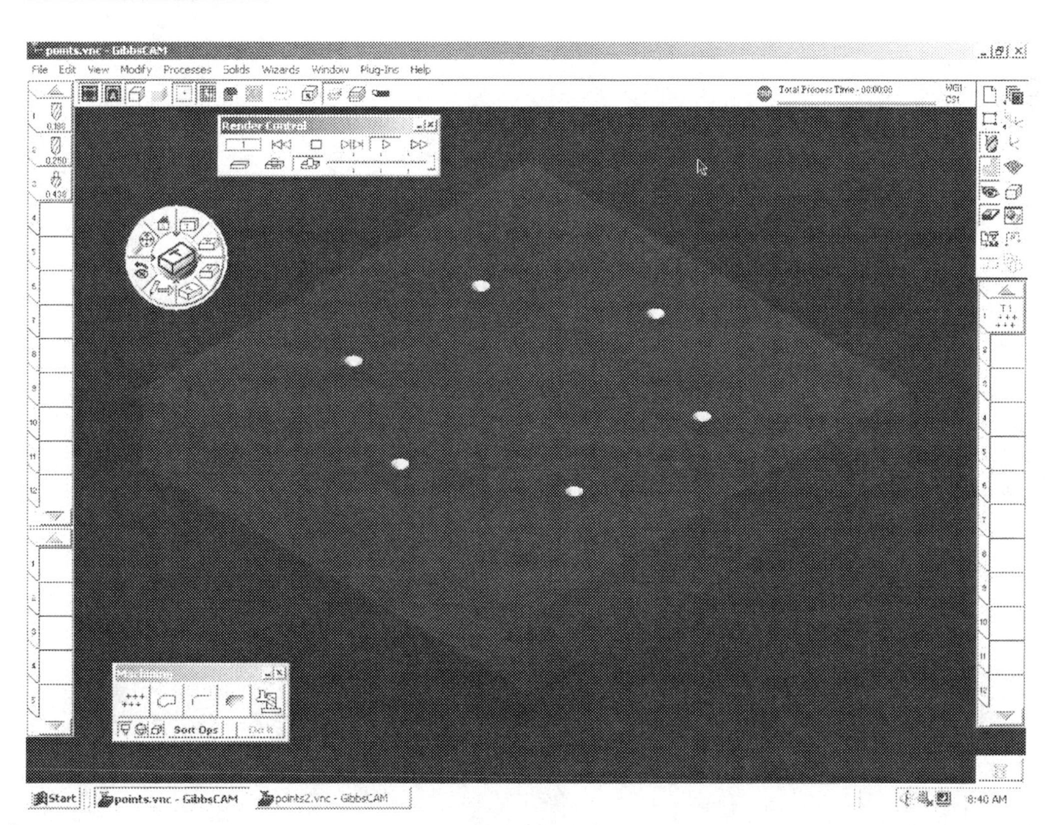


Close the Render Control dialog box.

Drag the Tool 2 from the Tool List to Tile 1 of the Process List.

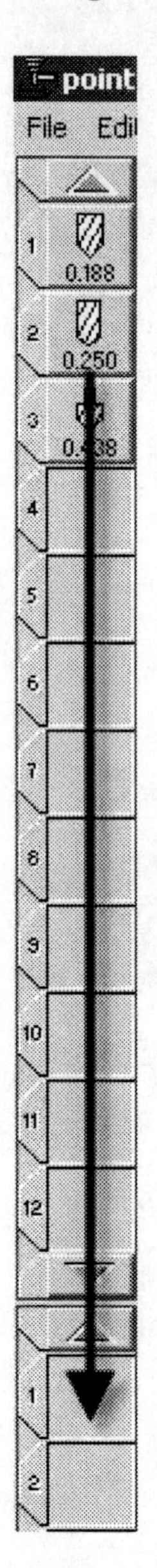


Drag the Drilling Icon from the Machining Palette to Tile 1 of the Process List.

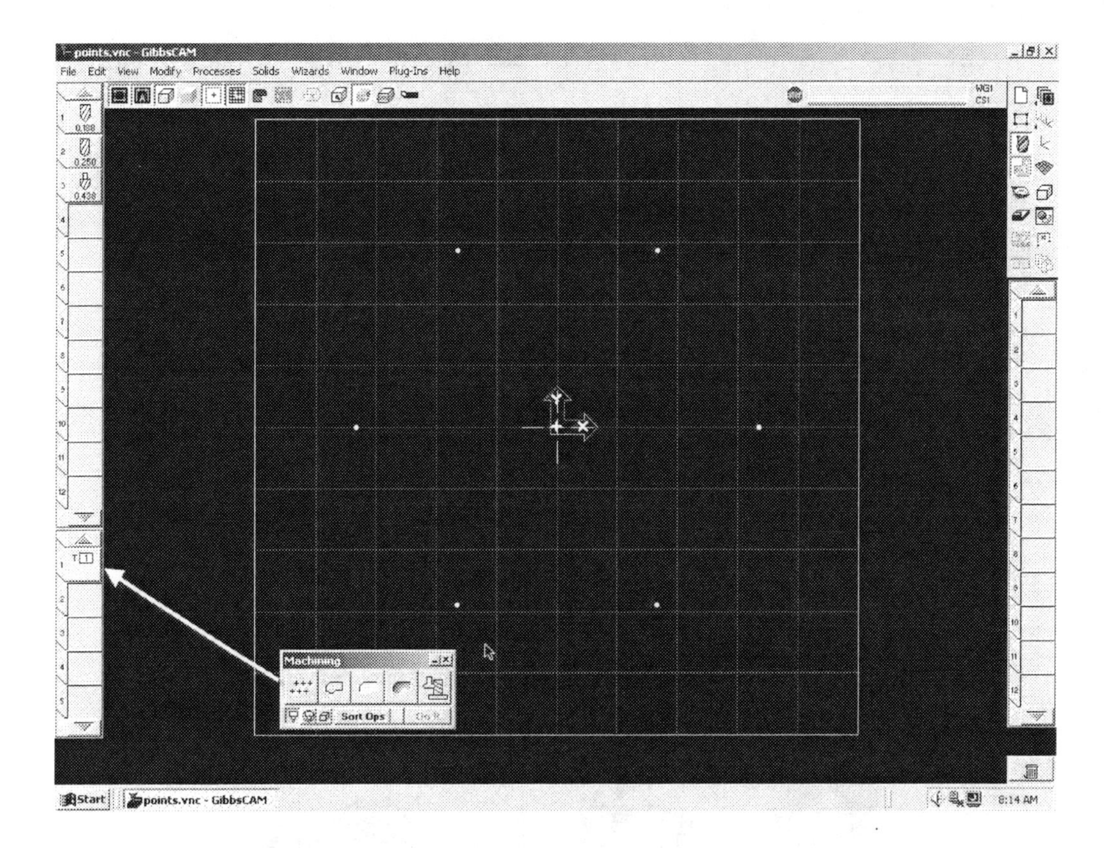


The Process Holes dialog box will appear. Enter the values as shown.

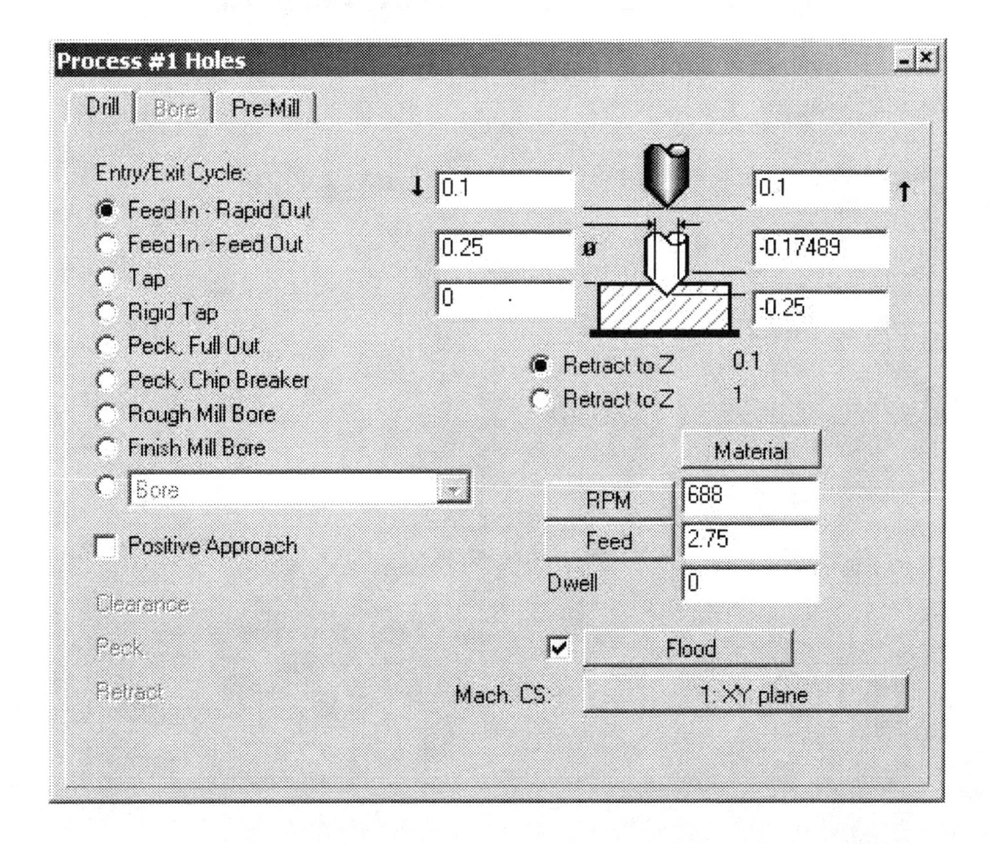

Close the dialog box. Hold down the shift key and drag a box around the points to select them.

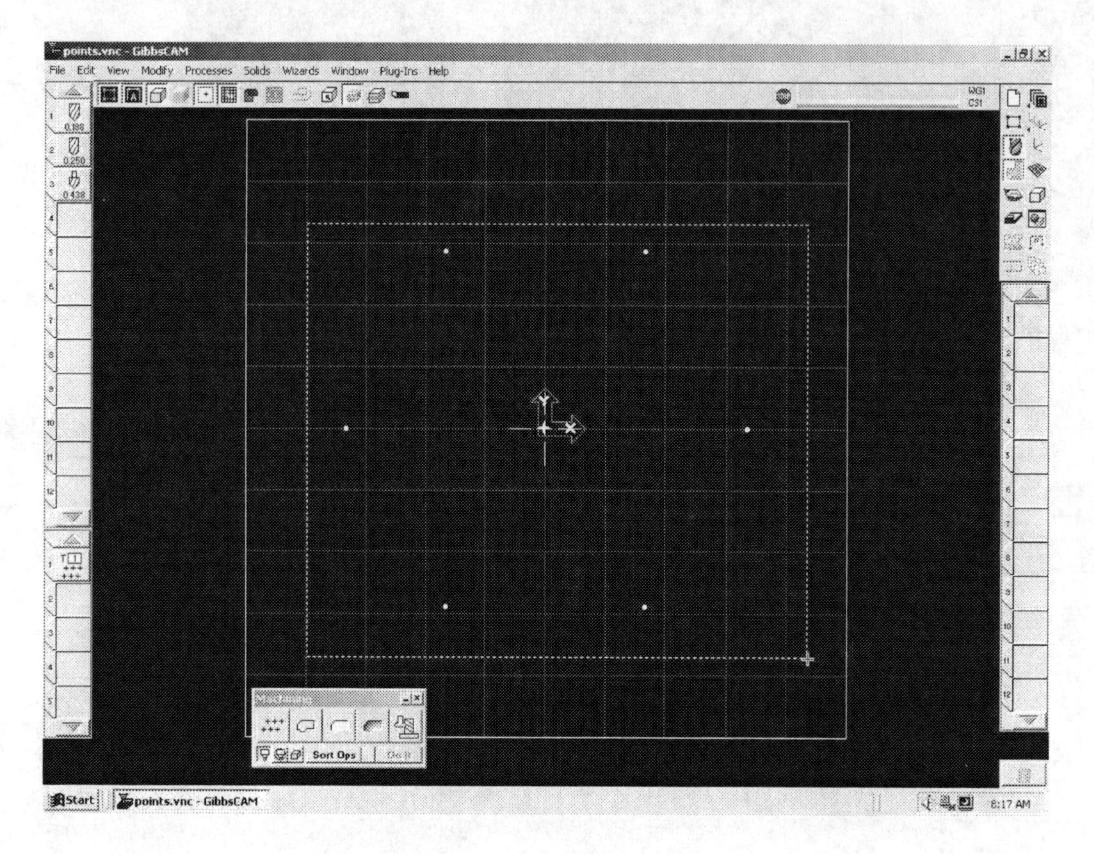


Click on Do It in the Machining Palette.

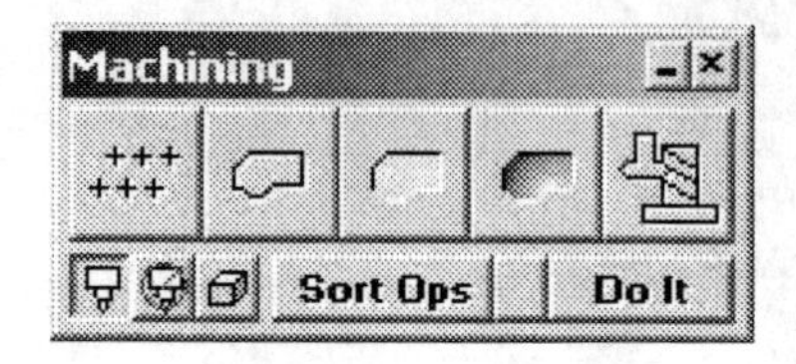


Render your part.

Select Tile 3 in the Operations List to begin the next operation.

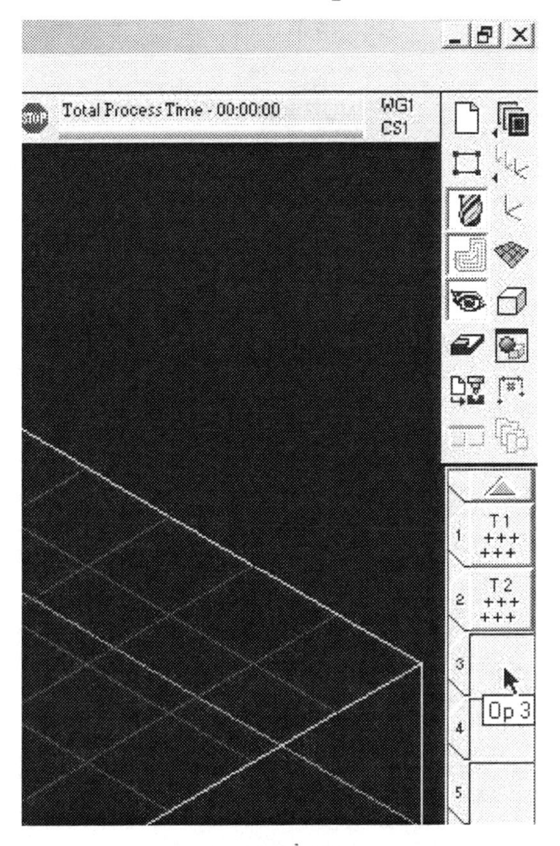


Drag Tile 1 from the Process List to the Trash Can.

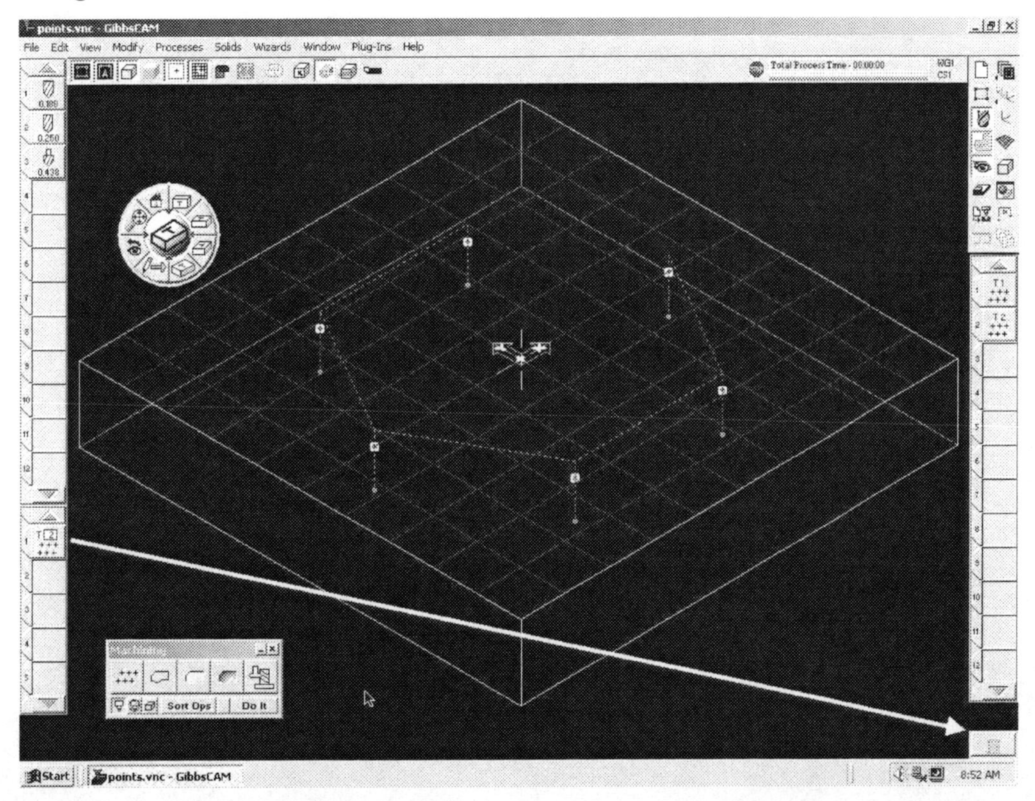

Drag Tool 3 from the Tool List to Tile 1 of the Process List. Drag the Holes icon from the Machining Palette to Tile 1 of the Process List.

The Process Holes dialog box will appear. Enter the values as shown.

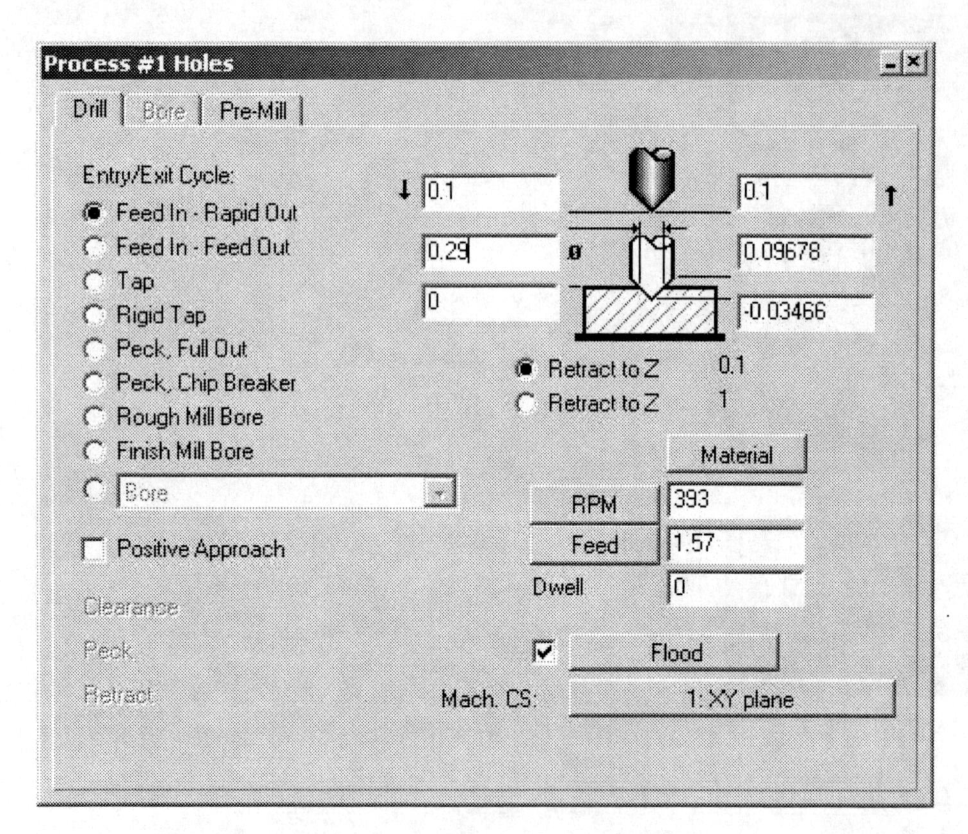


Close the dialog box. Hold down the shift key and drag a box around the points to select them. Click on Do It in the Machining Palette.

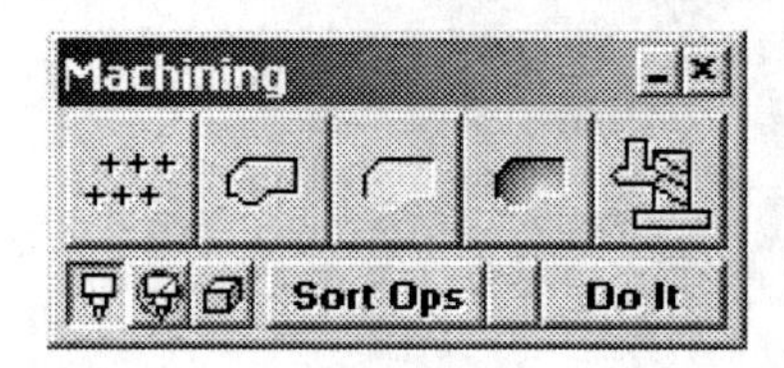


Render your part.

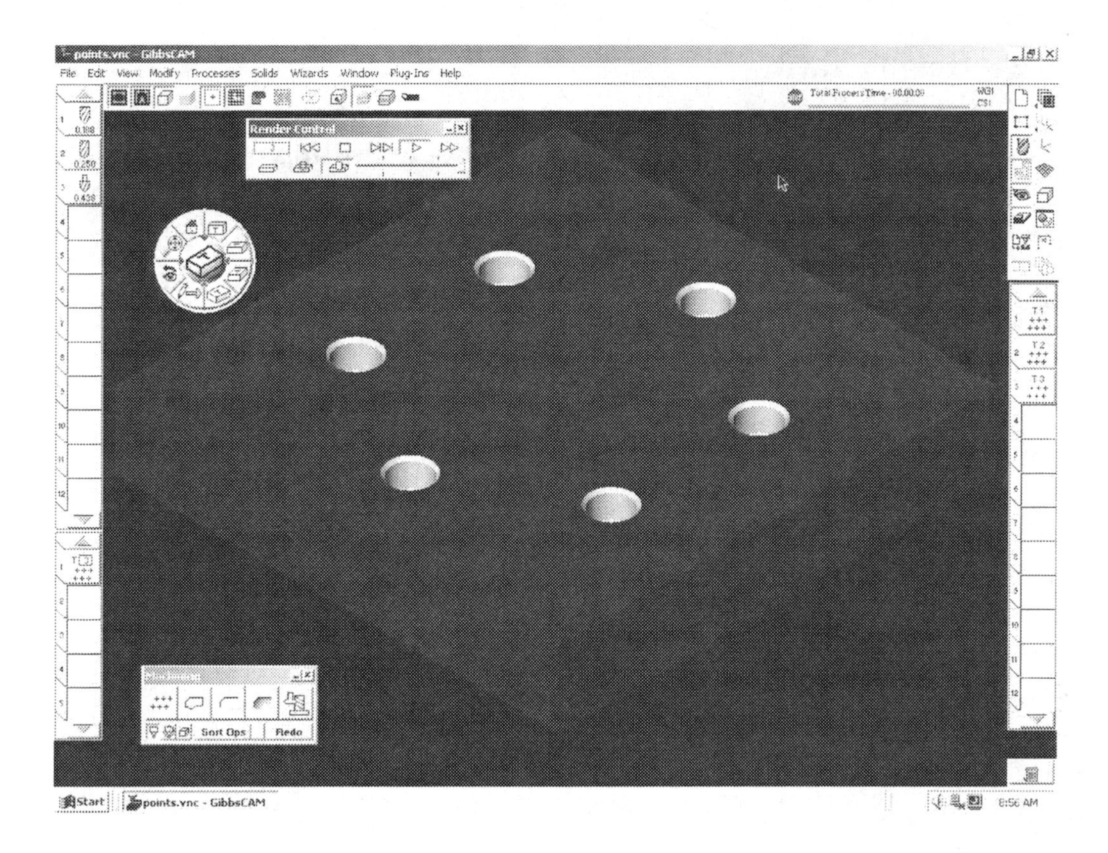


This completes the exercise. Save the file. Choose *File-Save* from the *Main Menu*.

## Exercise 2 – Hole Wizard

In this exercise you will use the hole wizard to create drilling operations.

Create a new part file. Select *File-New* from the *Main Menu*.

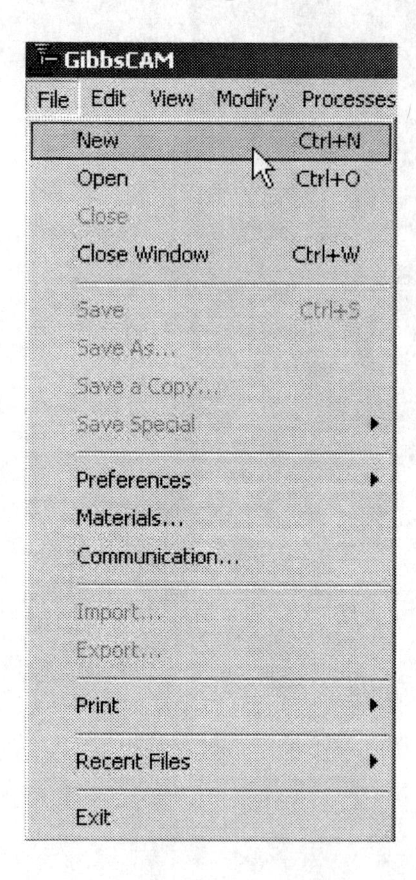


Save the new part file as HOLE_WIZARD.

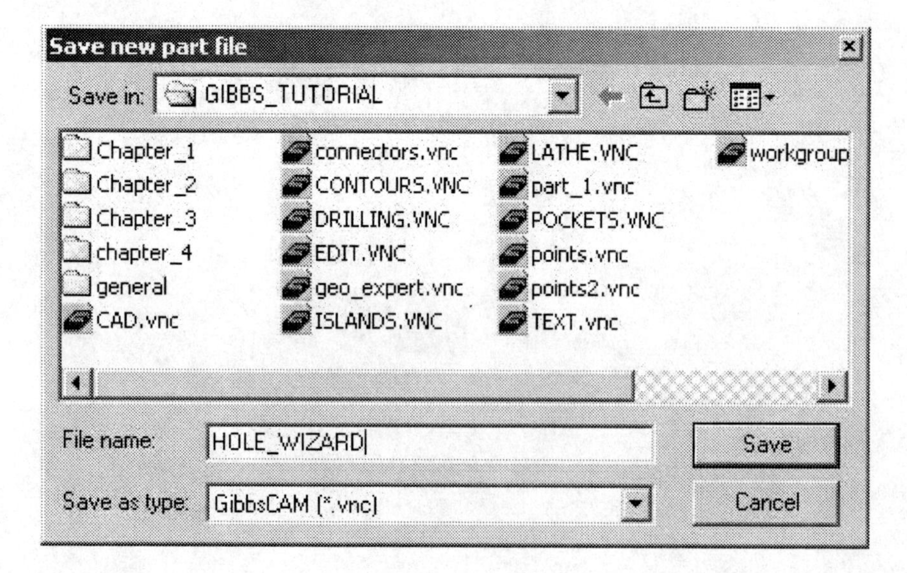

Select the Document Control icon from the Top Level Palette.

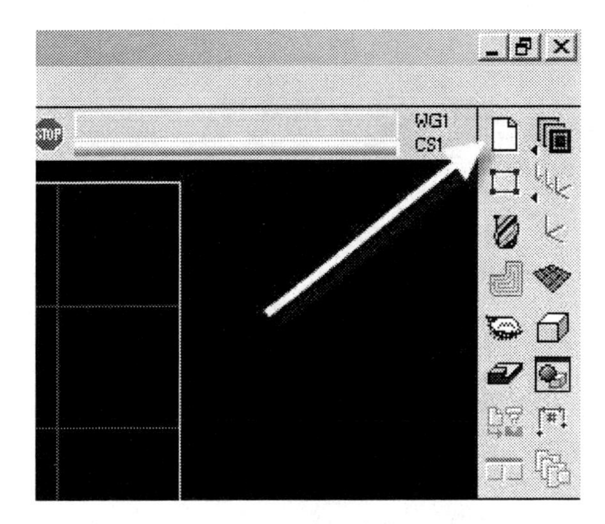


Select 3-Axis Vertical Mill as the machine type and enter the values for the stock size as shown.

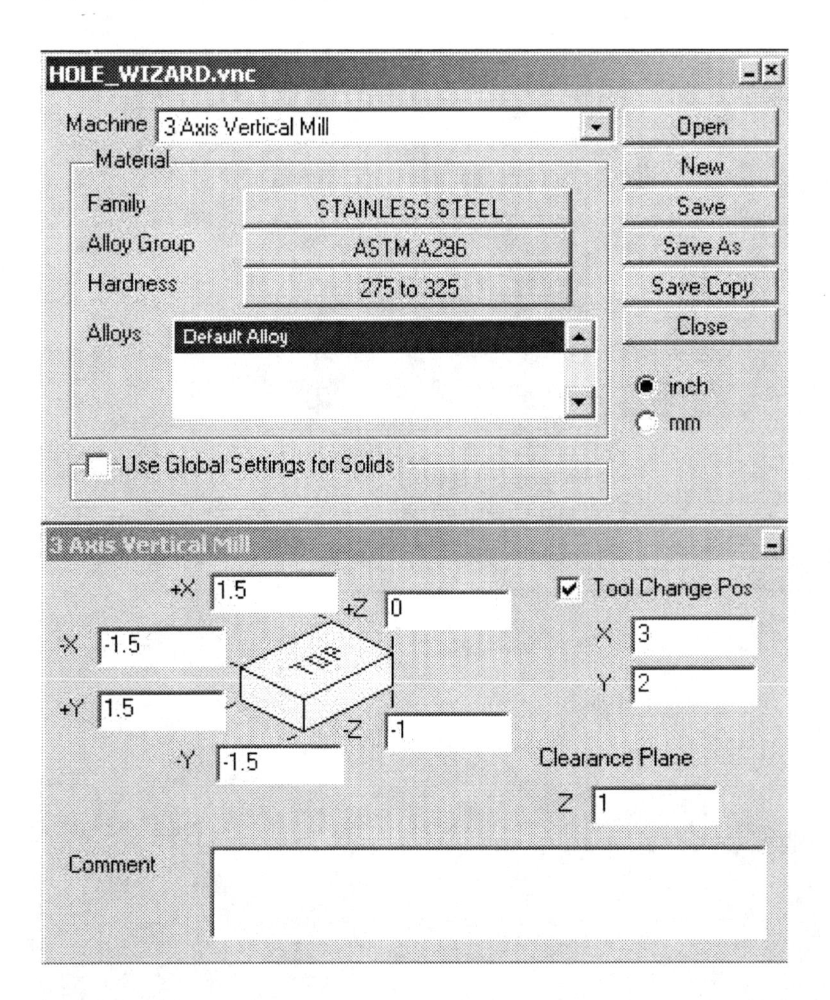

Close the dialog box.

Select *Wizards – Hole Wizard* from the *Main Menu*.

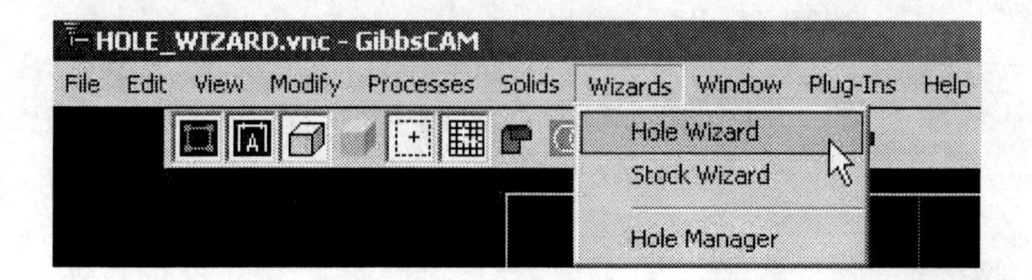


The Hole Wizard - Step One: Hole Shape dialog box will appear. Select a ream hole and then click on Next.

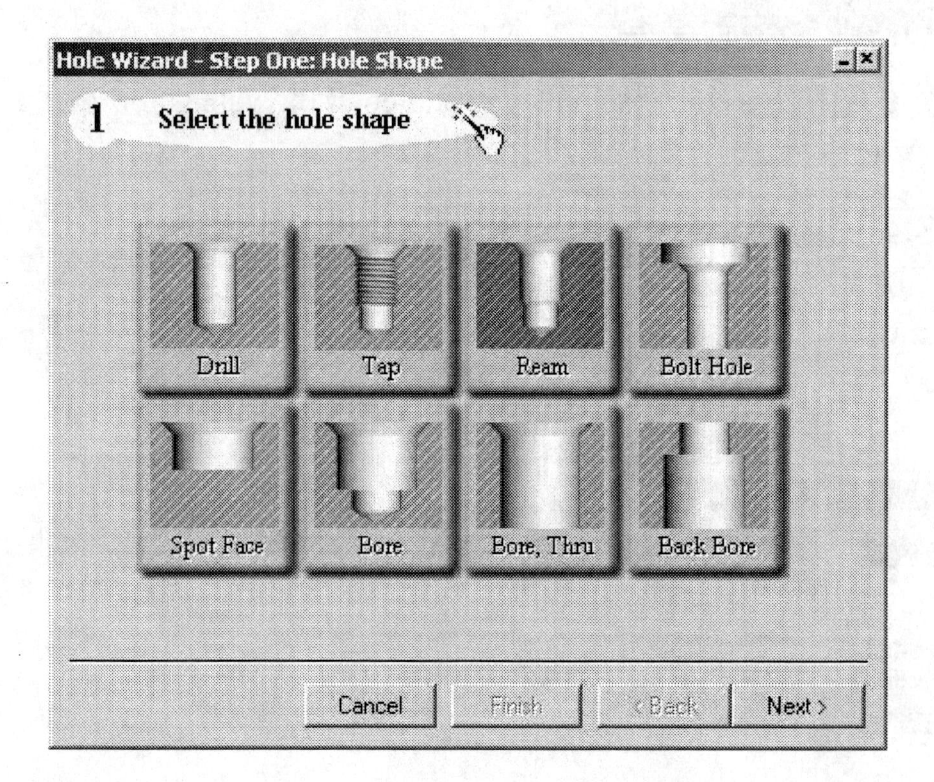

The Hole Wizard – Step Two: Ream Hole dialog box will appear. Enter the values as shown in the next graphic to define the size of the ream hole and click on Next.

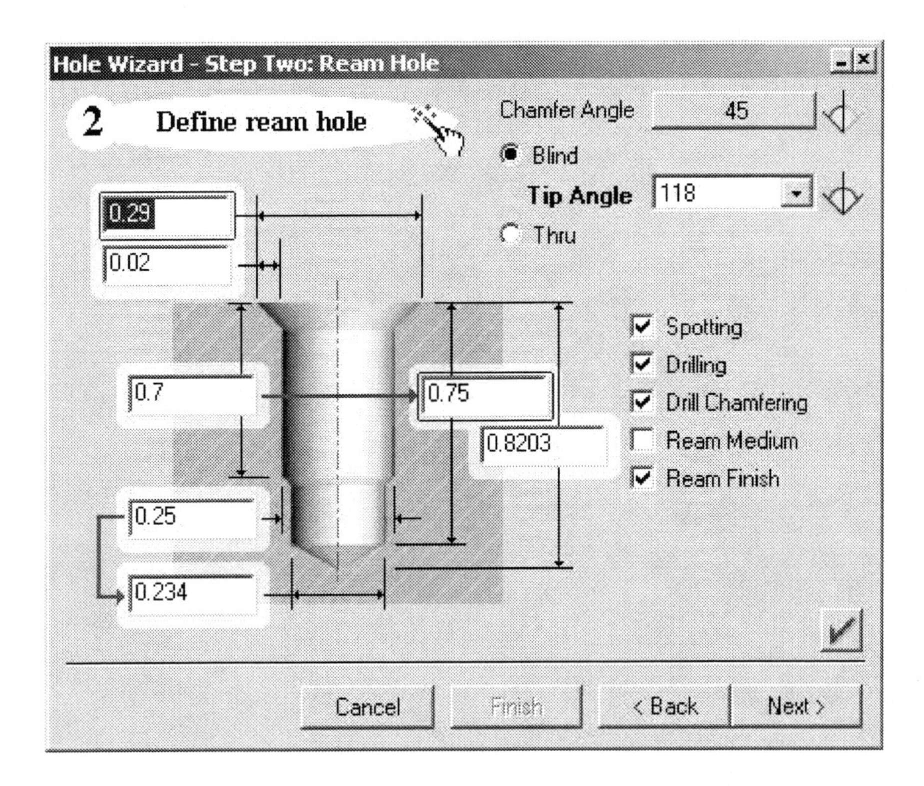


The Hole Wizard checker will appear and prompt you to create tools. Select Create Spot Drill Tool. You may need to use the scroll bar in the dialog box to find it.

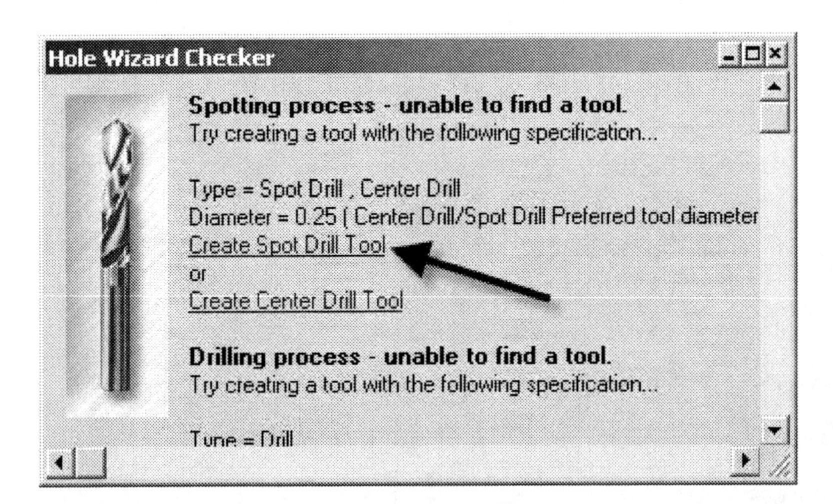


The Hole Wizard Checker will prompt you to create another tool. Select Create Drill Tool.

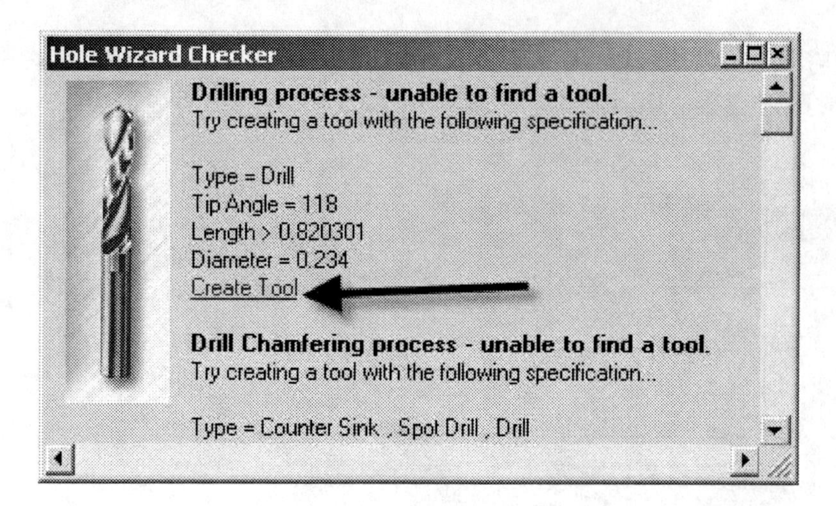


You will again be prompted to create another tool. Click on Create Counter Sink Tool.

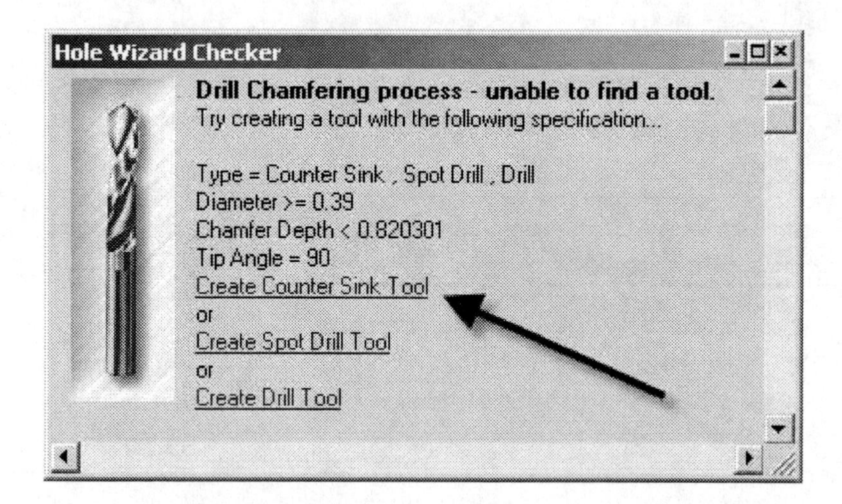


The Hole Wizard checker will prompt you to create the final tool. Click on Create Tool.

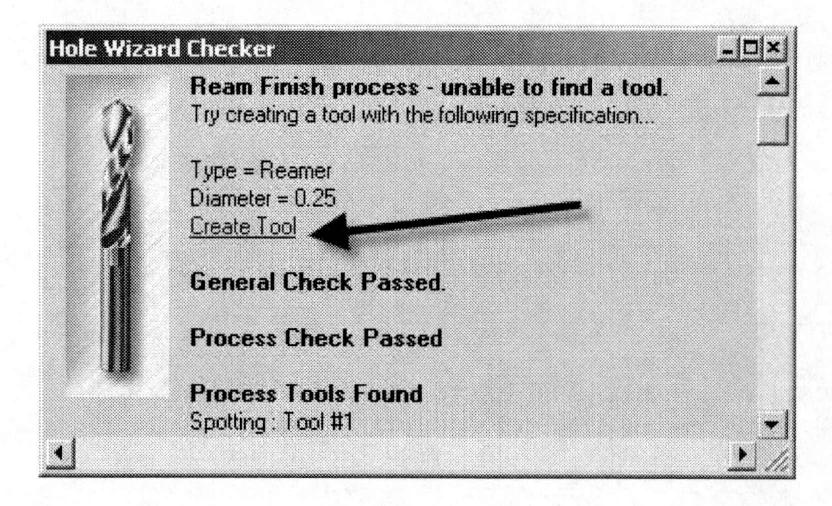

The Wizard Checker will now display a dialog box showing you that you have created all of the tools needed for the operations. Close this dialog box.

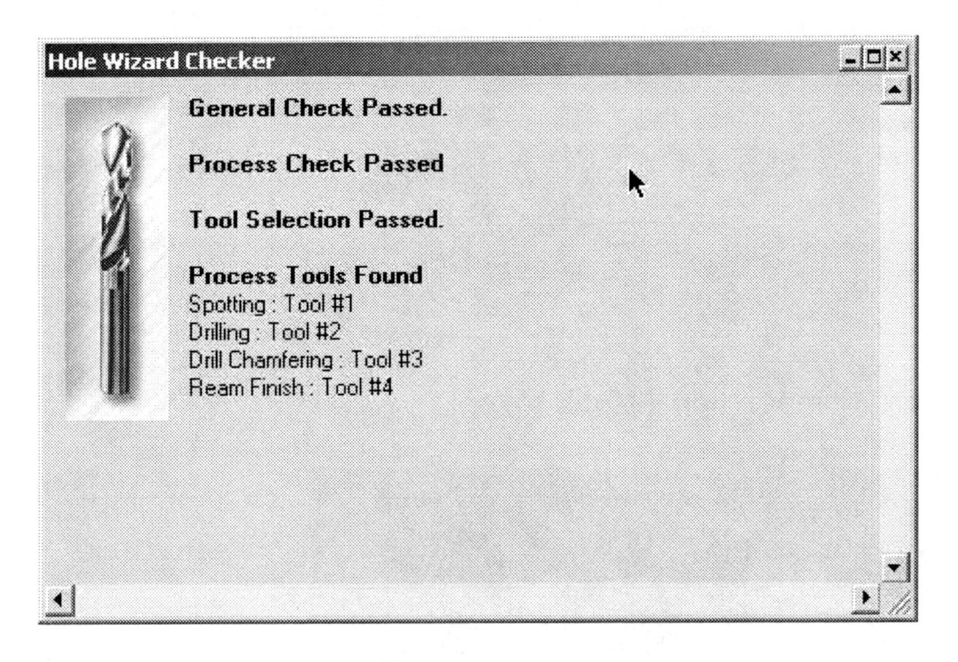


The Hole Wizard – Step Three: Pattern dialog box will appear. Click on the bolt circle icon as shown.

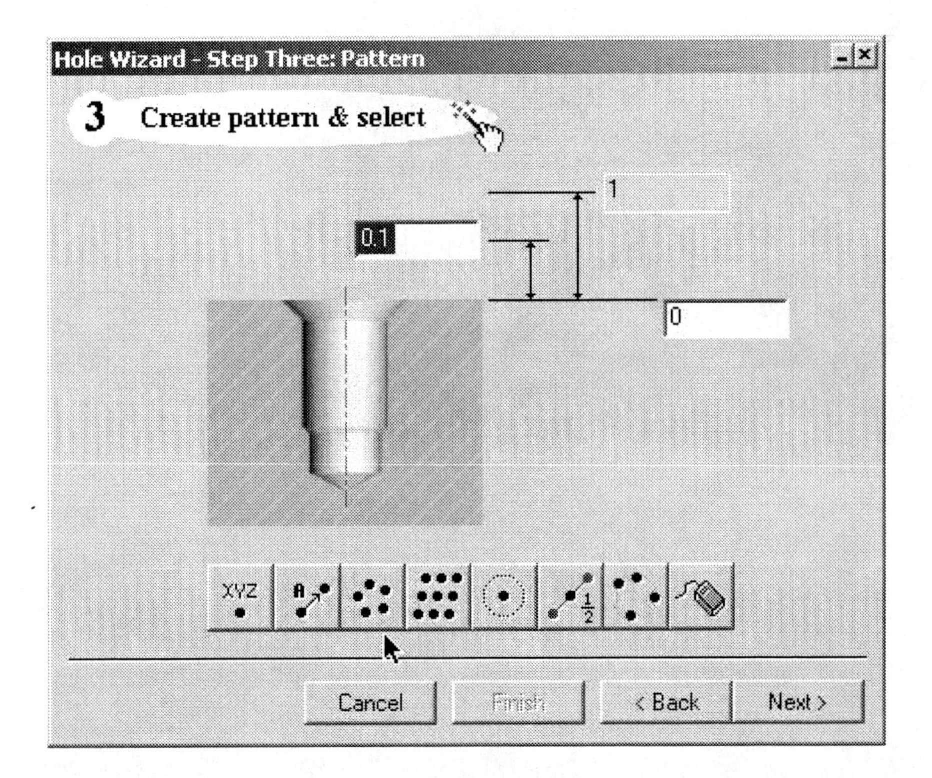

The Bolt Circle dialog box will appear. Enter the information as shown and click on Do It to create the pattern. Close the dialog box.

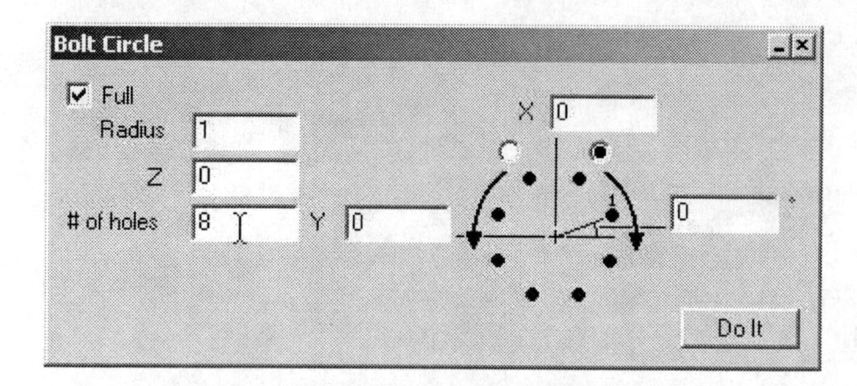


Hit the Return Icon to return to the Hole Wizard.

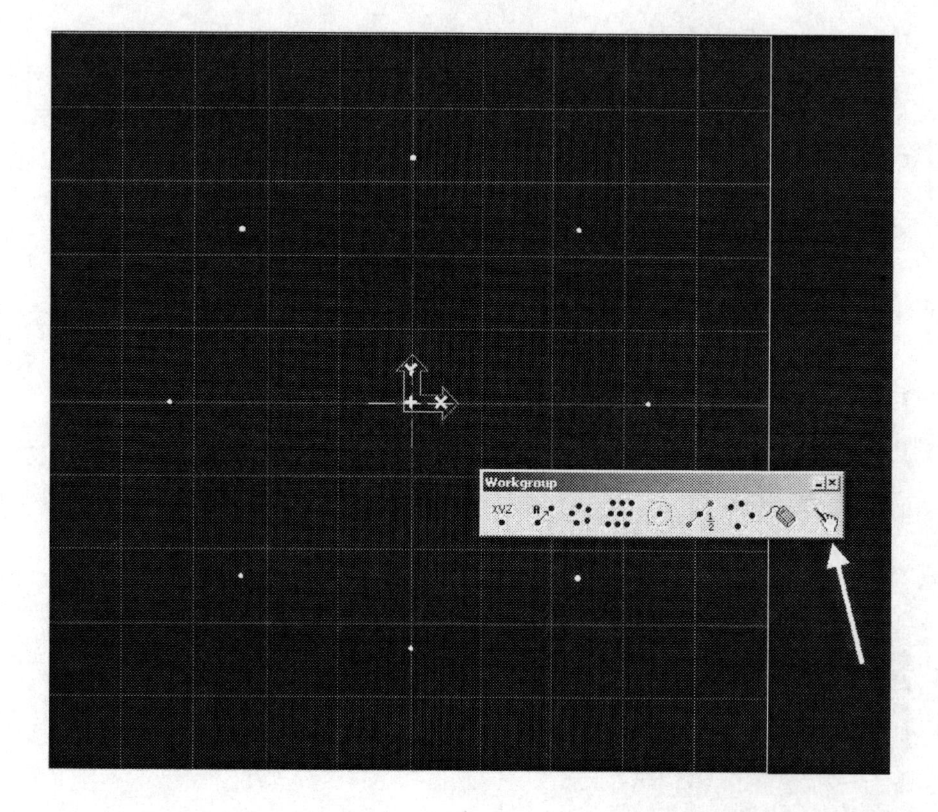

Hold down the control key and drag a box around the points as shown. Click on Next.

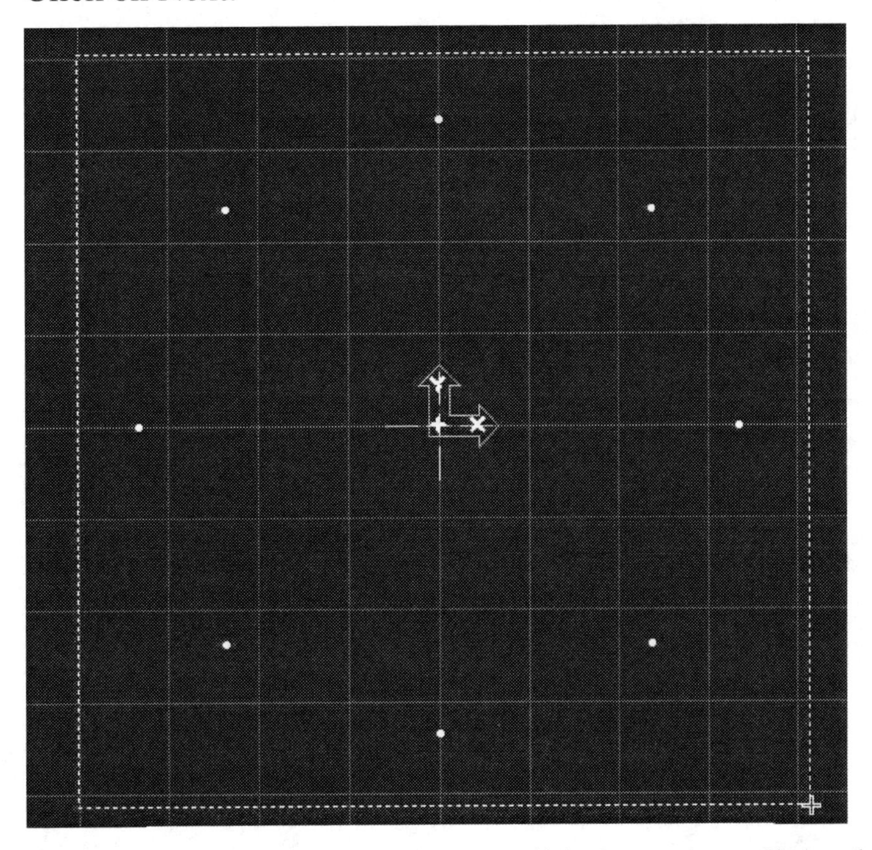

The Hole Wizard – Step Four: Build Operations dialog box will appear. Click on Build Processes.

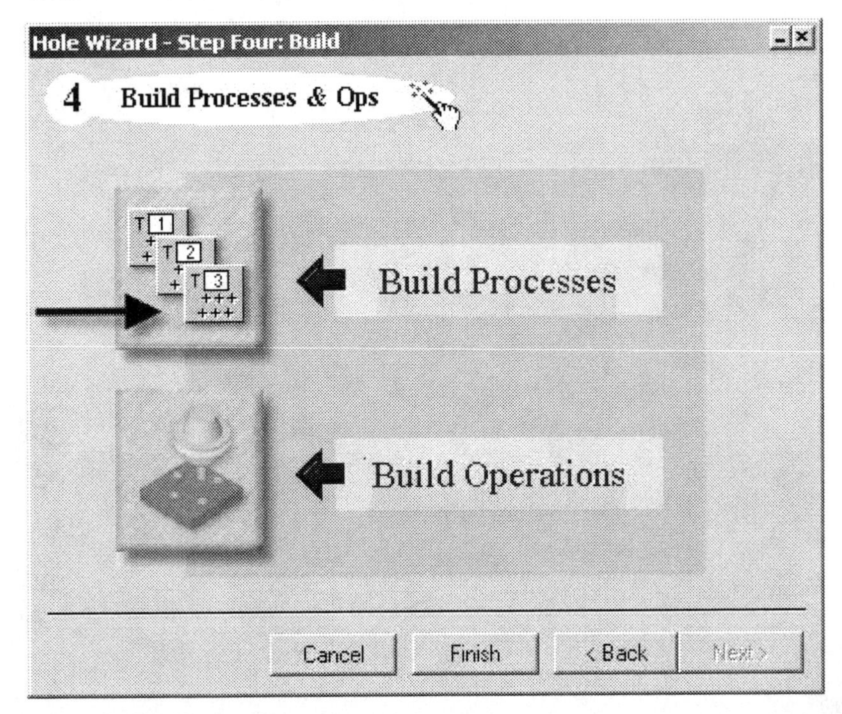

The machining processes will appear in the Process list. Click on the Build Operations icon.

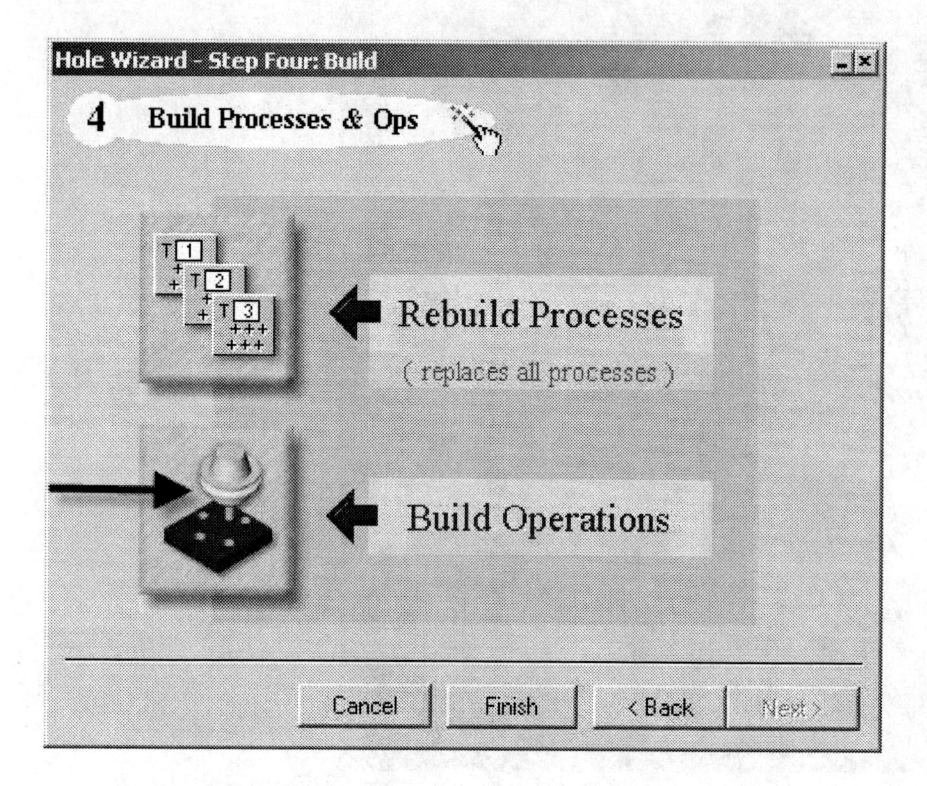


The operations will appear in the Operations List.
Render the part. Your screen should look like the next graphic.

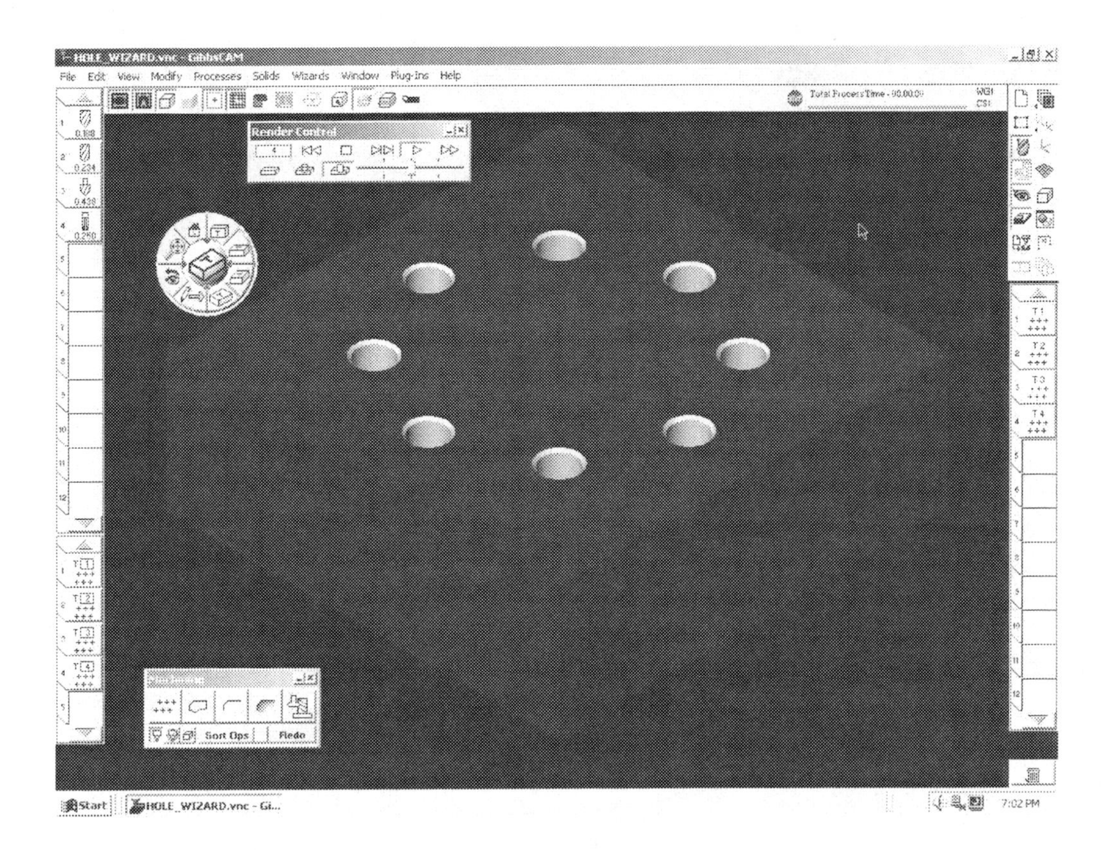


This completes the exercise. Save the file. Choose *File-Save* from the *Main Menu*.

## Exercise 3 – Hole Manager

In this exercise you will use the Automatic Feature Recognition to create machining operations for the hole data stored in a CAD file.

Note: You will need the SolidSurfacer option to complete this exercise.

Open the file HOLE_MGR.SLDPRT. Select *File-Open* from the *Main Menu*. *This file can be downloaded at www.schroff1.com.*

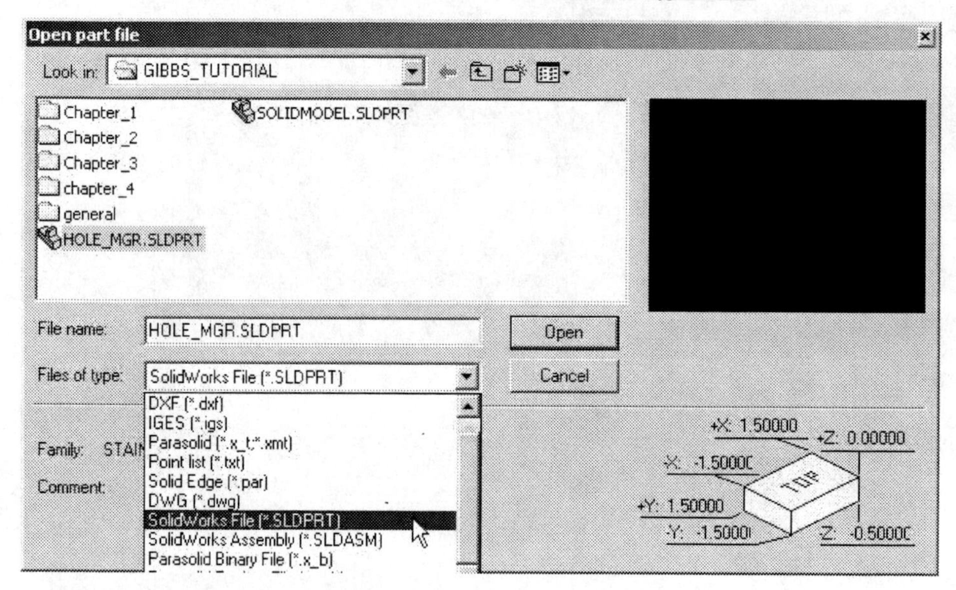


The Solid File Option dialog box will appear. Enter the values as shown.

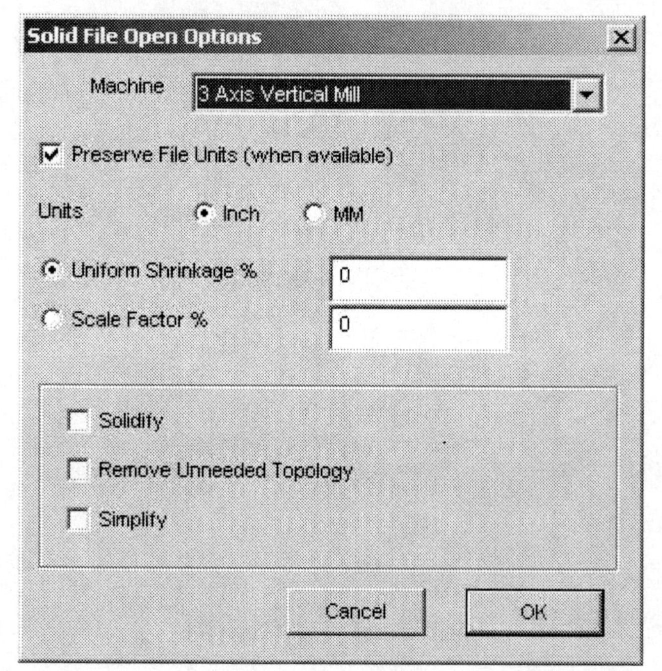

Click on OK. The Select Configuration dialog box will appear. This will show all available configurations of the CAD file. Select the default configuration and click on OK.

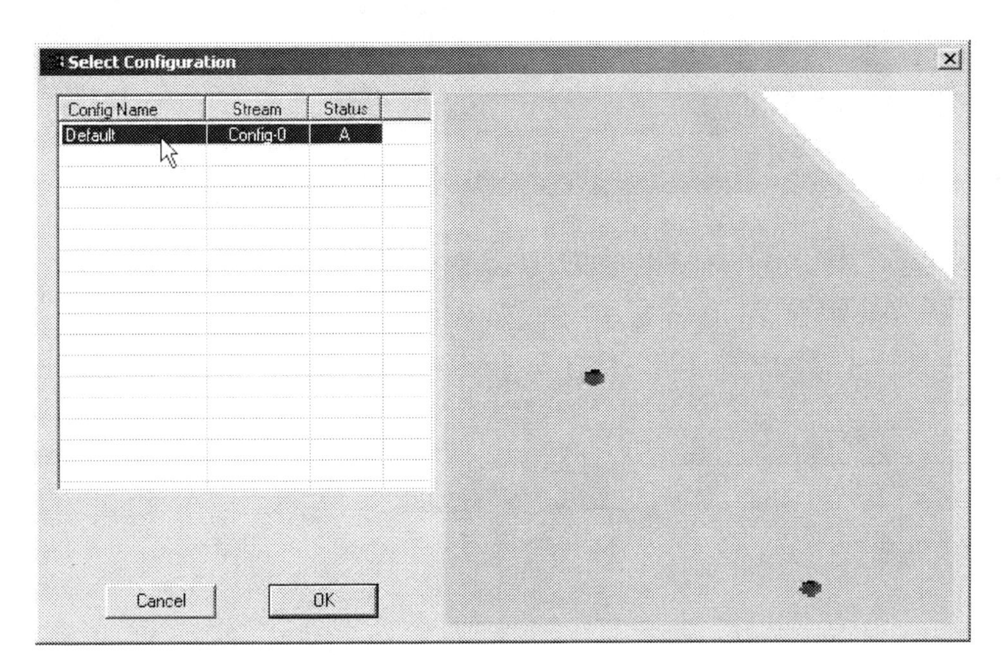


The part file will appear on your screen.

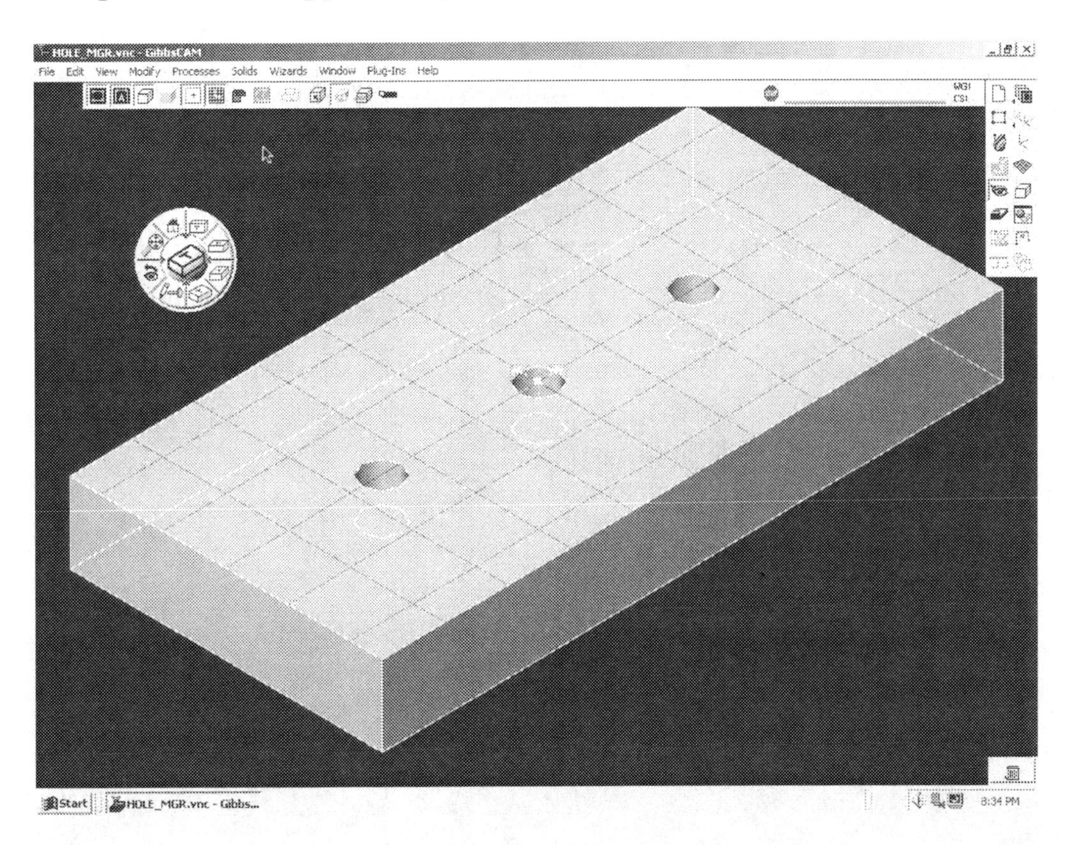

Select *Wizards – Hole Manager* from the *Main Menu*.

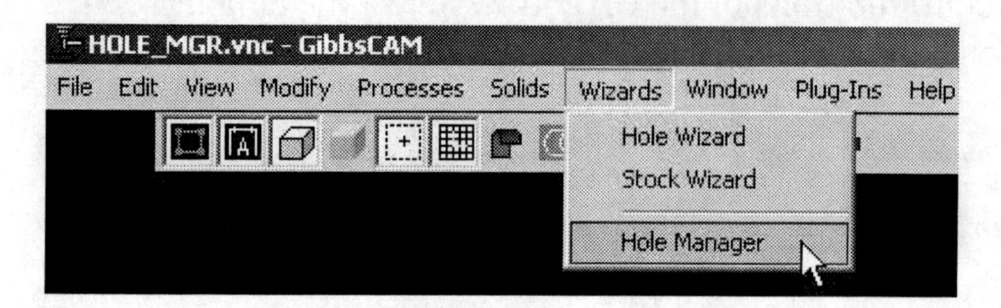


The Hole Manager dialog box will appear. Select the Run AFR (automated feature recognition) icon.

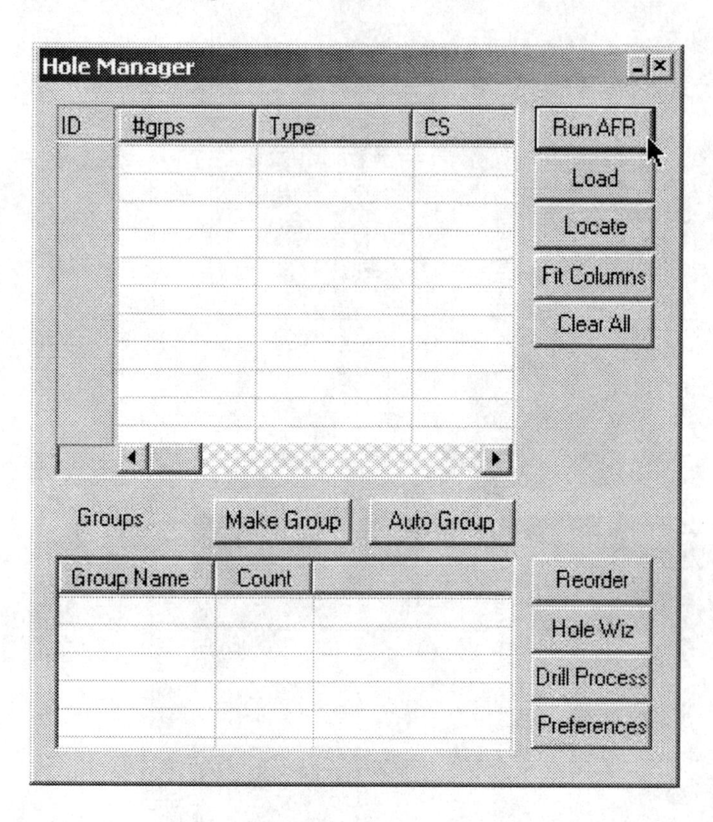

The AFR Update Option dialog box will appear. Enter the values as shown and click on OK.

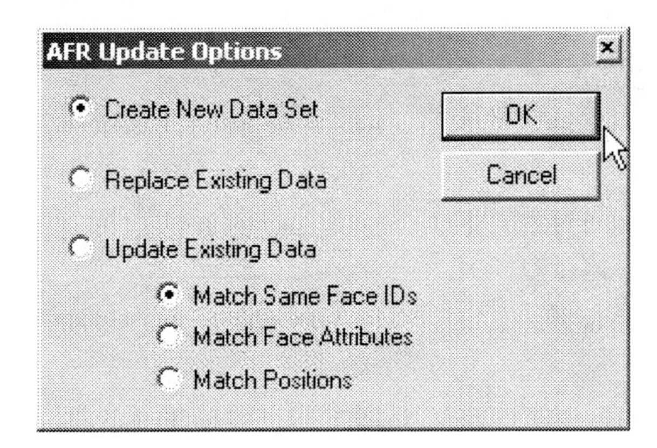


The AFR Options dialog box will appear. Enter HOLES for the WorkGroup name and select the options as shown and click on OK.

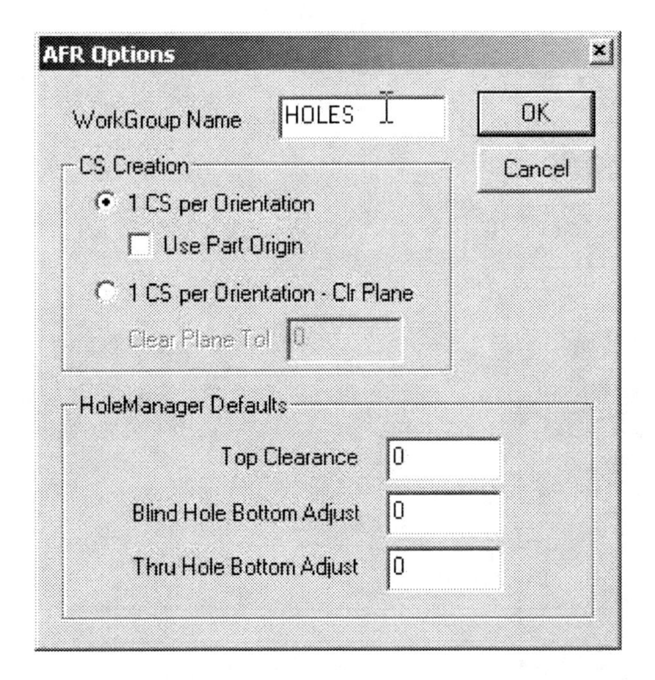

The hole information from the CAD file now appears in the Hole Manager.

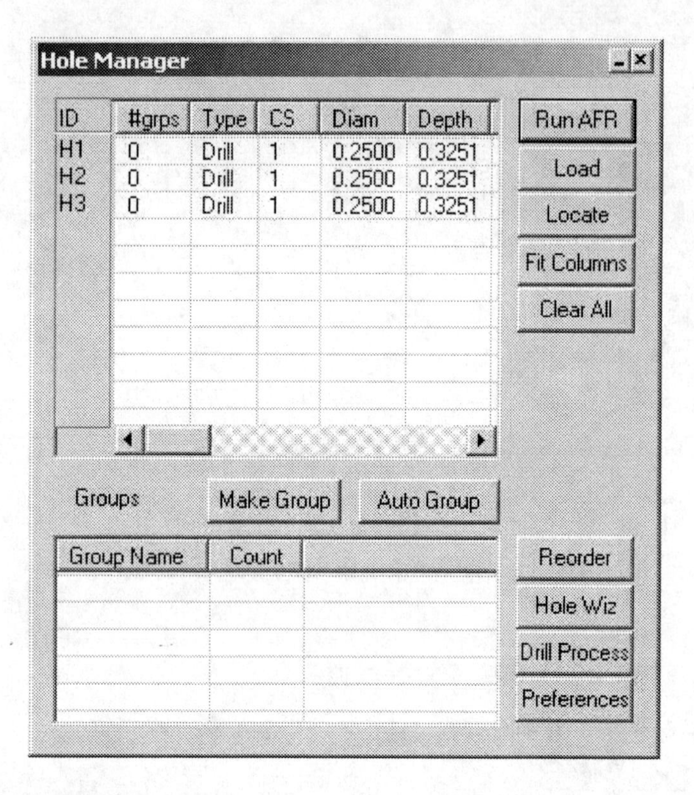


Hold down the control key and select the 3 holes. Select Make Group as shown.

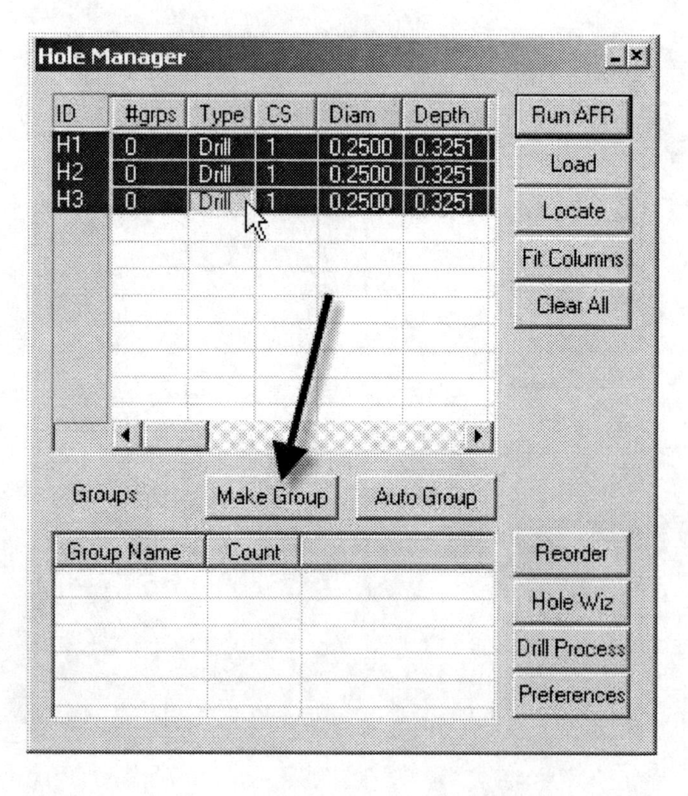

Group 1 will now appear. Select it and click on Hole Wizard.

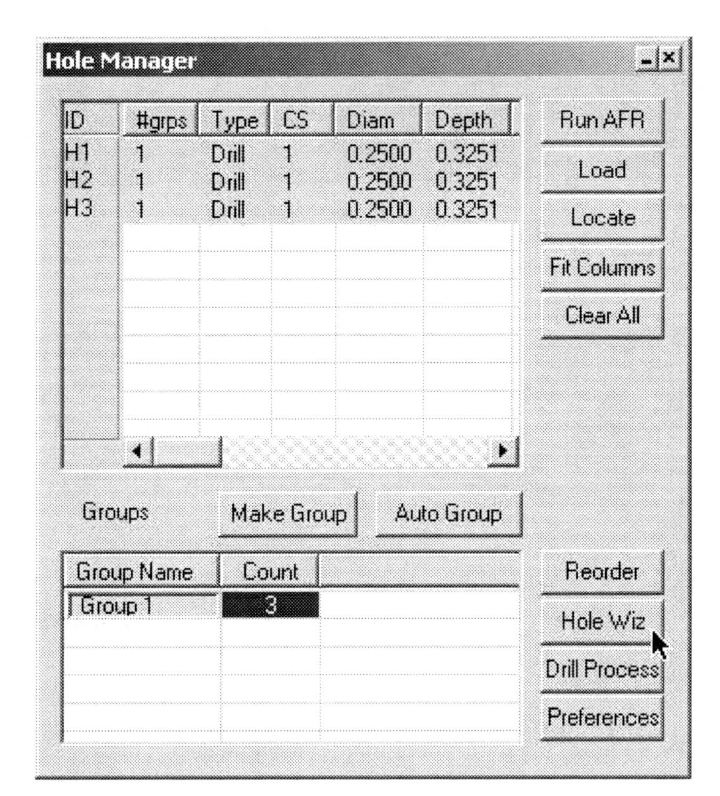


The Hole Wizard - Step Two: Drill Hole dialog box will appear. The size of the hole is defined as shown. Click on Next.

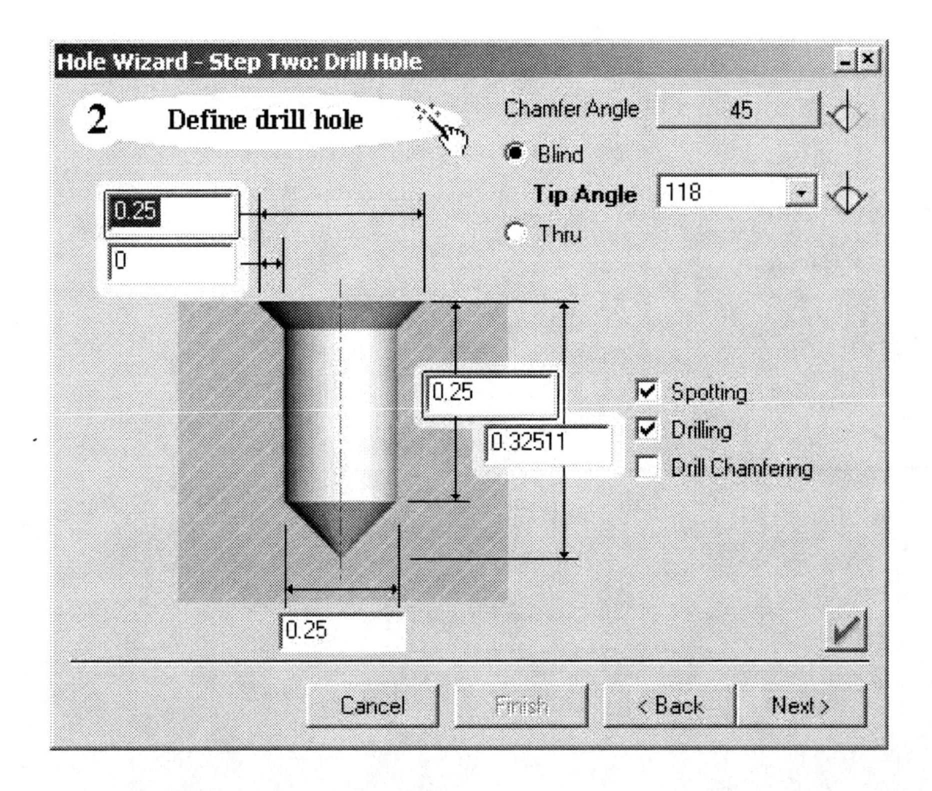

You will be prompted by the Hole Wizard Checker to create tools. Select Create Spot Drill Tool.

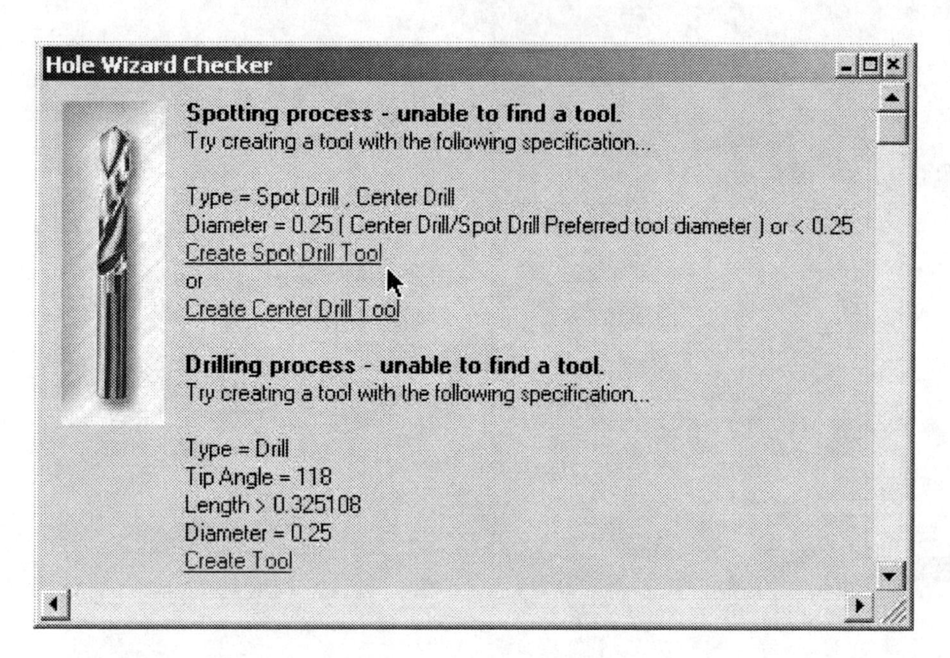


You will be prompted to create another tool. Select Create Tool.

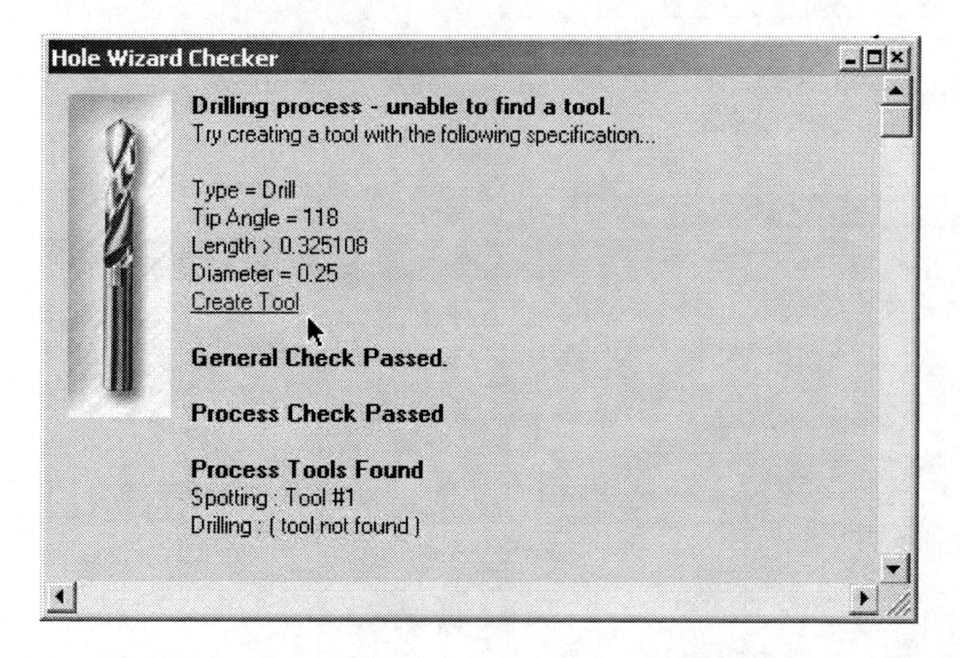

The Hole Wizard Checker will verify that all tools have been created.

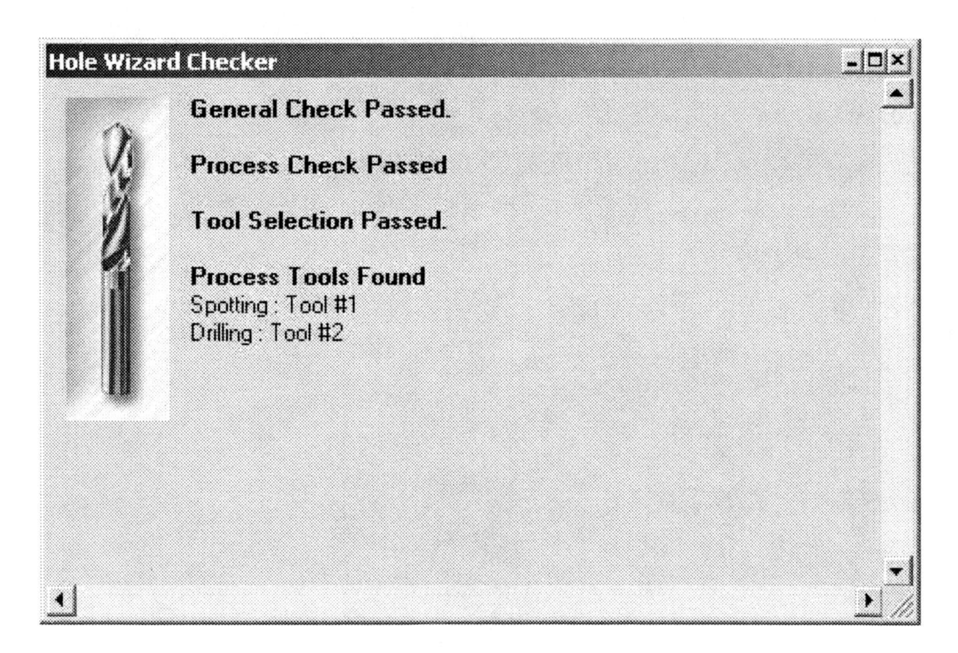


Close the Hole Wizard Checker dialog box. Click on Next in the Hole Wizard – Step Three: Pattern dialog box.

Click on Build Processes.

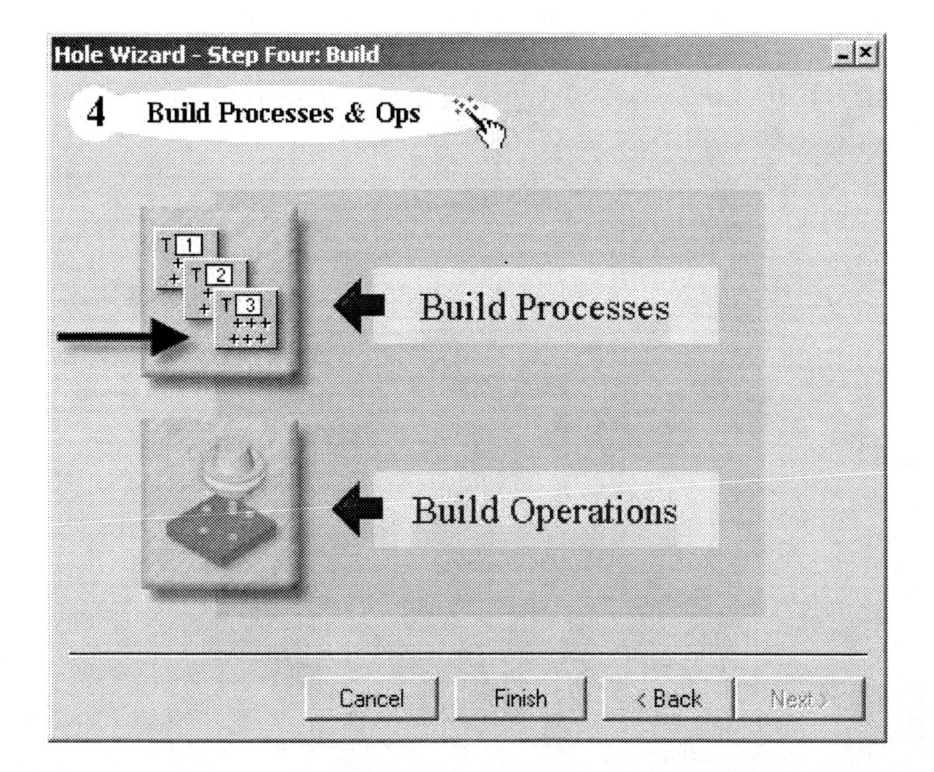

Click on Build Operations.

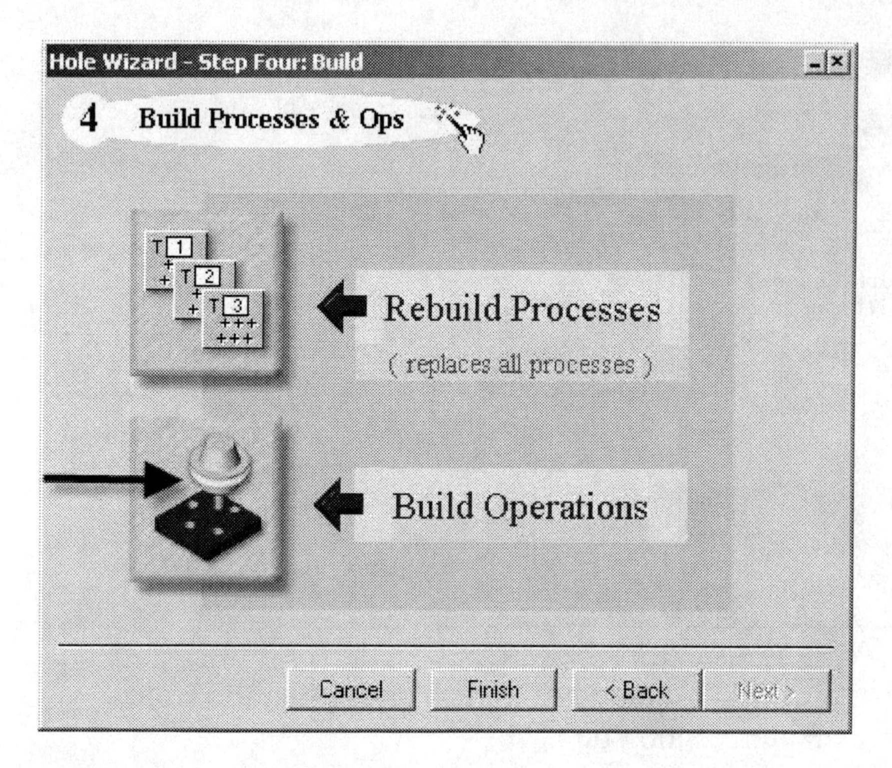


Your screen should look like this.

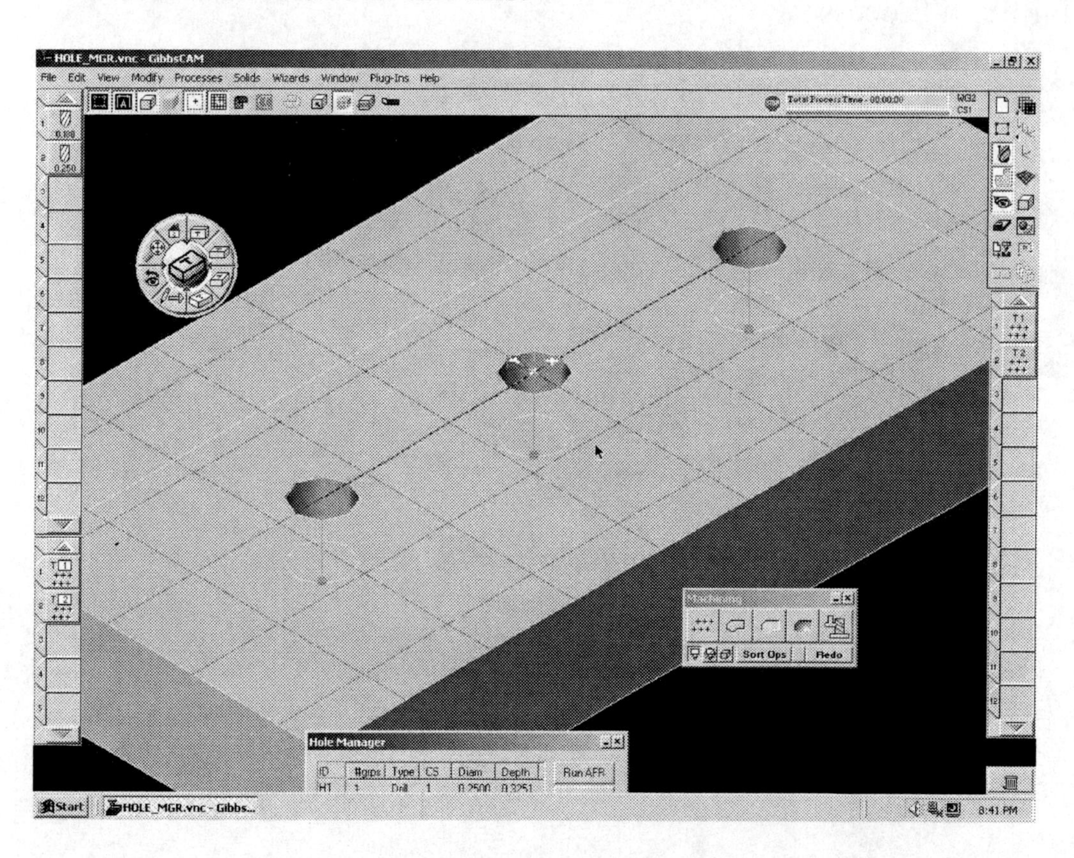

Render the part.

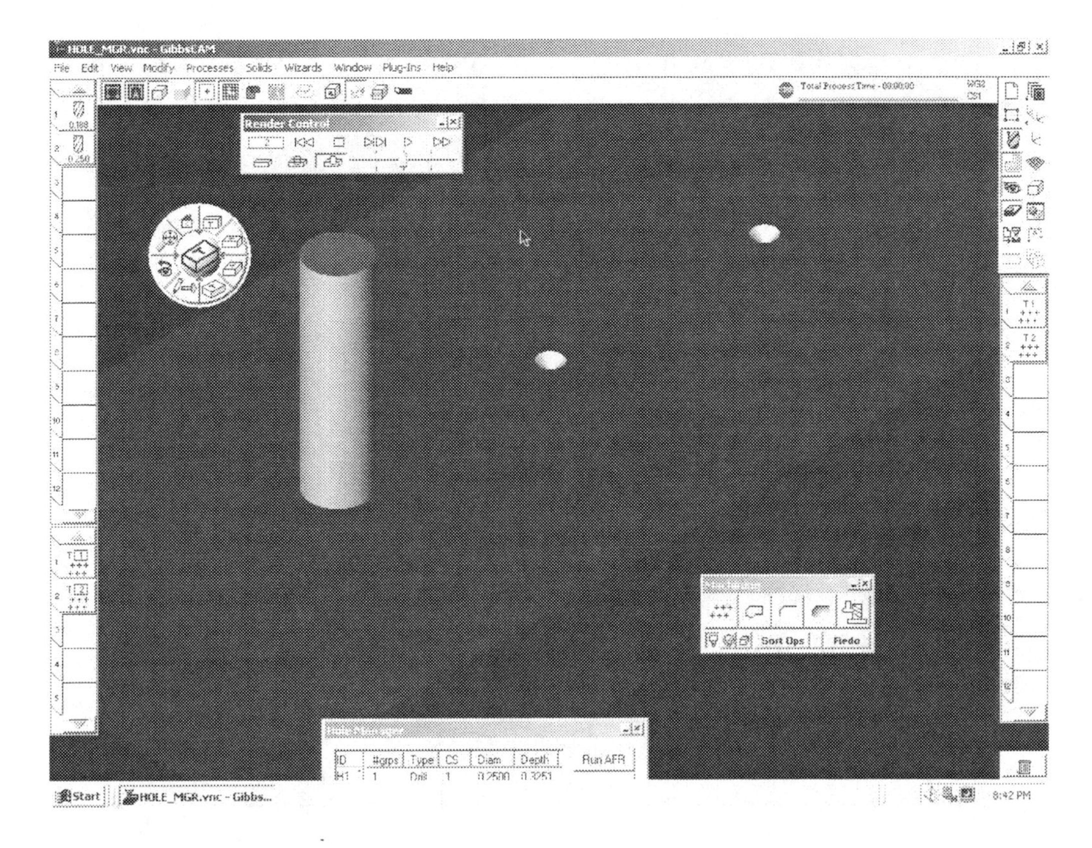

This completes the exercise. Save the file. Select *File-Save* from the *Main Menu.*

## Exercise 4 – Roughing and Contours

In this exercise you will create roughing and contour machining operations to machine 3 pockets in a part.

Open the file pockets.vnc. *This file can be downloaded at www.schroff1.com.*

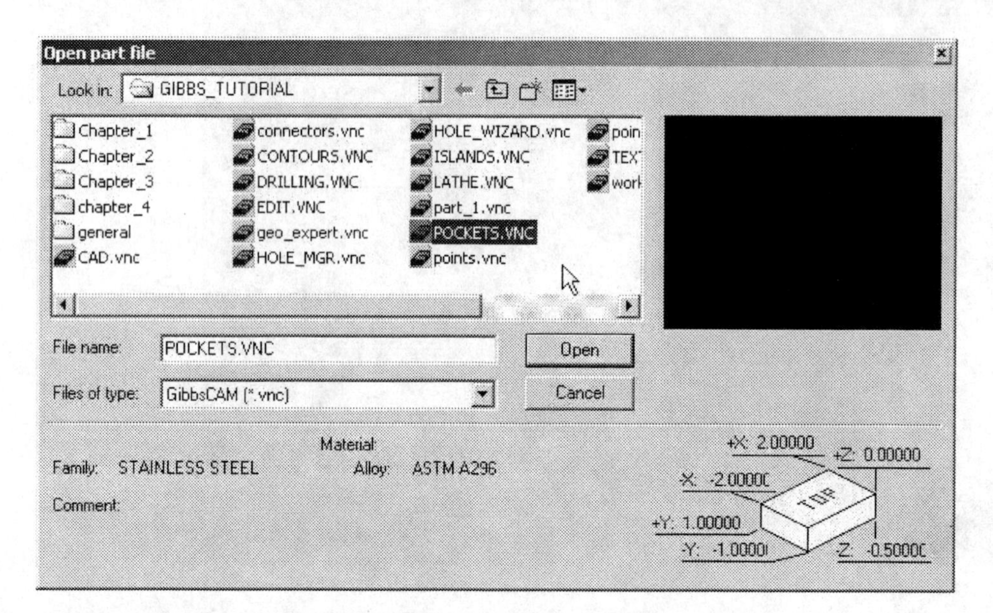


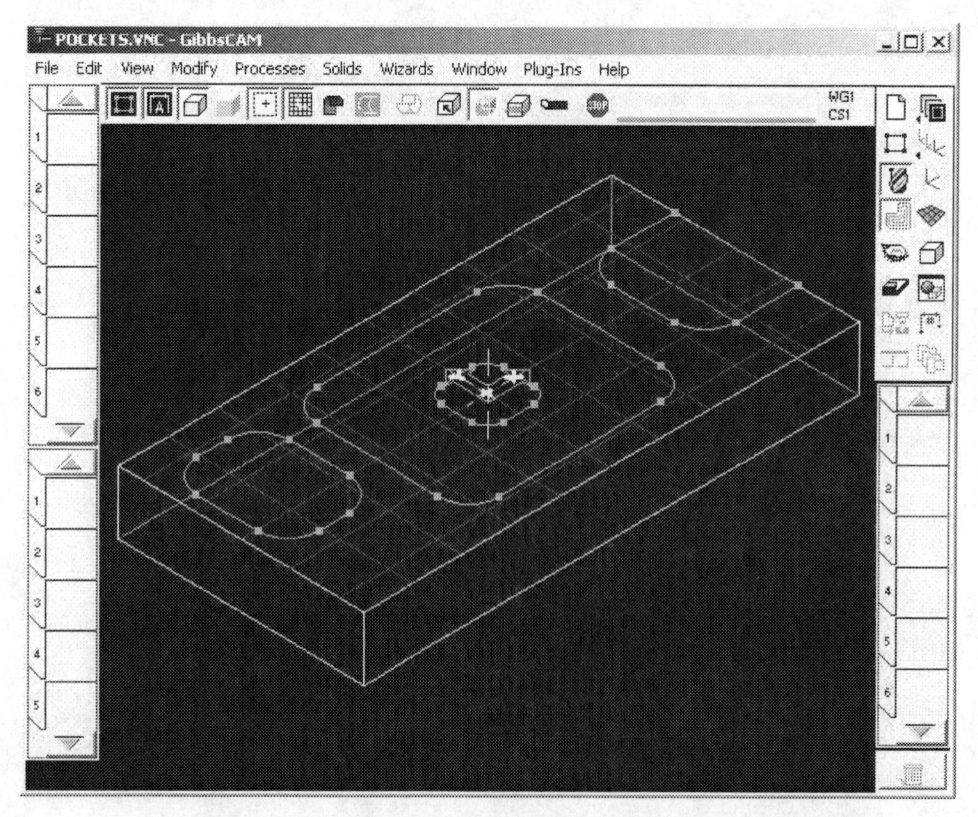

Select the Tools icon from the Top Level Palette.

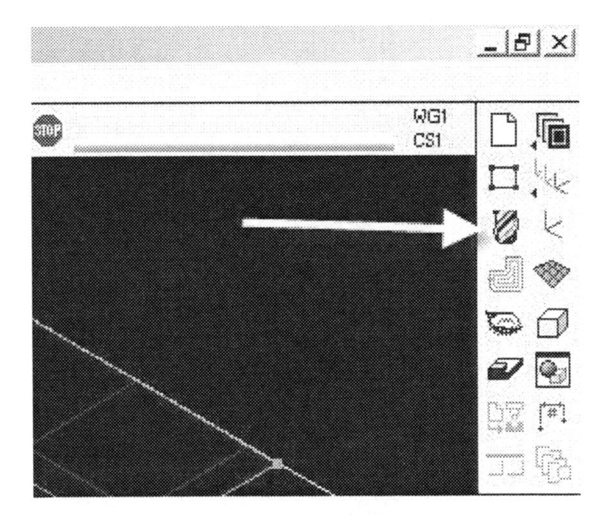


Double-click on Tile 1 in the Tool List.

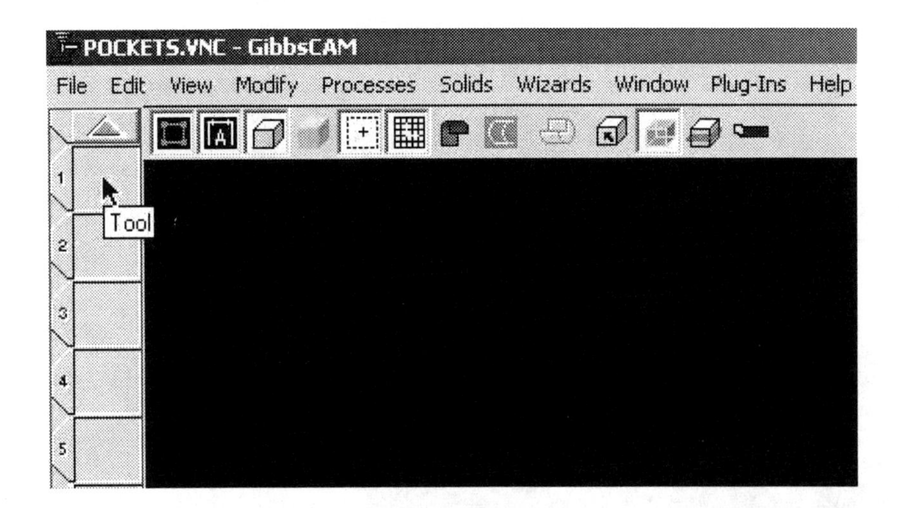


Create Tool 1. Select Rough End Mill as the tool type and enter the values as shown in the next graphic to define the size of the tool. Close the dialog box.

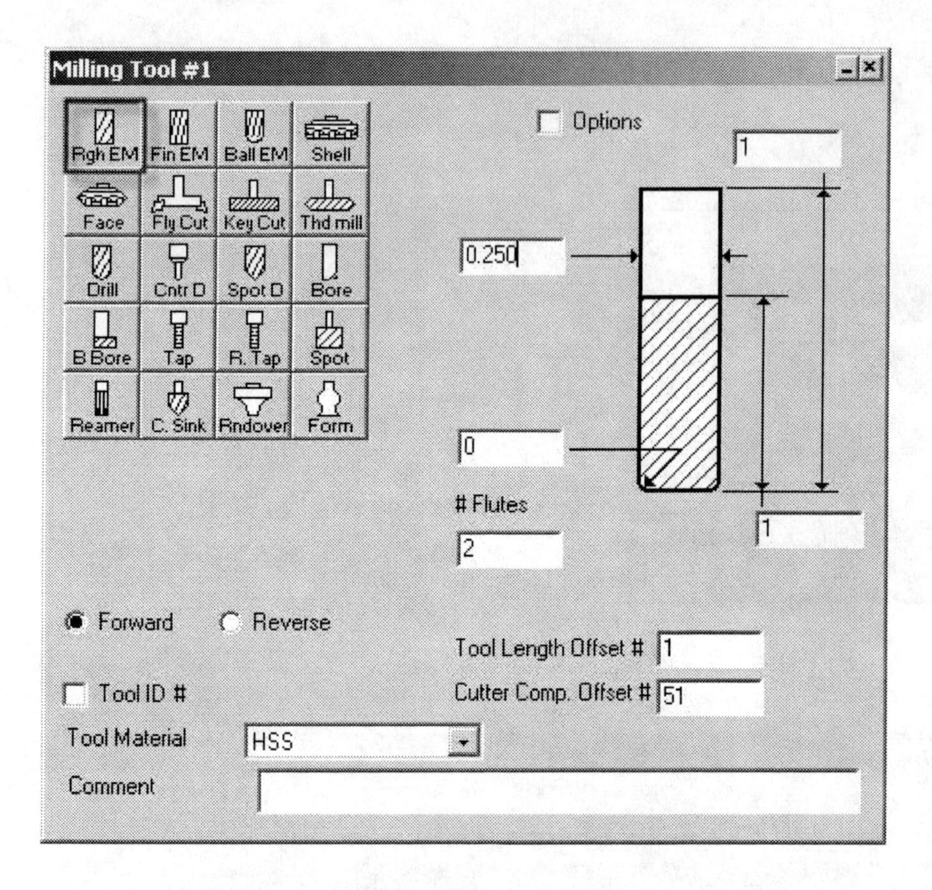


Tool 1 will appear in the Tool List.

Double-click on Tile 2 in the Tool List.

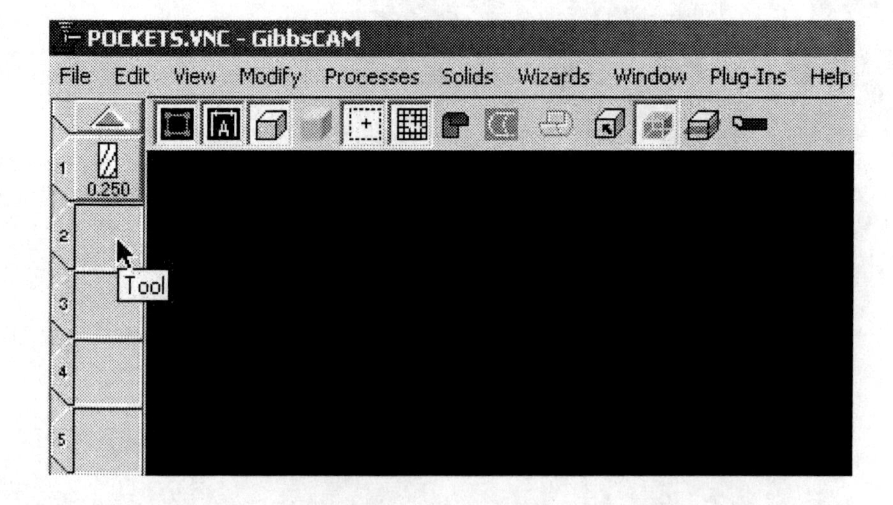

Tool 2 will be a .375 dia. Rough End Mill. Enter the values as shown to define the size of the tool. Close the dialog box.

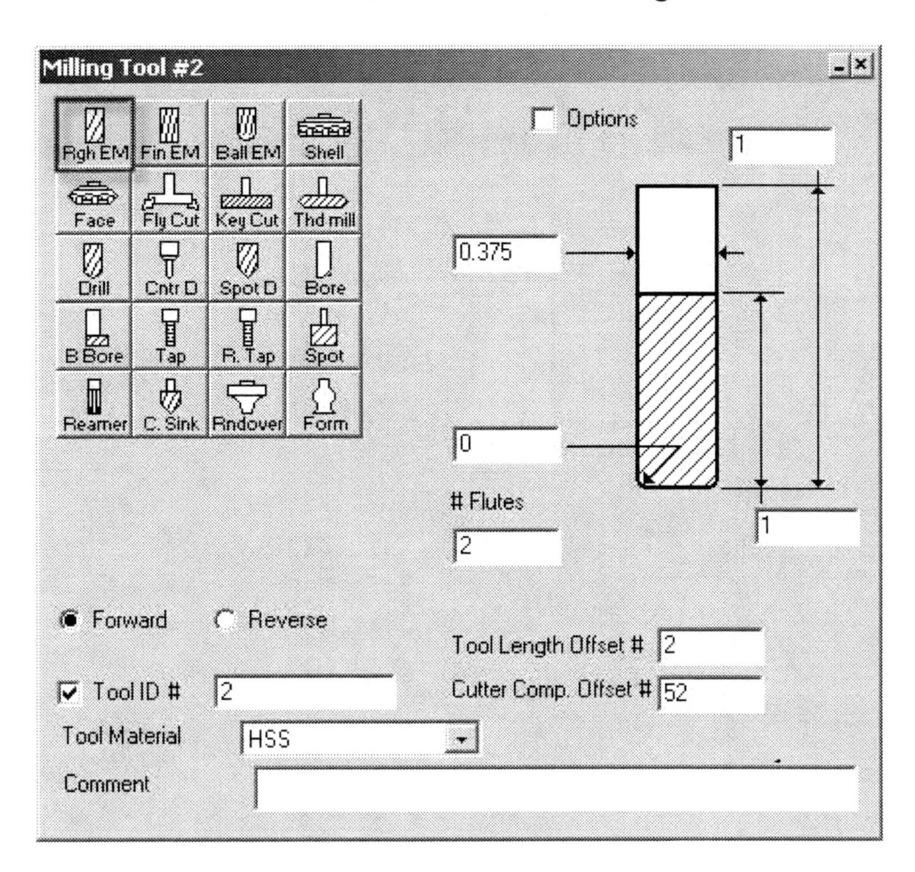

Double-click on Tile 3 in the Tool List to create a .250 dia. Finish End Mill as shown.

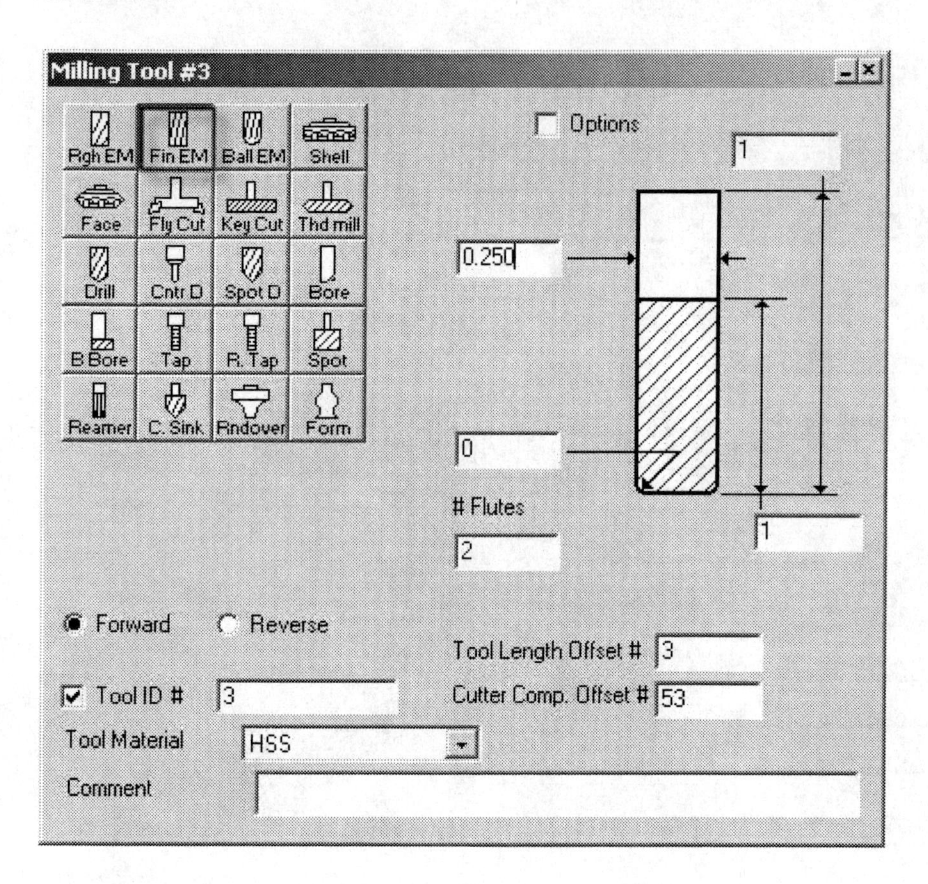


Click on the CAM icon on the Top Level Palette.

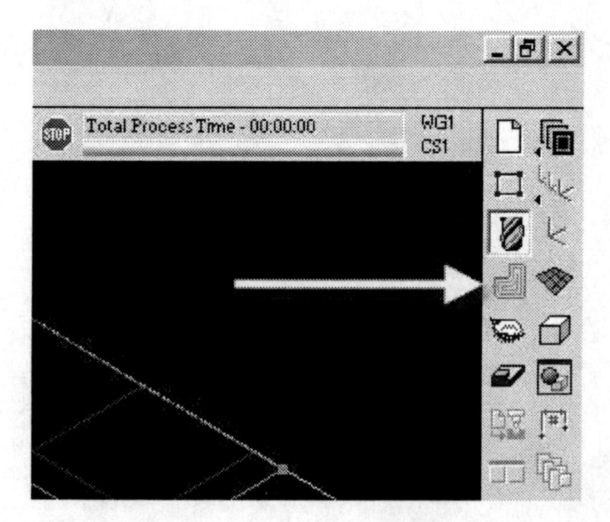

The Machining Palette will appear.

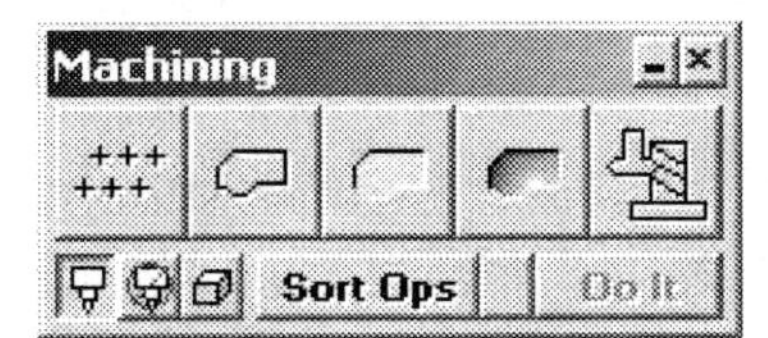


Drag Tool 1 from the Tool List to Tile 1 of the Process List.

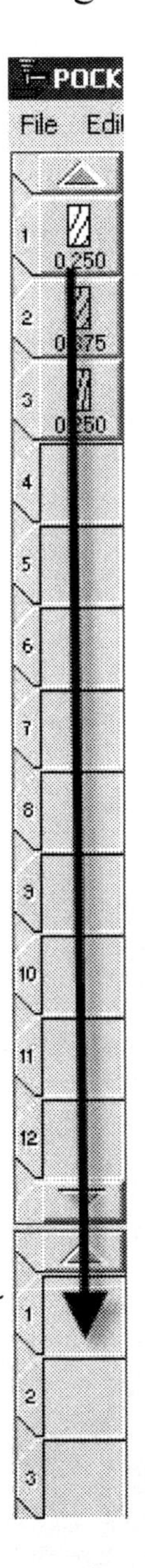

Drag the Rough Icon from the Machining Palette to Tile 1 of the Process List as shown in the following graphic.

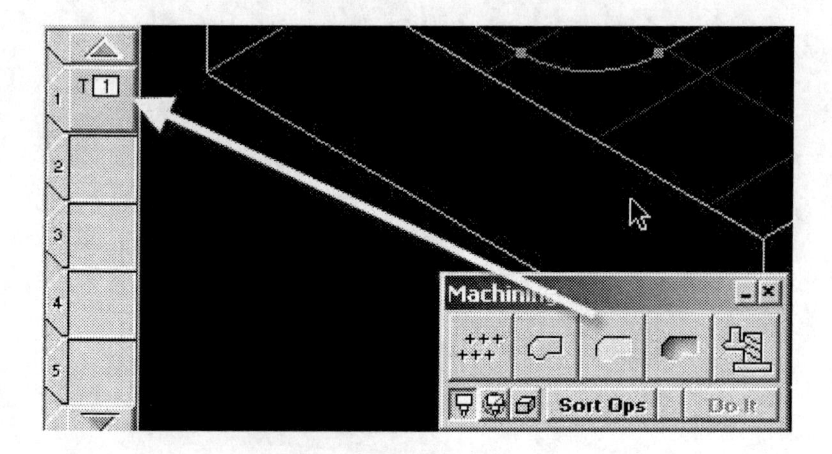


The Roughing dialog box will appear. Enter the values as shown. Close the dialog box.

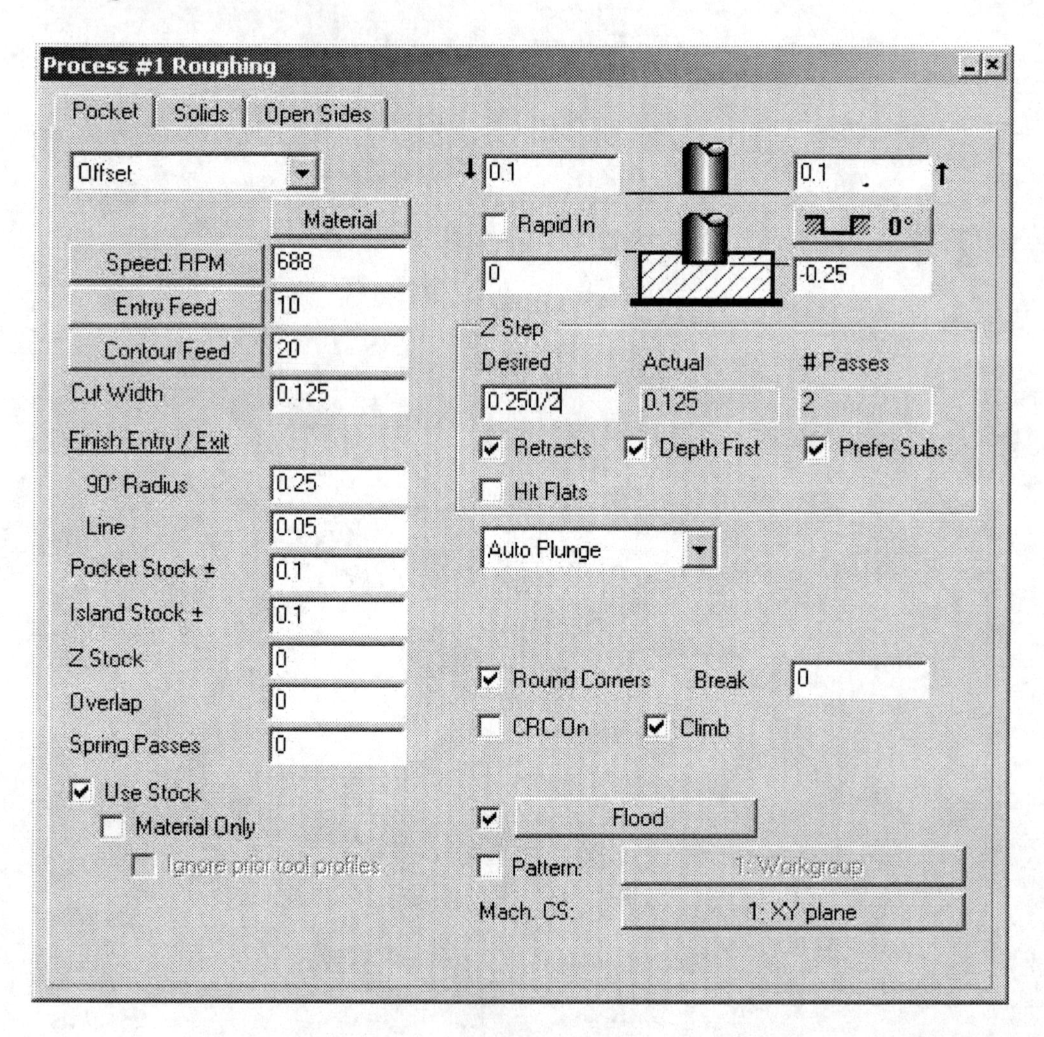

Hold down the control key and double-click on both profiles as shown to select them.

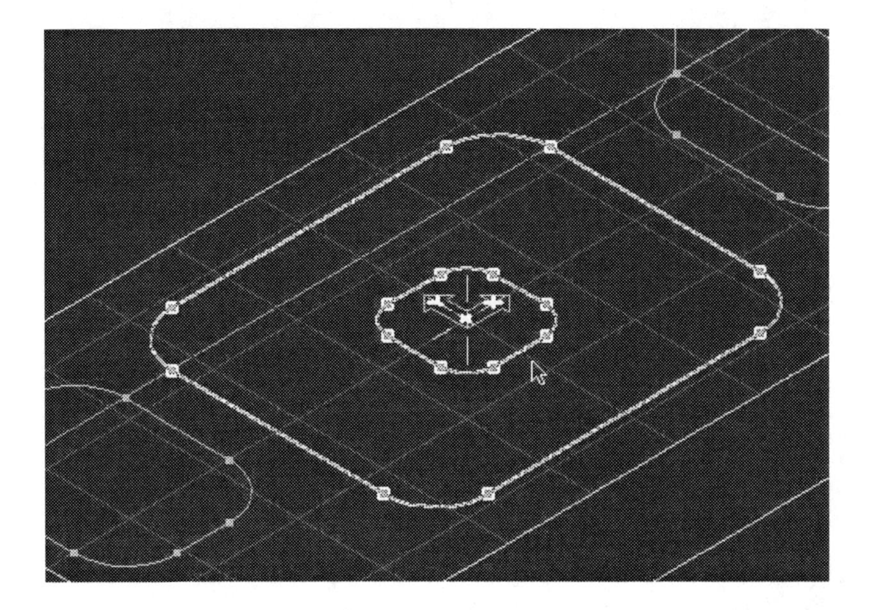

Click on Do It on the Machining Palette.

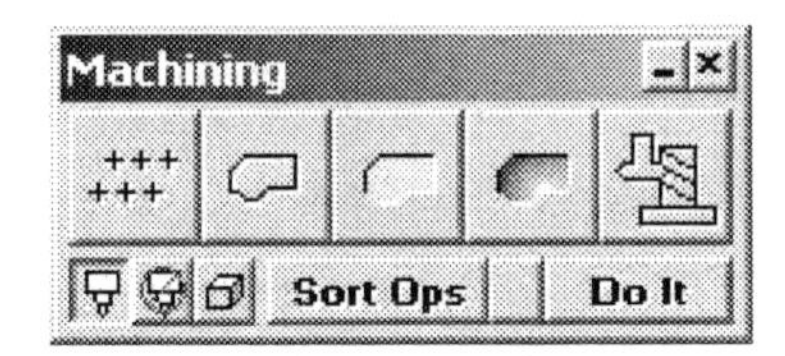


Your screen should look like the following graphic.

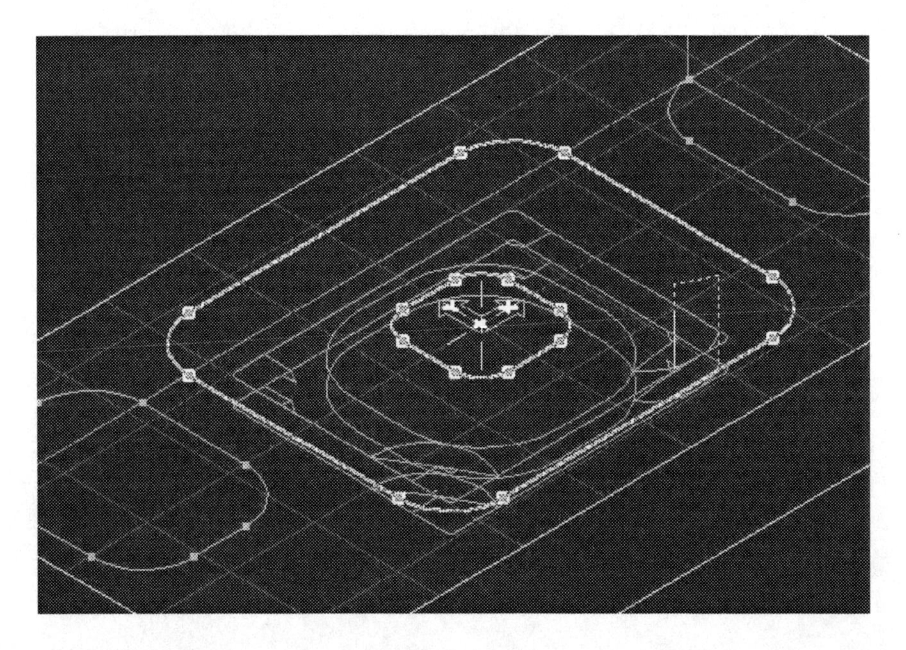

Render the part.

To begin the next operation, select Tile 2 in the Operations List.

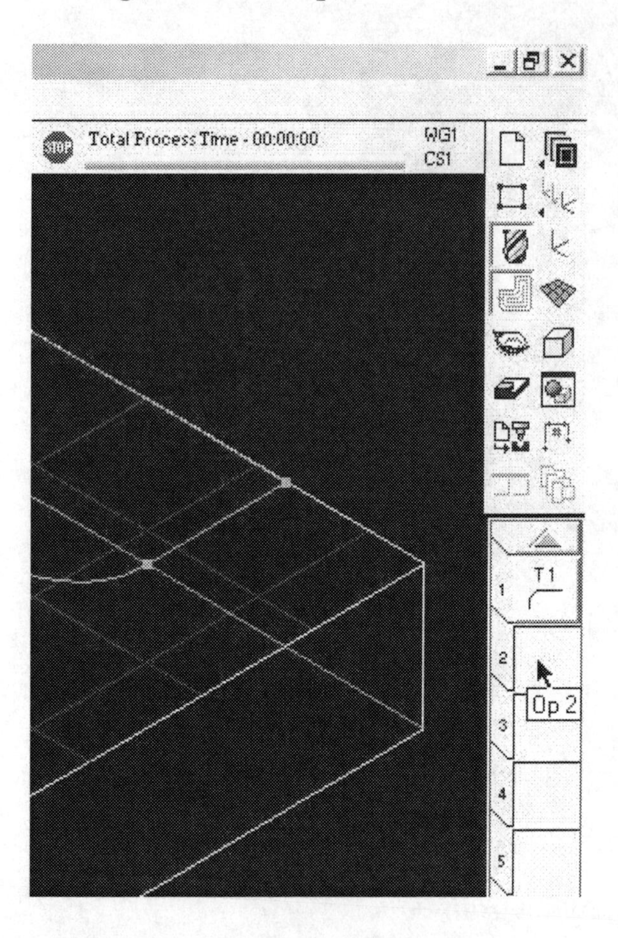


Drag Tile 1 from the Process List to the Trash Can.

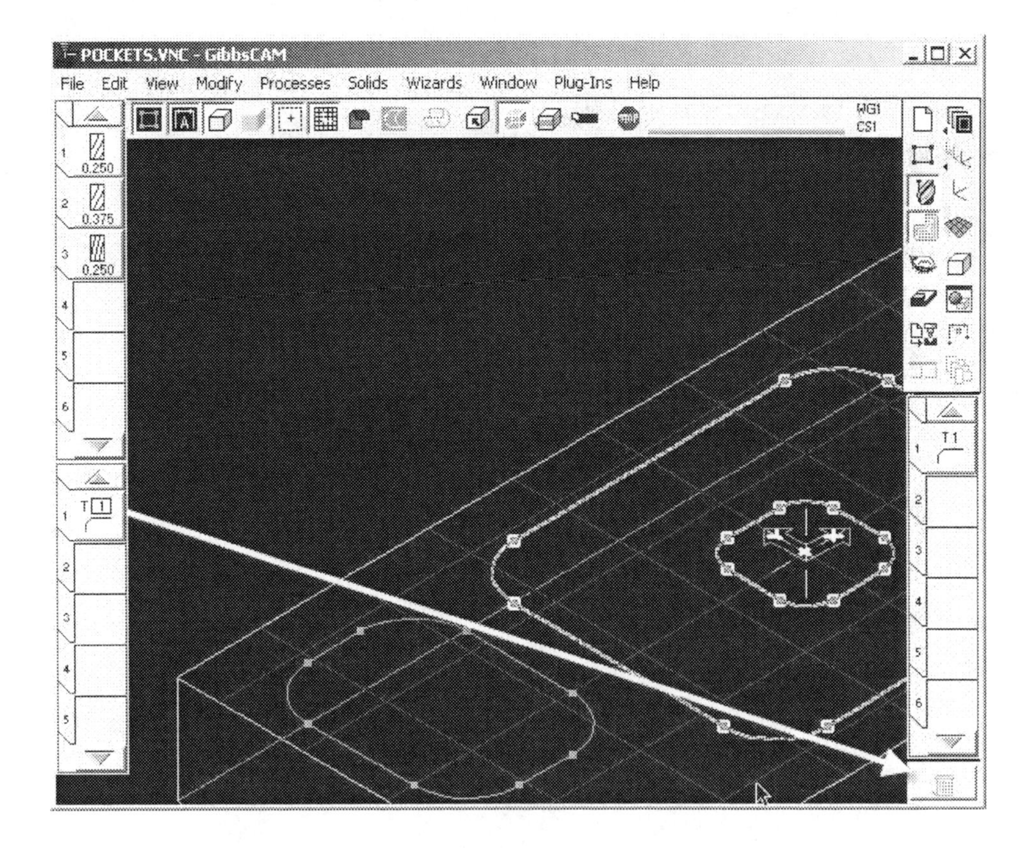


Before you begin creating the next machining operation, you will need to change one of the lines of one of the pockets to an air wall. This is because one of the sides of the pocket will be not be surrounded by material.

Select the line as shown with the right-mouse button. Select Change feature to from "Wall to Air."

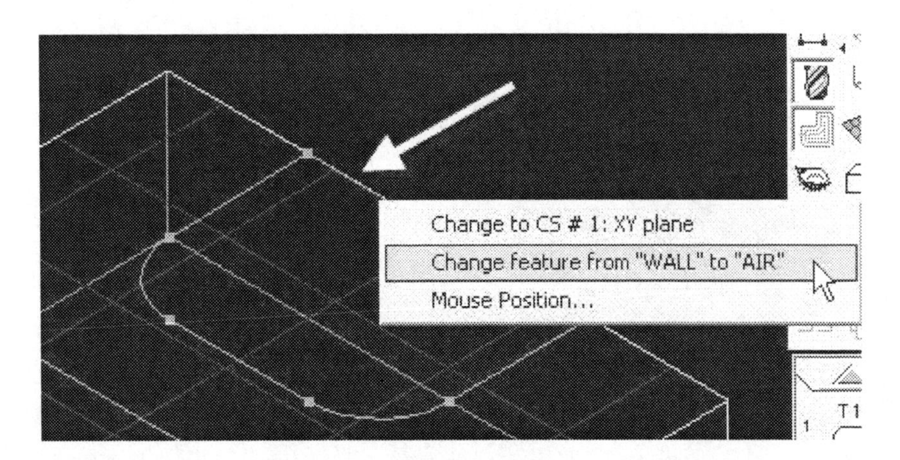

Because the 2 profiles from the previous operation are still selected, click anywhere on the model to deselect them.

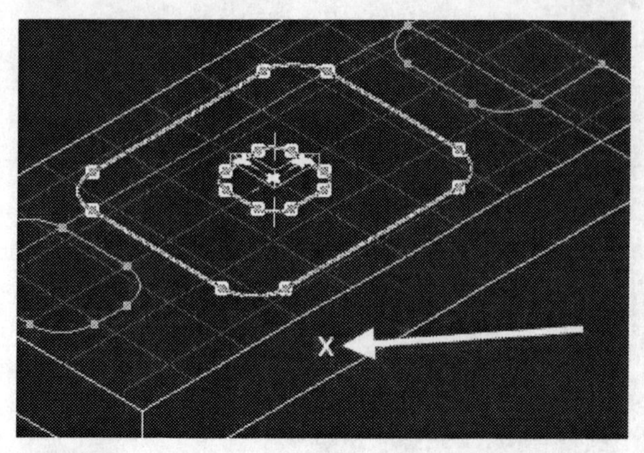


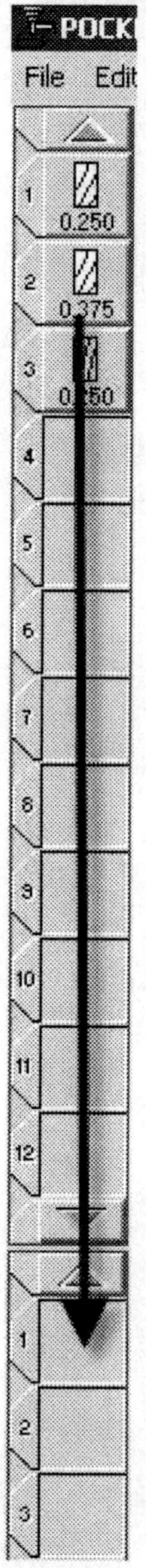


Drag Tool 2 from the Tool List to Tile 1 of the Process List.

Drag the Rough Icon from the Machining Palette to Tile 1 of the Process List.

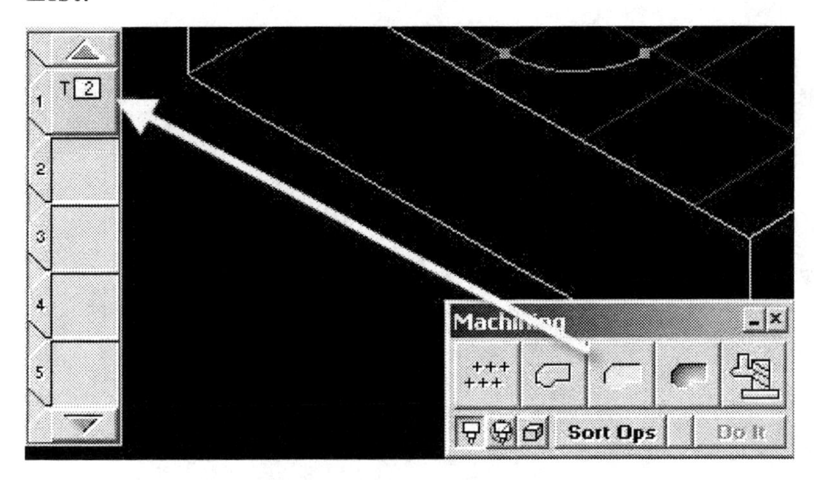


The Roughing dialog box will appear. Enter the values as shown and close the dialog box.

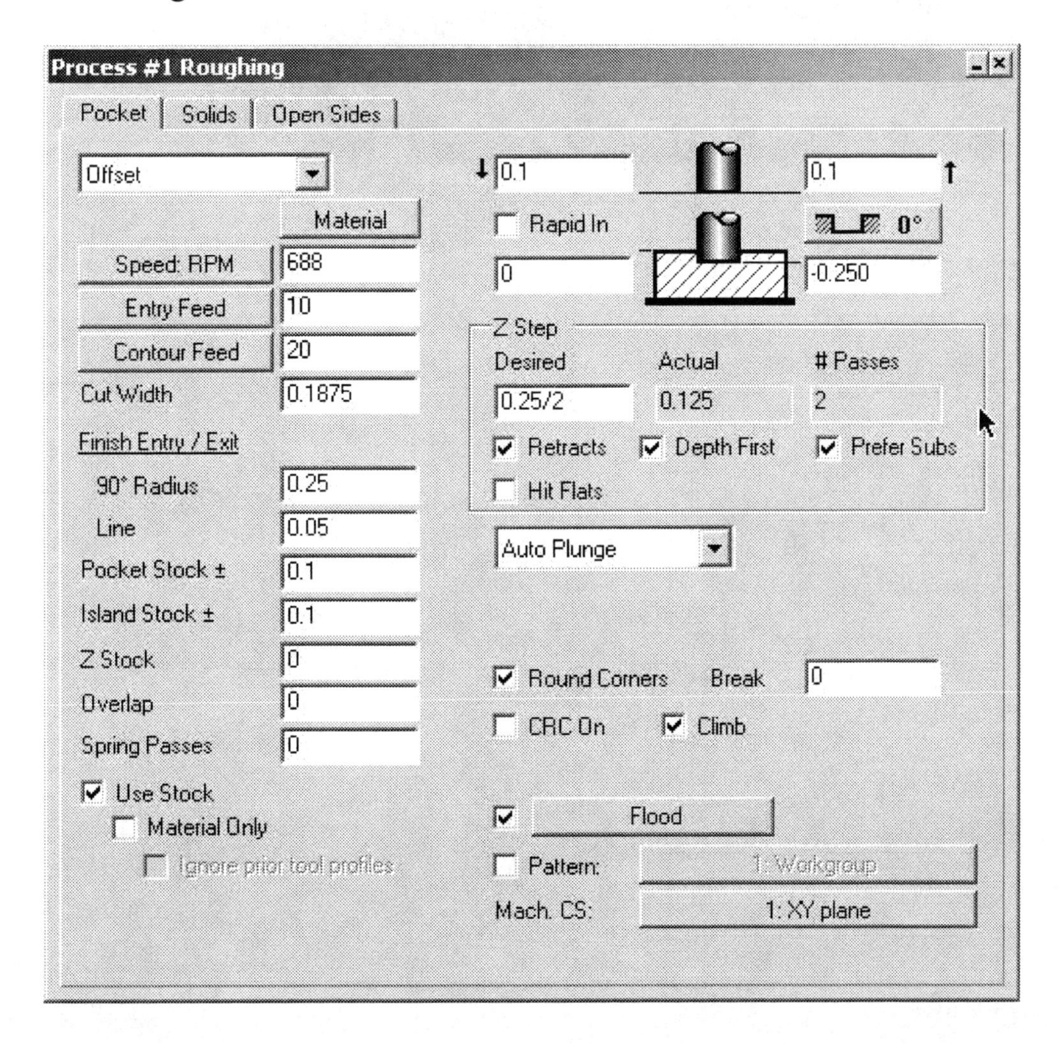

Hold down the control key and select the 2 profiles as shown.

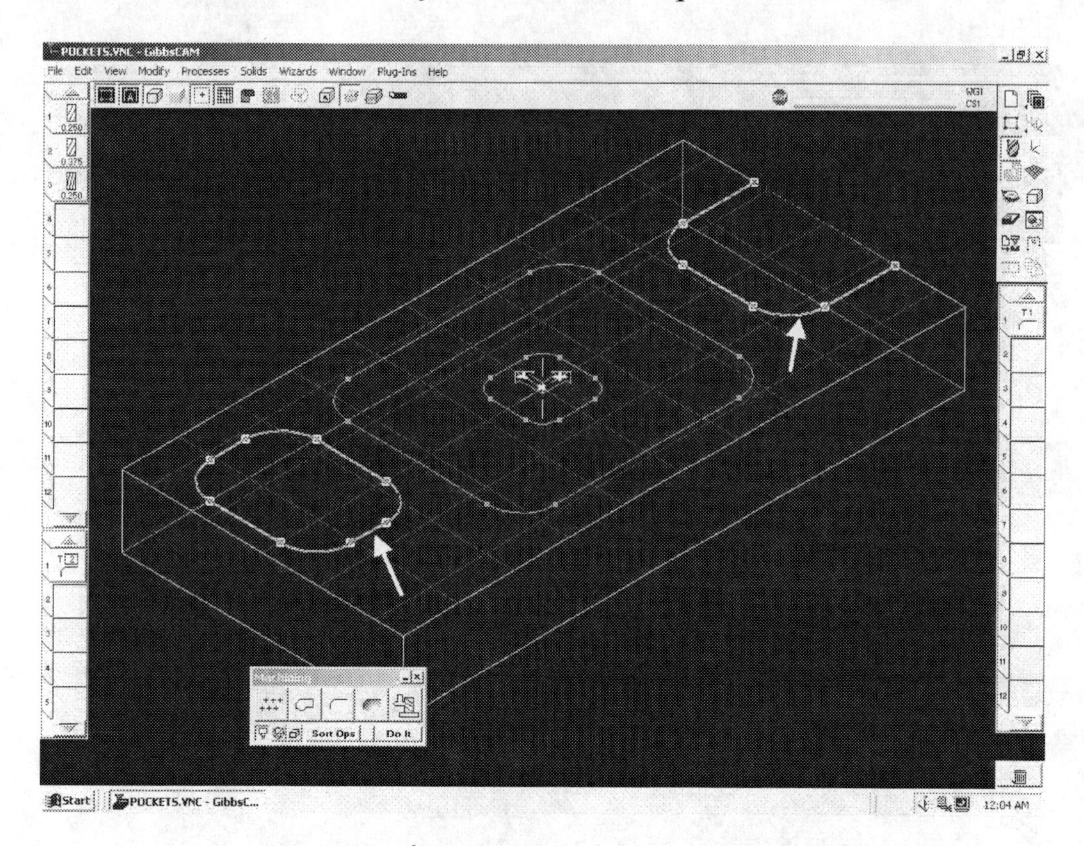


Click on Do It on the Machining Palette.

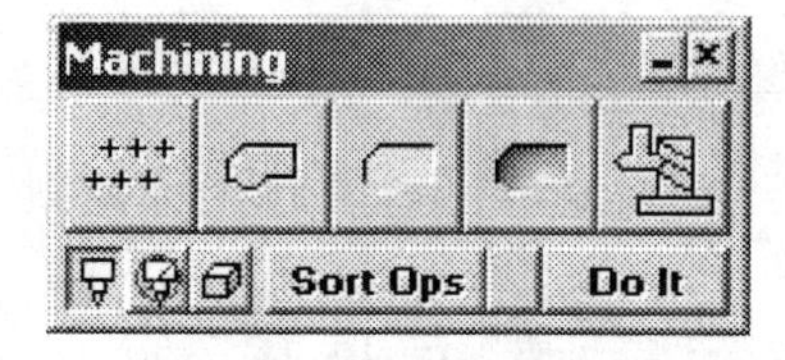


Your screen should look like the next graphic.

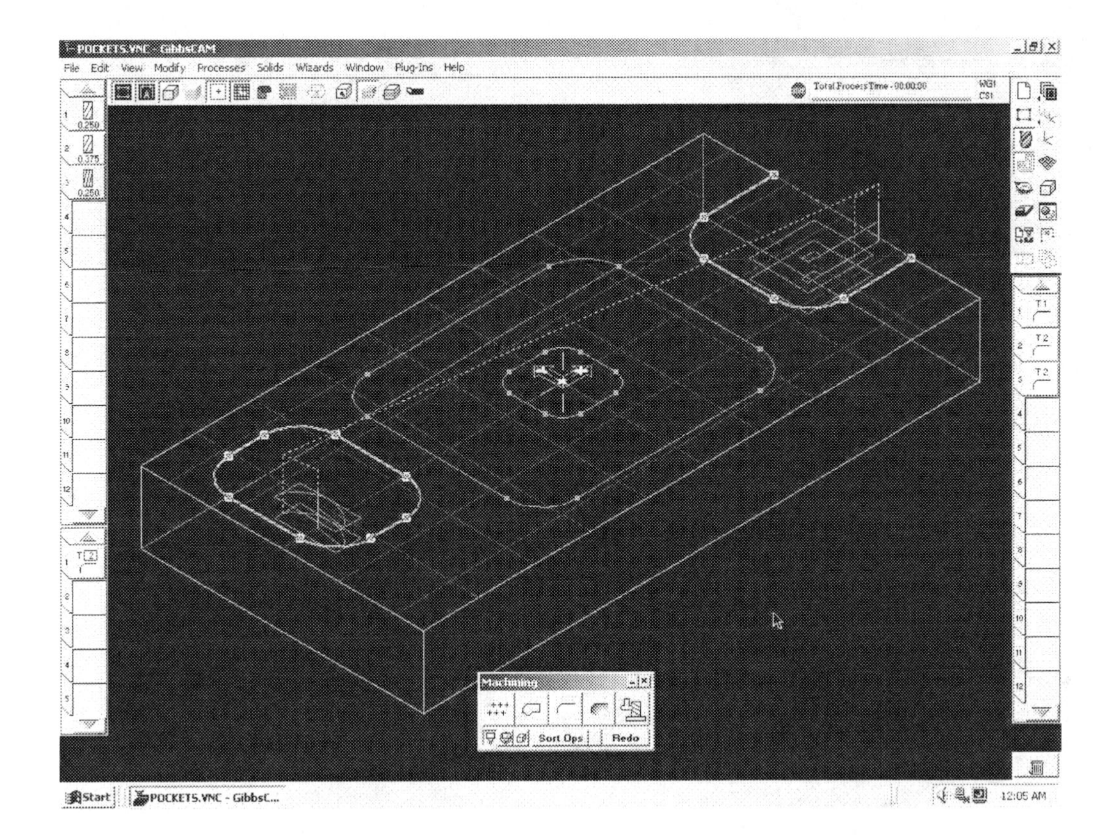


Render the part and then close the Render Control dialog box.

Click on Tile 4 in the Operations List.

Drag Tile 1 from the Process List to the Trash Can.

Drag Tool 3 from the Tool List to Tile 1 of the Process List.

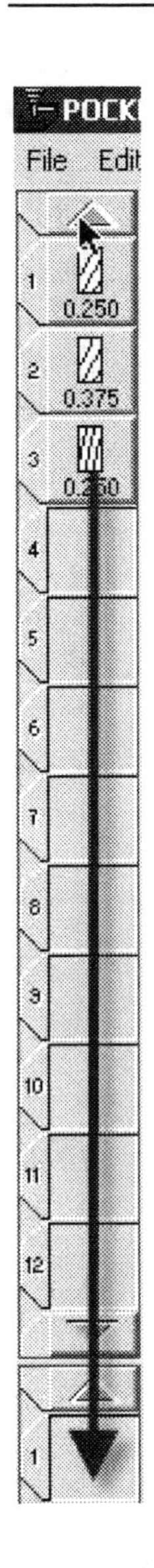


Drag the Contour icon from the Machining Palette to Tile 1 of the Process List.

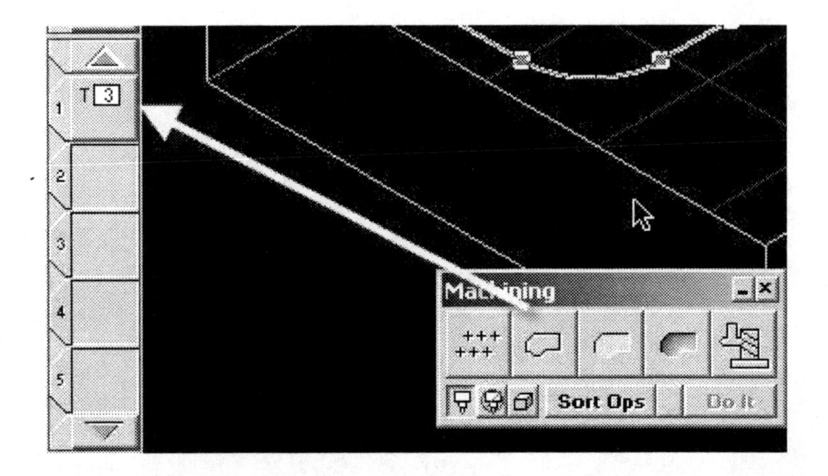

The Contour dialog box will appear. Enter the values as shown and close the dialog box.

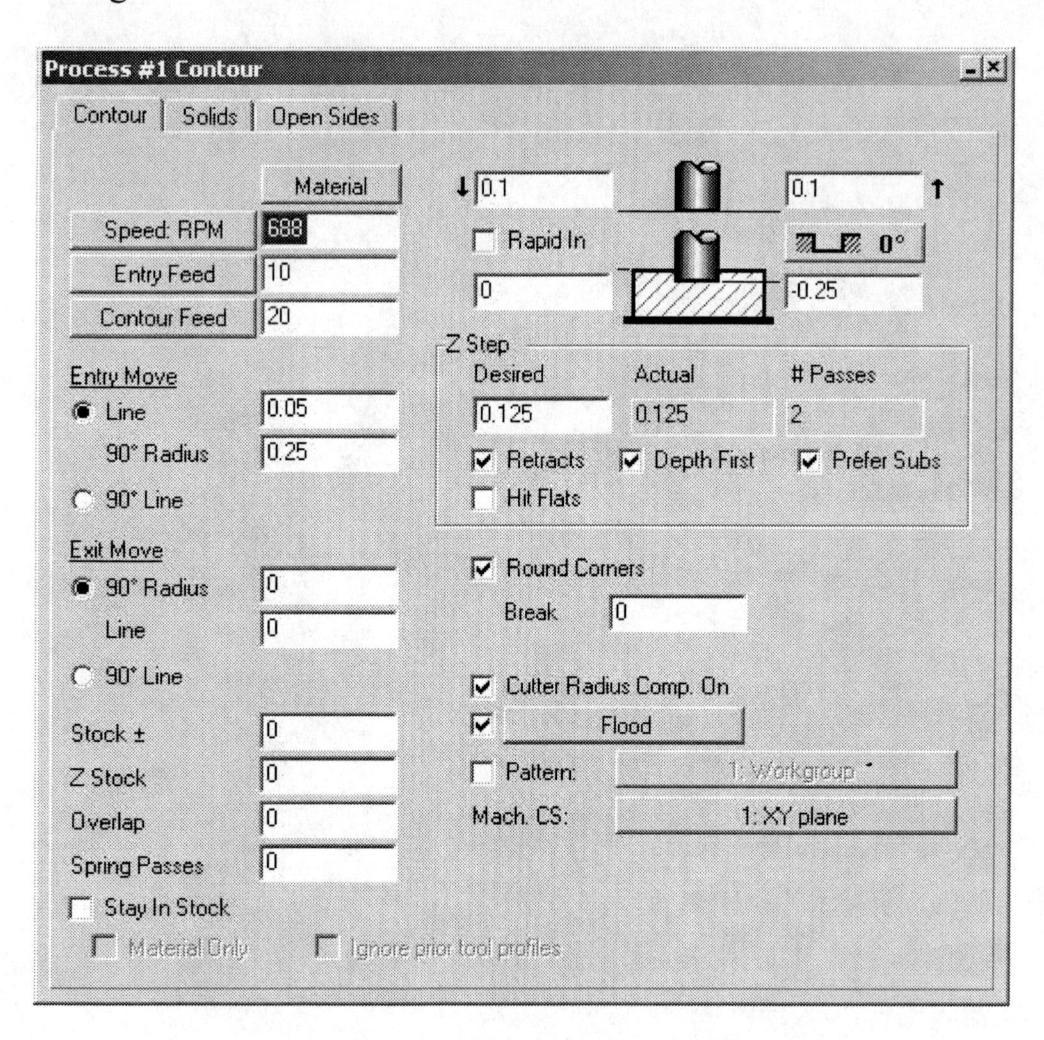

Select the pocket as shown. Click on the circle that designates the inside tool position as shown. The blue arrow designates the direction of the cutting tool. Select the arrow as shown.

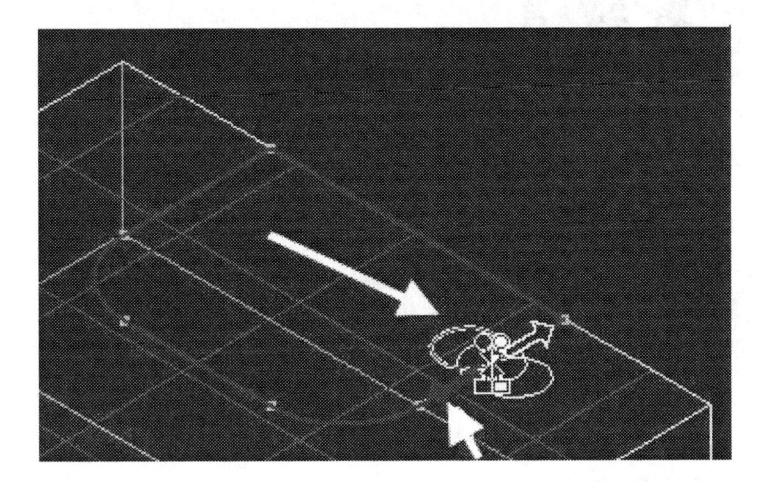

Notes:

1. The system places 3 circles on the profile. You can place the cutting tool on the inside, centerline, or outside edge of the profile by selecting the appropriate circle.
2. The blue arrow indicates the tool direction. You can change the tool direction by clicking on the other arrow.
3. The white square (Start Feature) indicates the feature that the tool will start cutting on.
4. The white dot (Start Point) is the point on the Start Feature where the tool will begin cutting.
5. The black square (End Feature) indicates the feature that the tool will stop cutting on.
6. The black dot (End Point) is the point on the End Feature where the tool will stop cutting.
7. You can change the position of any of these markers by clicking and dragging them.

Click and drag the black square (end feature) to the line segment as shown.

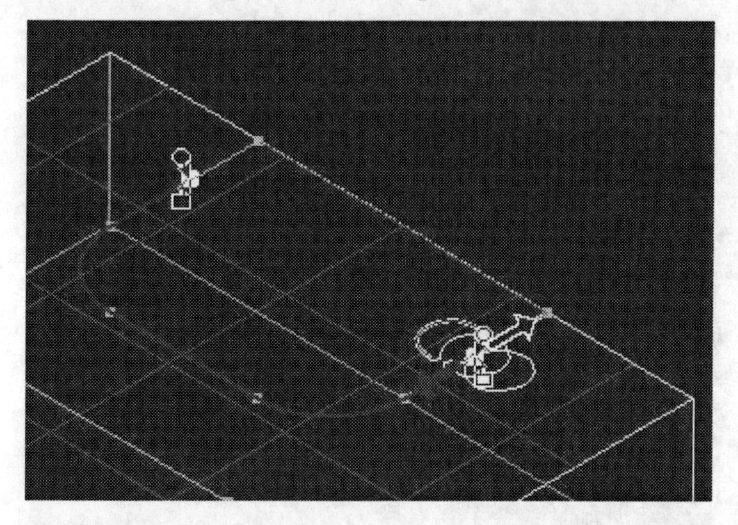

Click and drag the black circle (end point) out past the stock as shown.

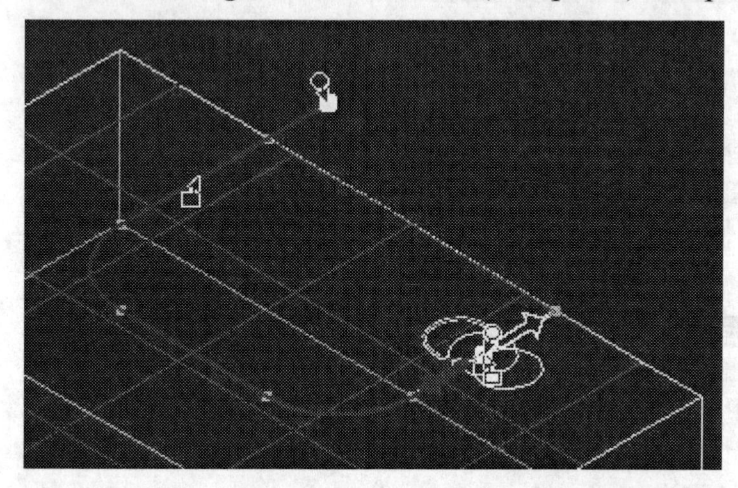

Click and drag the white dot (start point) out past the stock as shown.

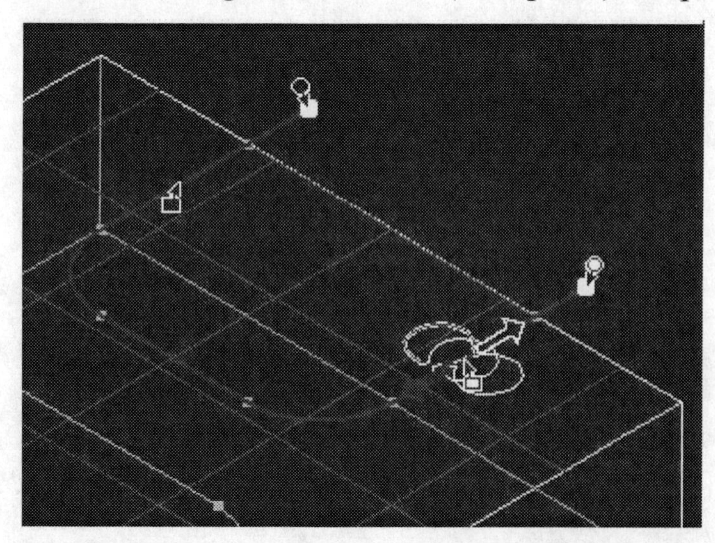

Click on Do It on the Machining Palette.

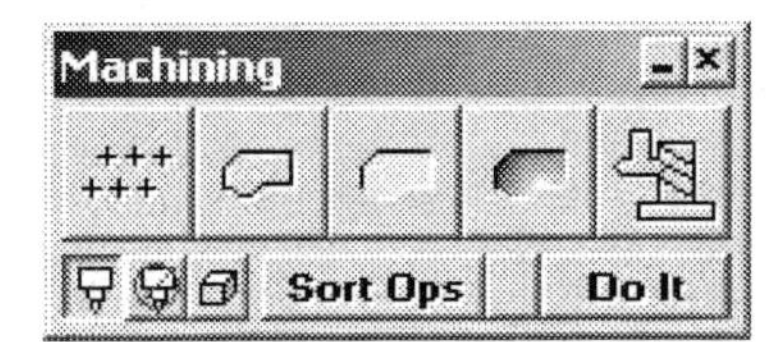


Your screen should look like this.

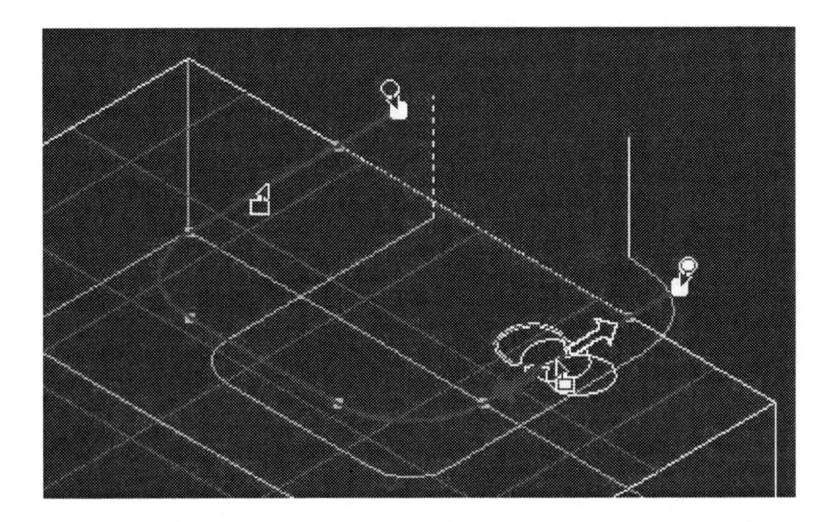

Render the part.

## On your own:

Using Tool 3, add a finishing pass (contour operation) on the remaining 3 profiles. Use a cutting depth of .25 inches.

This completes the exercise. Save the File. Select *File-Save* from the *Main Menu.*

## Exercise 5 – Machining from a 3D CAD Solid Model

In this exercise you will rough out a pocket using the Profiler and determine cutting depths from the model itself.

Open the file SOLIDMODEL.VNC that you created in Chapter 3 – Exercise 2.

Create a .250 dia. Rough End Mill tool.

Activate the Profiler and move it so the profiles of the pocket are visible (green).

Drag Tool 1 from the Tool List to Tile 1 of the Process List. Drag the Roughing icon from the Machining palette to Tile 1 of the Process List.

The Roughing dialog box will appear. To determine the top Z surface of material for this operation, place your cursor in the dialog box that is highlighted with a black frame in the graphic below. Hold down the Alt key and pick the top face of the part as shown. This value will be generated by your selection on the model.

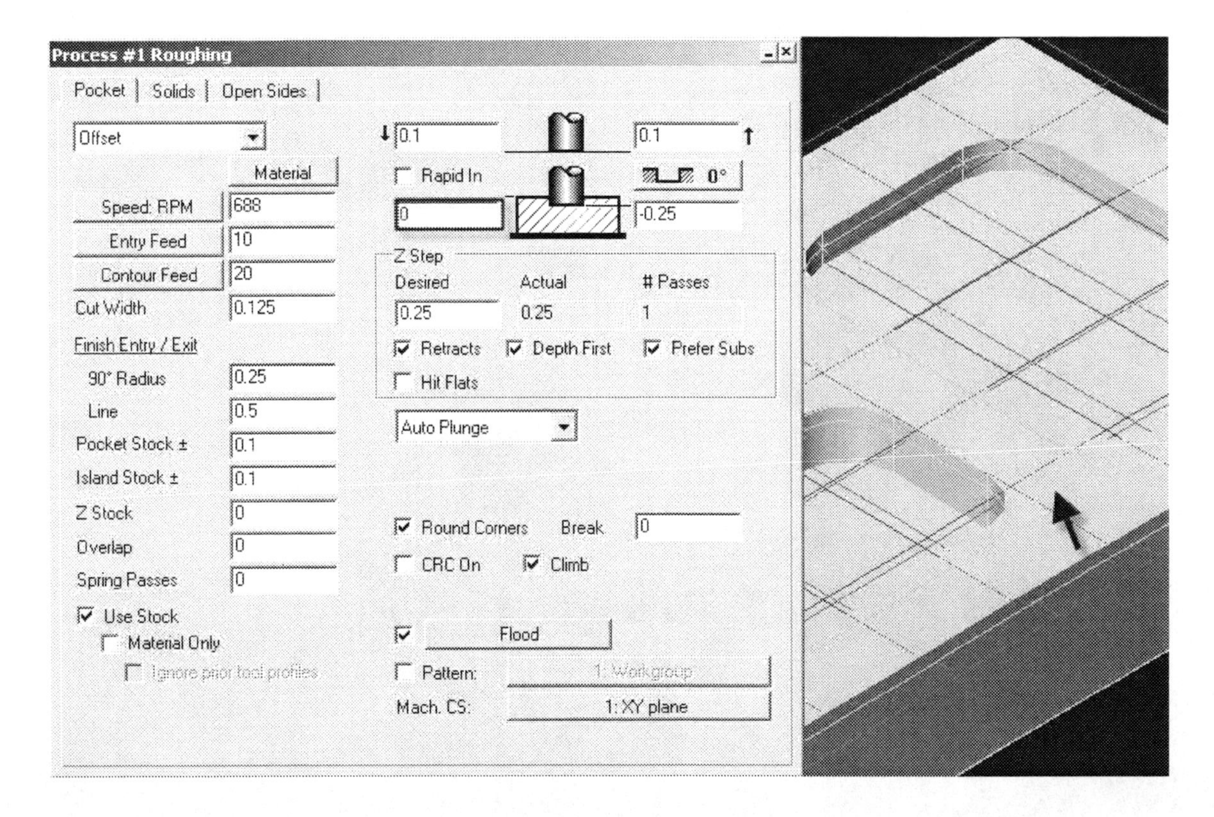

To determine the cutting Z depth, place your cursor in the dialog box that is highlighted with a black frame in the graphic below as shown. Hold down the Alt key and select the bottom face of the pocket as shown. Enter the remaining values as shown

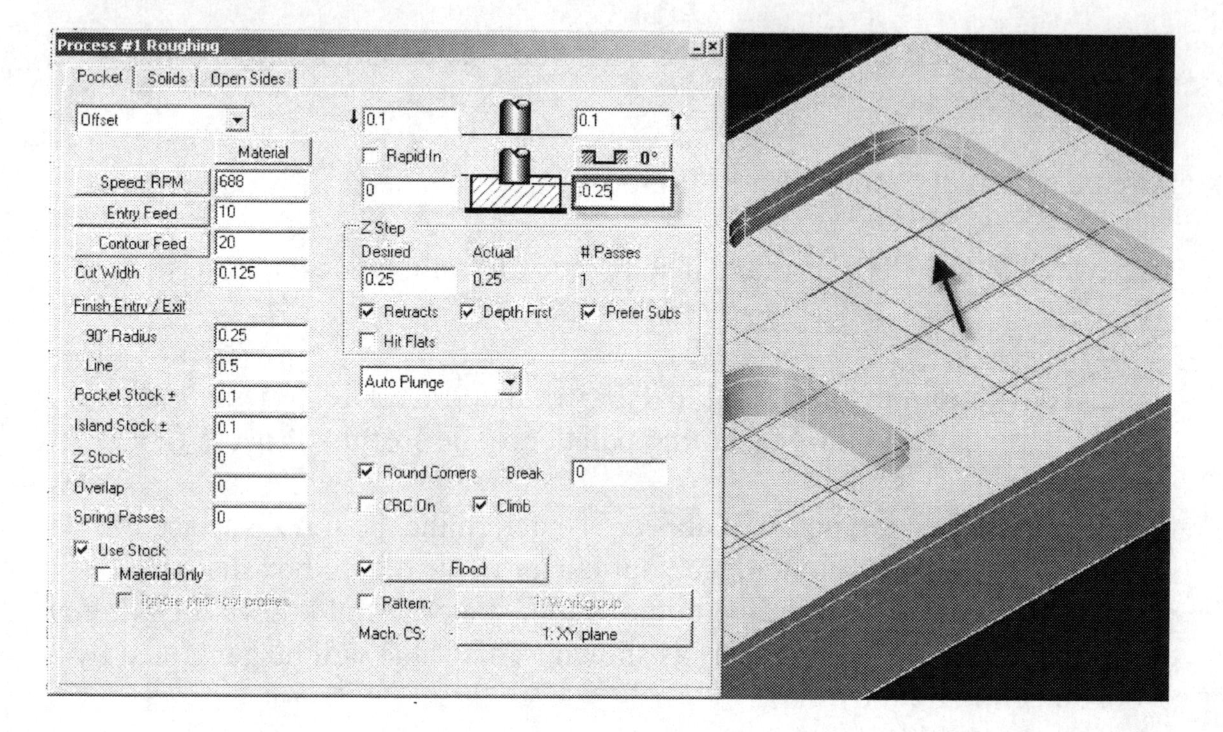

Close the dialog box. Double-click on the green profile as shown and click on Do It on the Machining Palette.

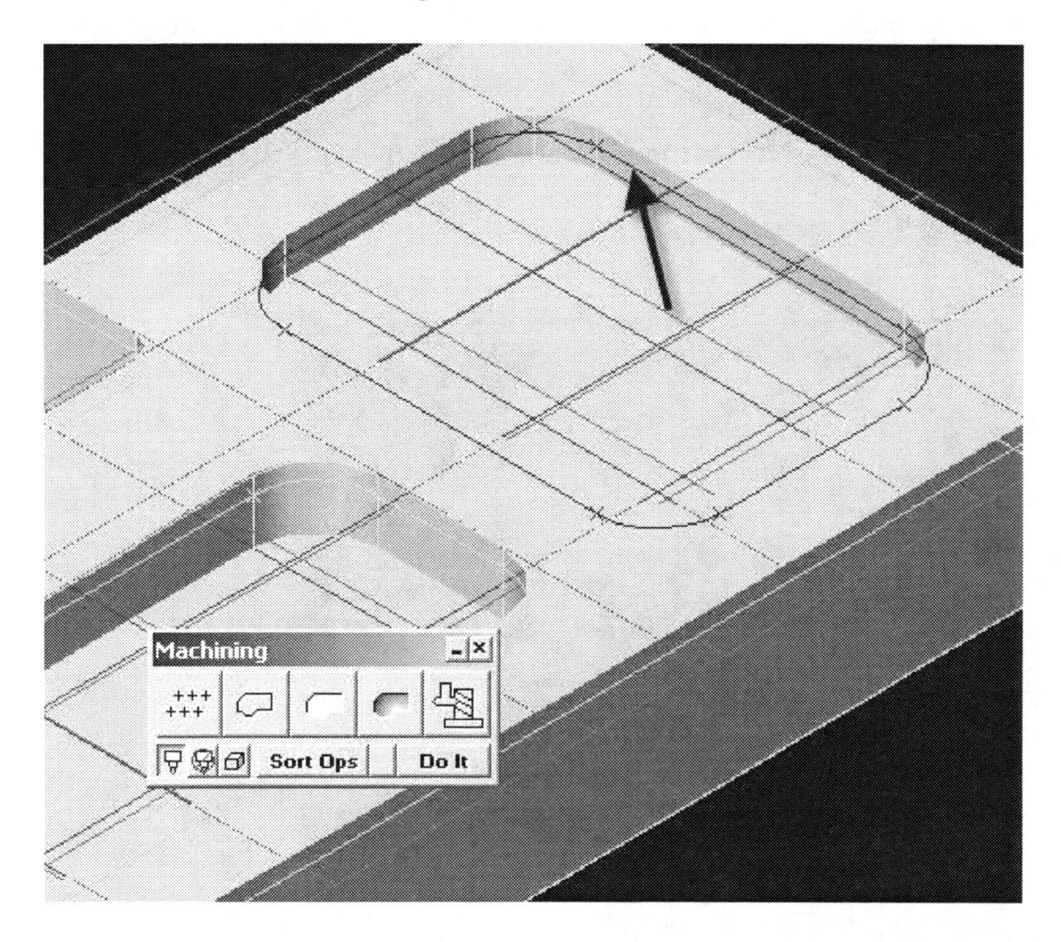


Render your part.

**On your own:**

Rough out the other pocket using the profiler and Tool 1.
Create Tool 2 - .250 dia. Finish End Mill
Add a finishing pass (contour operation) to both pockets.

This completes the exercise. Save your file. Choose *File-Save* from the *Main Menu*.

## Exercise 6 – Editing Machining Operations

In this exercise you will edit a machining operation that you created in Exercise 1 of this chapter.

Open the file POINTS.VNC that you created in Chapter 4 – Exercise 1

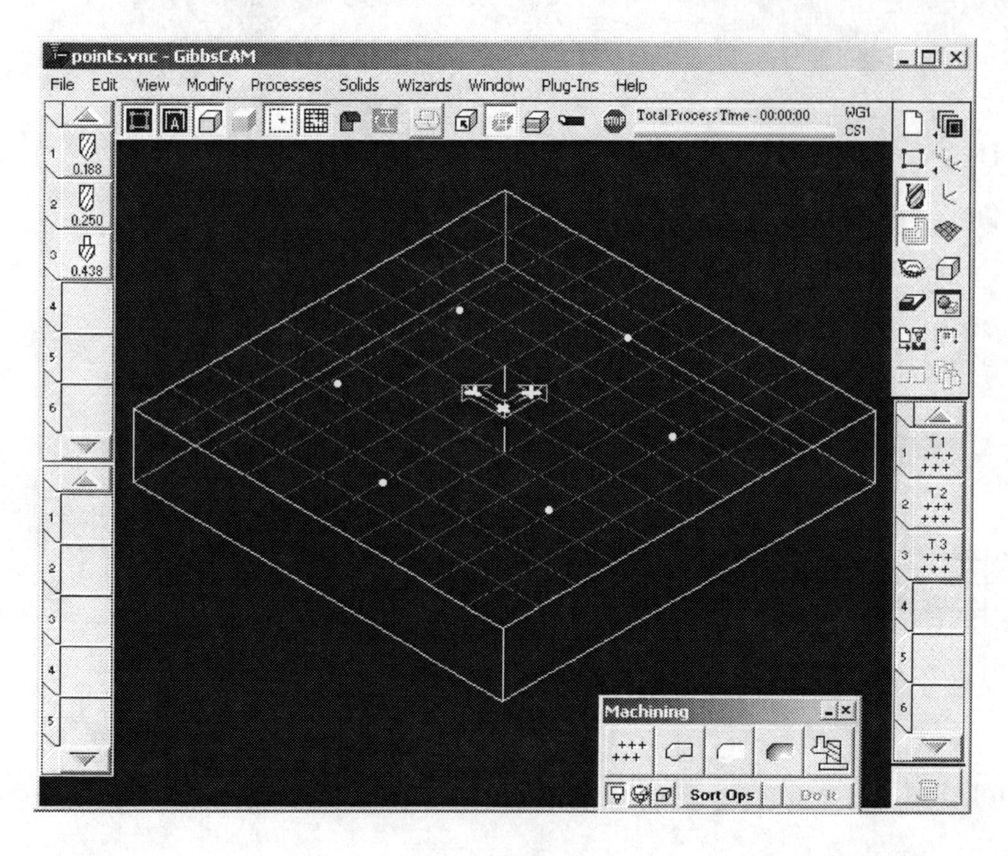


Double-click on Tile 2 in the Operations List as shown in the next graphic.

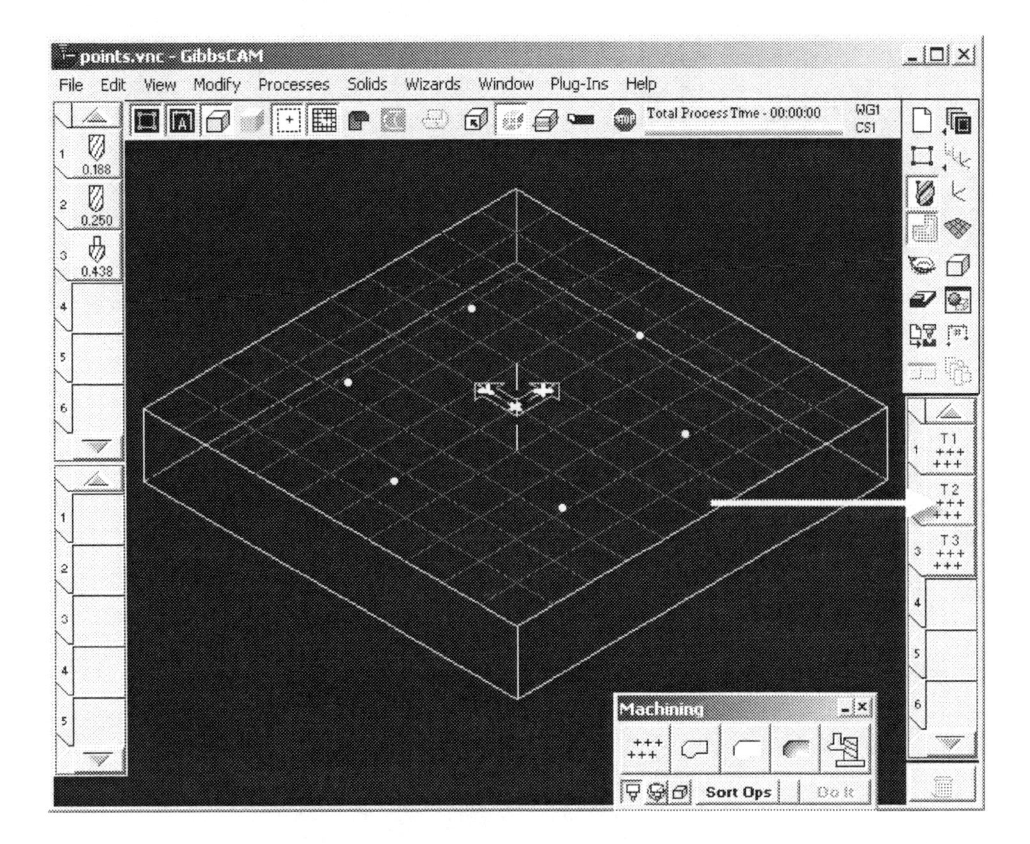


This will activate Tile 1 that you created in the Process List.

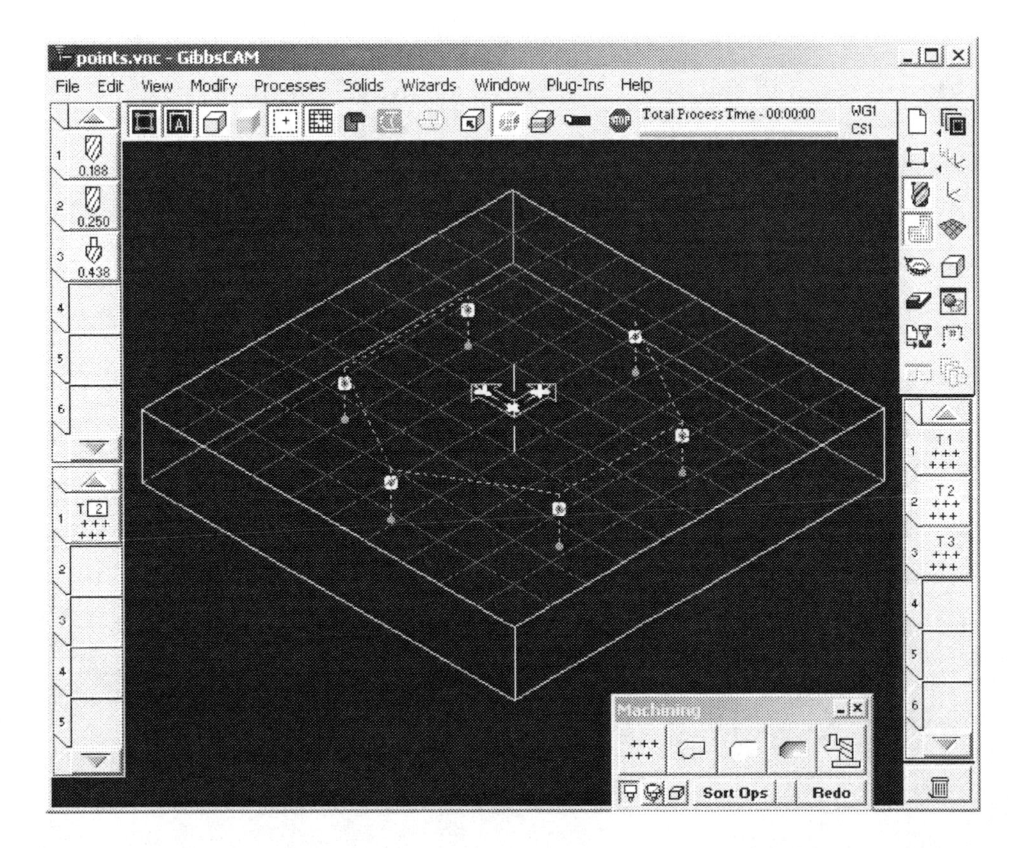

Double-click on Tile 1 in the Process List. The Process dialog box will appear. Edit the Z depth to -.350 as shown.

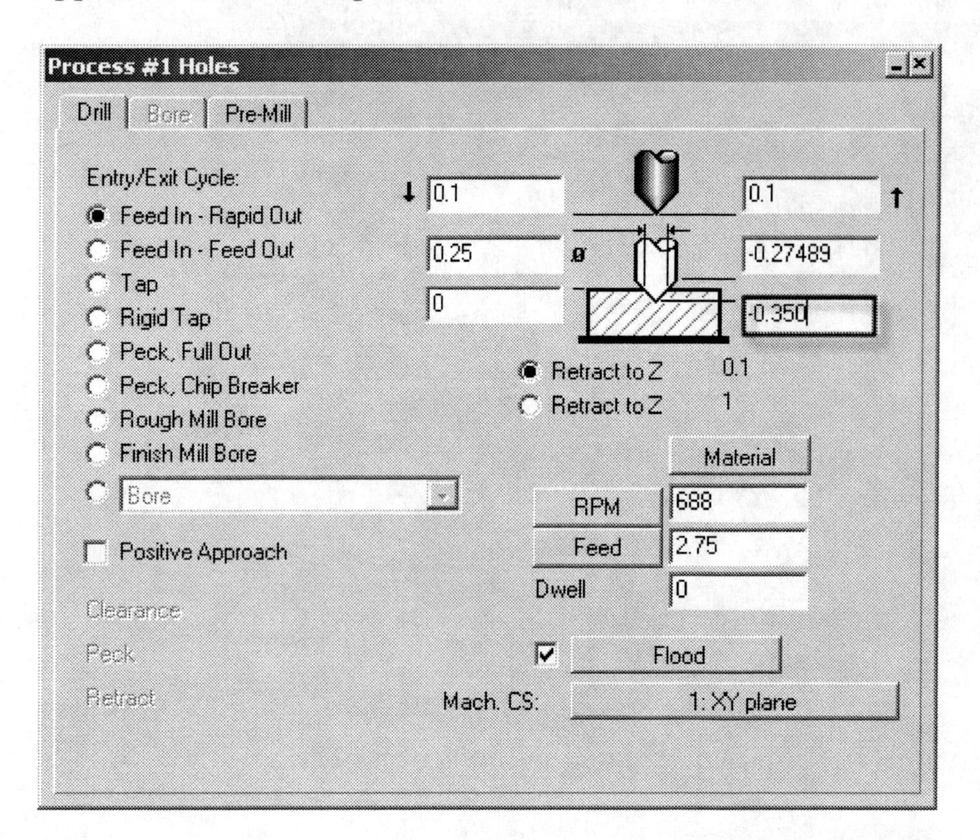


Close the dialog box. Click on Redo on the Machining Palette.

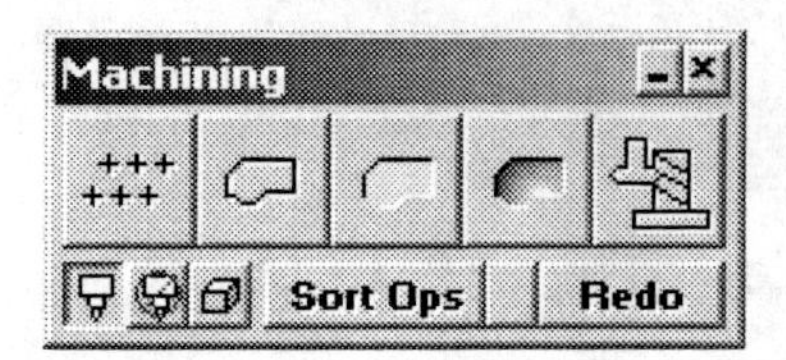


The machining operation has been edited. Render your part.

This completes the exercise. Save your file. Choose *File-Save* from the *Main Menu*.

Chapter 5

# Lathe

## Exercise 1 – Lathe

## Introduction

It is recommended that you complete Chapters 1-4 before attempting this exercise.

Moving from the mill to the lathe is an easy transition in GibbsCAM. The geometry creation tools for the Lathe are the same as the production mill. The CAM process for the Lathe module is also the same as the production mill. Once CAD geometry has been created or imported, you set up tools, create machining operations, simulate the machining operations, and post process the output into CNC code.

Note: The machine type, units, stock size, and CAD geometry for this part have been set up. This exercise uses inch units.

**Note: For the exercises in this chapter, activate balloons (found under *Help* in the *Main Menu).* This will allow you to view a full explanation of all of the data entry fields you will see in the tool creation and machining dialog boxes as you complete the exercises.**

Choose *File-Open* from the *Main Menu*. Select the file LATHE.vnc and click on Open. *This file can be downloaded at www.schroff1.com.*

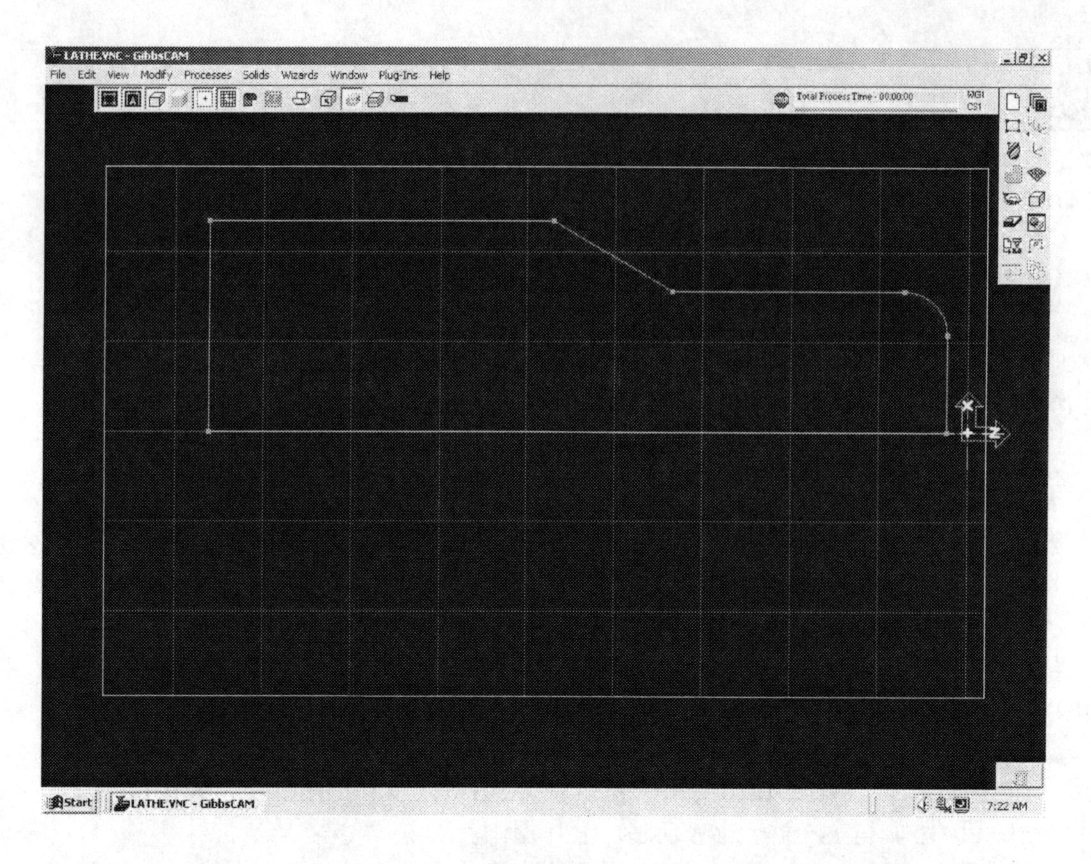


Select the Document icon on the Top Level Palette. The.Machine Type has been set to 2 Axis Lathe – 0.75 Shank and the stock size as shown in the following image.

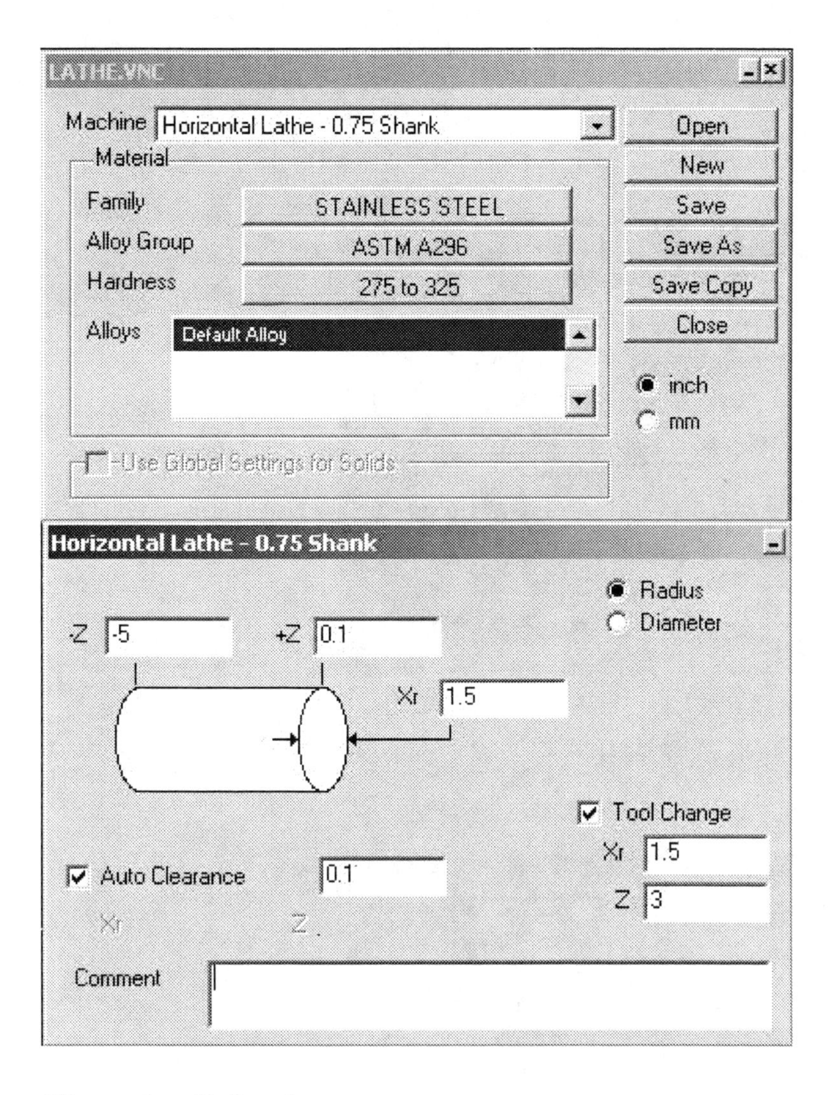


Close the dialog box.

To create the tools needed for this exercise, select the Tools icon in the Top Level Palette.

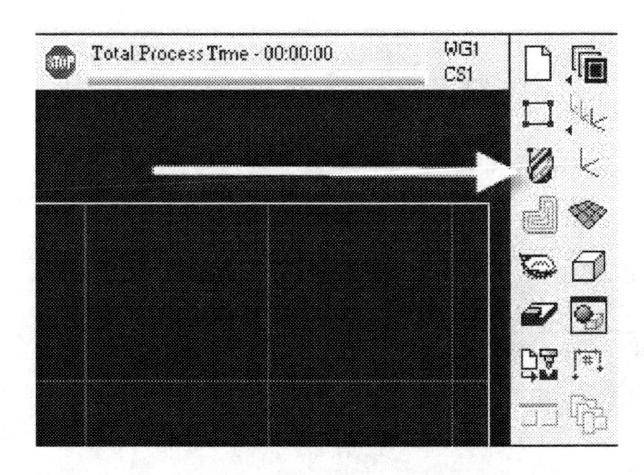

The Tool List will appear.

Double click on Tile 1 in the Tool List to create the first tool.

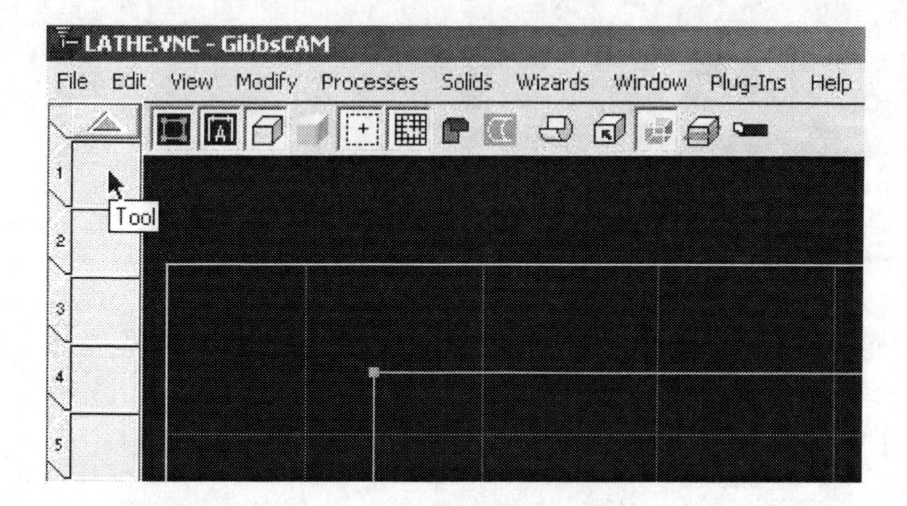


The Tool Creation dialog box will appear. Select the 80° C insert as the tool type.

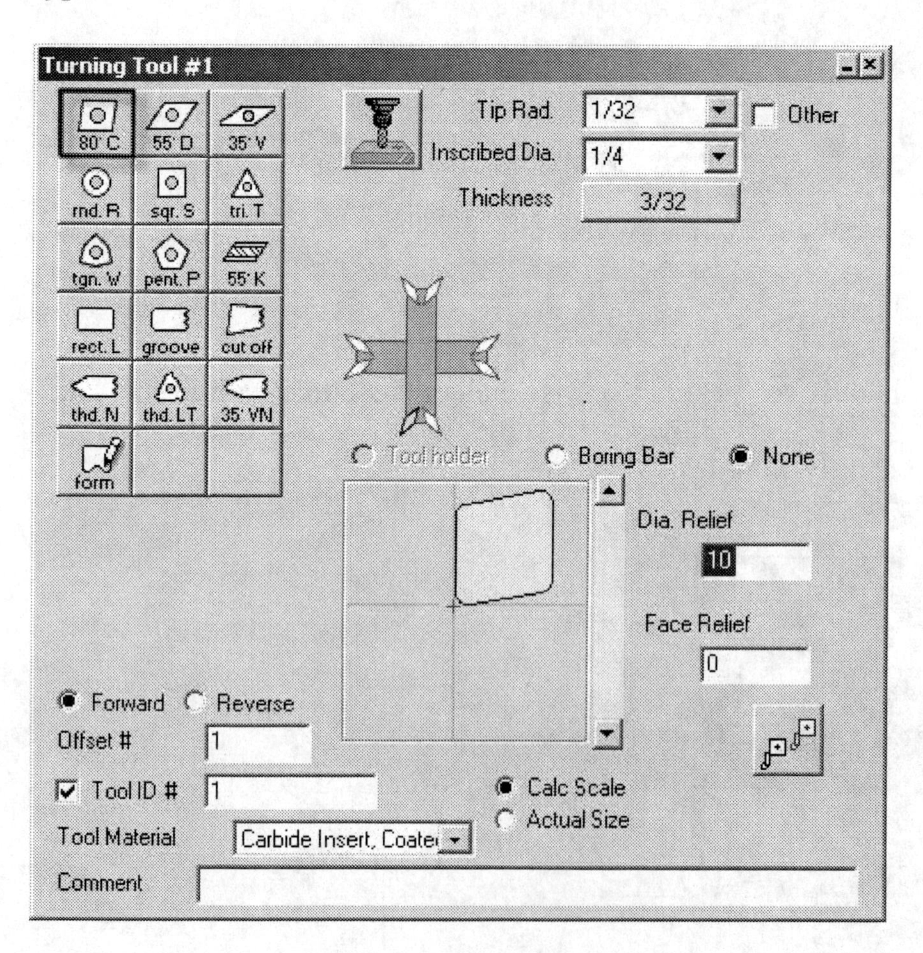

Define the size of the insert by selecting the values as shown from the pull-down menus.

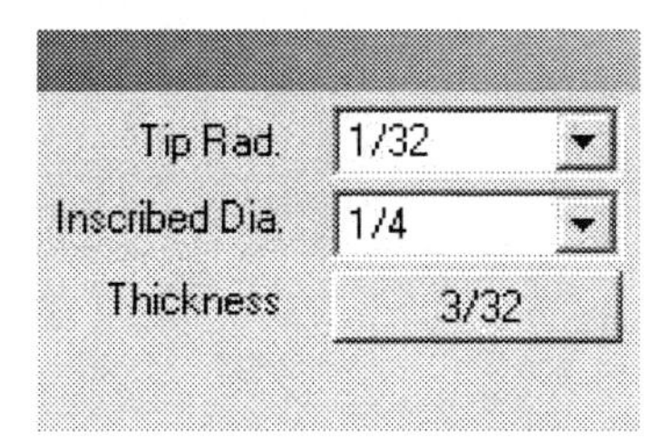


The Insert Orientation diagram determines the position of the tool insert. Select the position as shown.

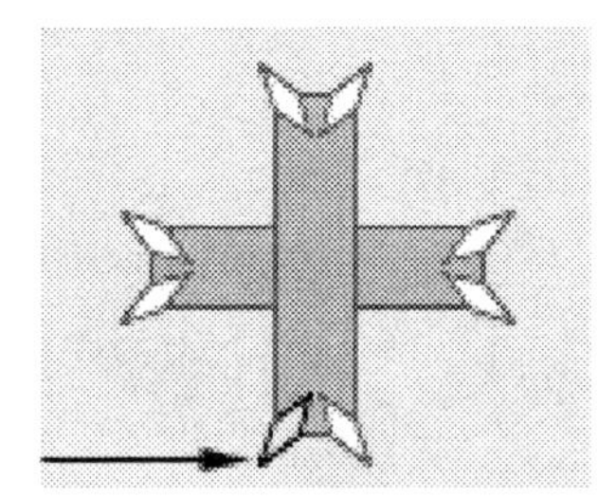

Click on the Tool Holder option as shown in the image below. A diagram of the tool holder will appear. The red circle represents the touch-off point for the tool. You can use the scroll bar to view and select more options in the list to the right of the tool holder diagram.

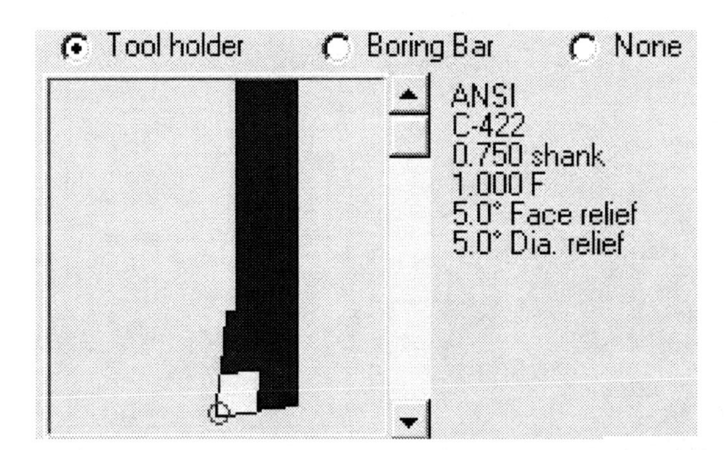


Close the Tool Creation dialog box. The tool will appear in the tool list.

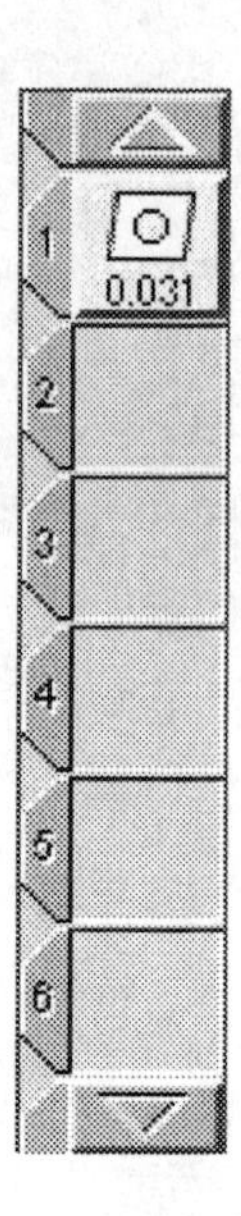


Double click on Tile 2 in the Tool List to create the second tool. The Tool Creation dialog box will appear. Select the 55° D insert. Enter the values as shown for the size and position of the insert. Close the Tool Creation dialog box.

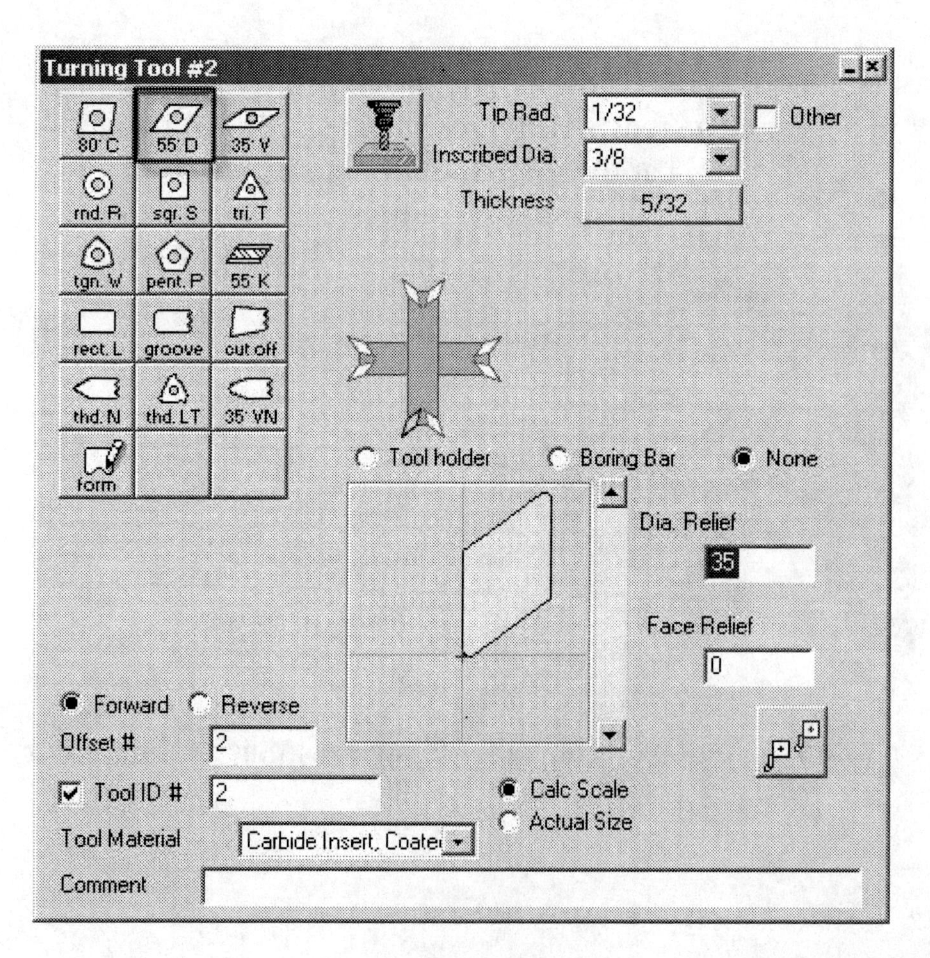

The tool will appear in the Tool List.

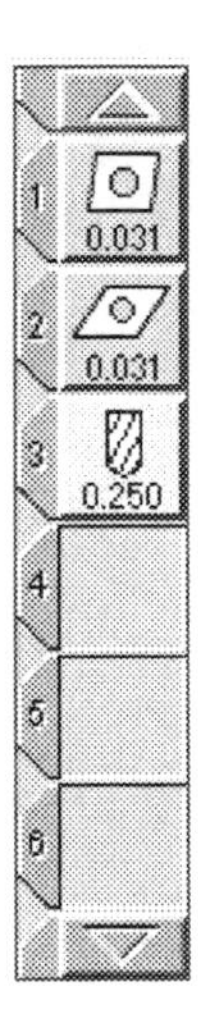


Select the CAM icon from the Top-Level Palette.

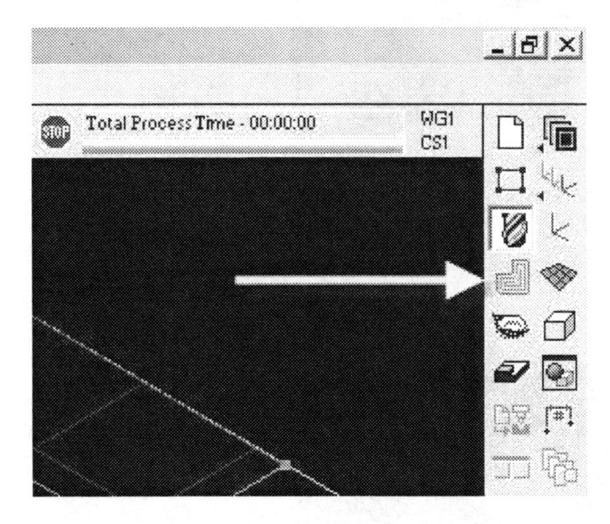


The Machining Palette, Process List, and Operations List will appear.

Machining operations are created by dragging a tool from the Tool List and a Machining Function from the Machining Palette on to the Process List.

Start the first machining operation by dragging Tool 1 from the Tool List to Tile 1 on the Process List.

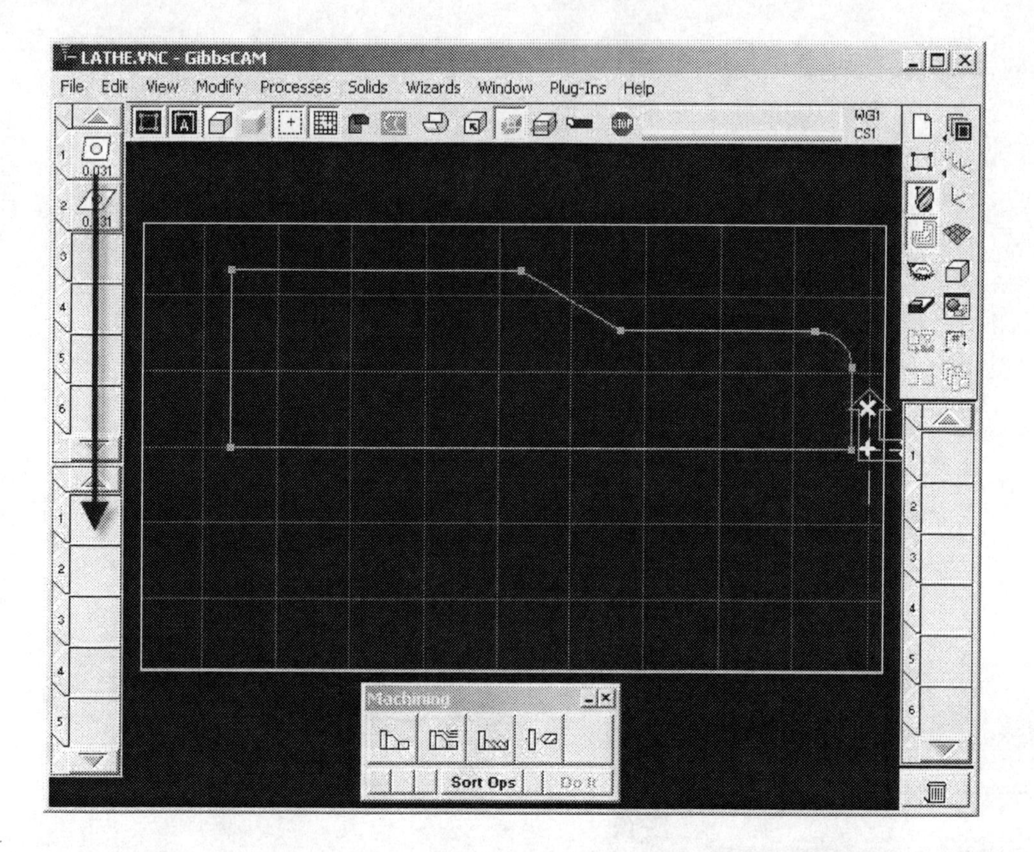


Drag the Contour icon to Tile 1 on the Process List.

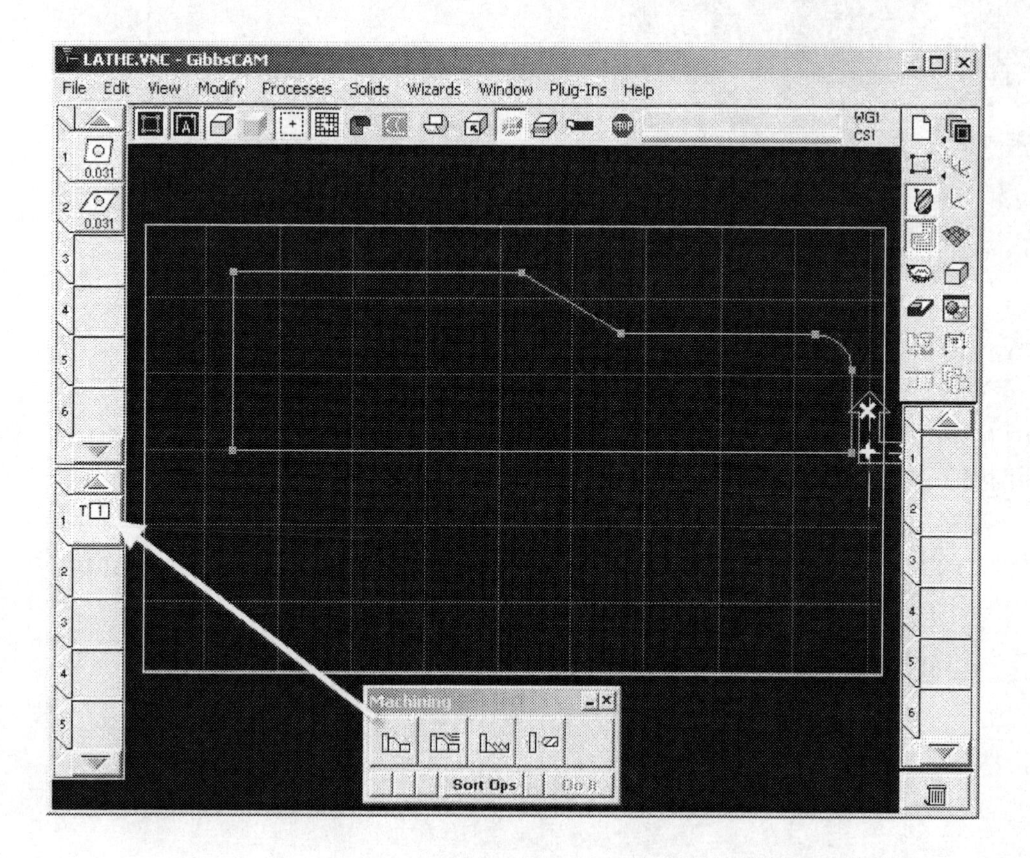

The Process dialog box will appear. Enter the values as shown.

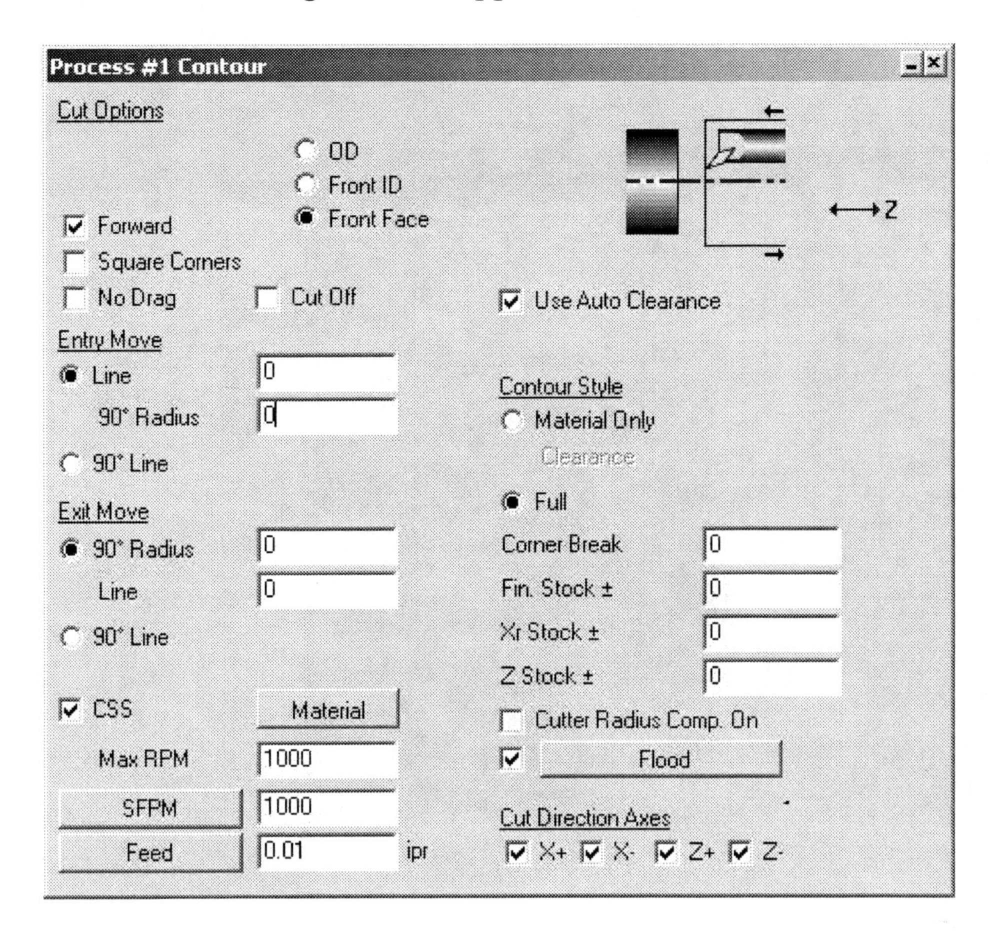


Close the Process dialog box.

Select the line as shown.

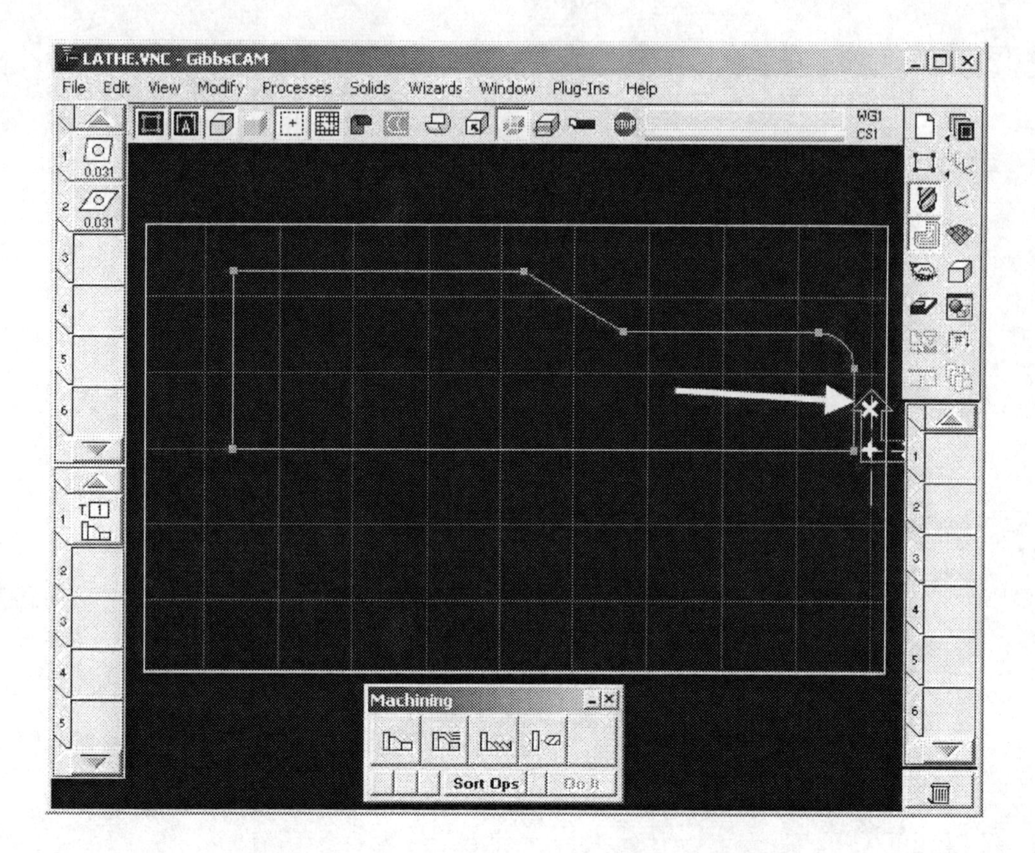


Notes:

1. The system places 3 circles on the profile. You can place the cutting tool on the inside, centerline, or outside edge of the profile by selecting the appropriate circle.
2. The blue arrow indicates the tool direction. You can change the tool direction by clicking on the other arrow.
3. The white square (Start Feature) indicates the feature that the tool will start cutting on.
4. The white dot (Start Point) is the point on the Start Feature where the tool will begin cutting.
5. The black square (End Feature) indicates the feature that the tool will stop cutting on.
6. The black dot (End Point) is the point on the End Feature where the tool will stop cutting.
7. You can change the position of any of these markers by clicking and dragging them.

Position the tool along the outside edge as shown in the image below. Make sure the tool direction is pointing down. Click and drag the white dot (Start Point) and the black dot (End Point) to the approximate locations as shown below.

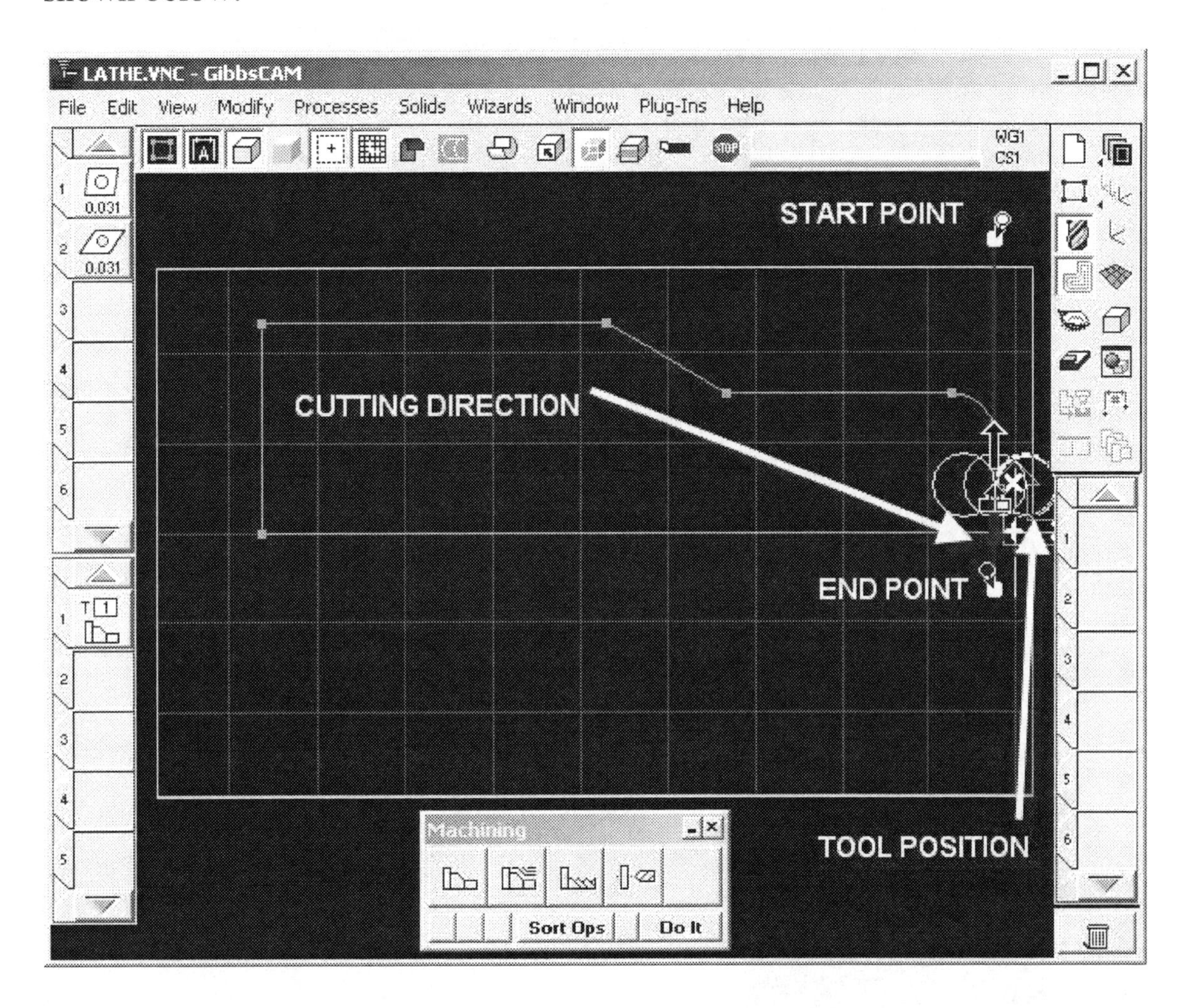


Click **on Do It**.

The orange line represents the tool path.

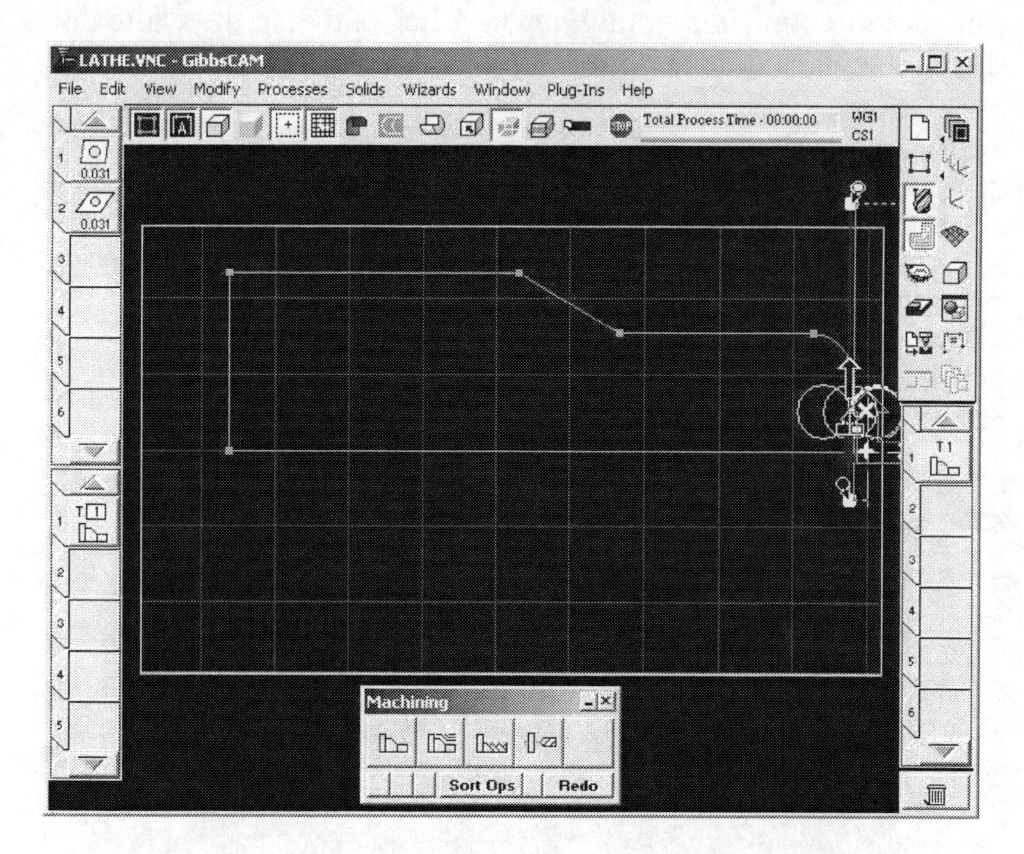


Begin the second operation by selecting Tile 2 on the Operations List.

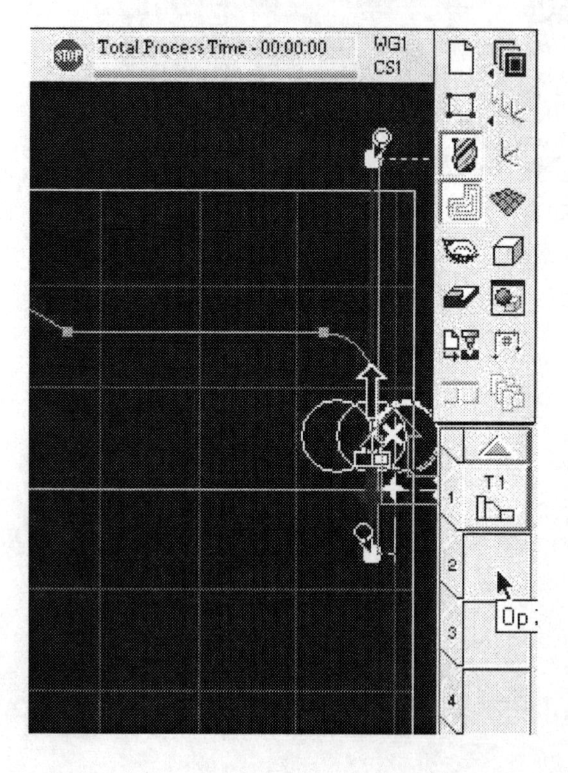

Click and drag Tile 1 on the Process List to the Trash Can.

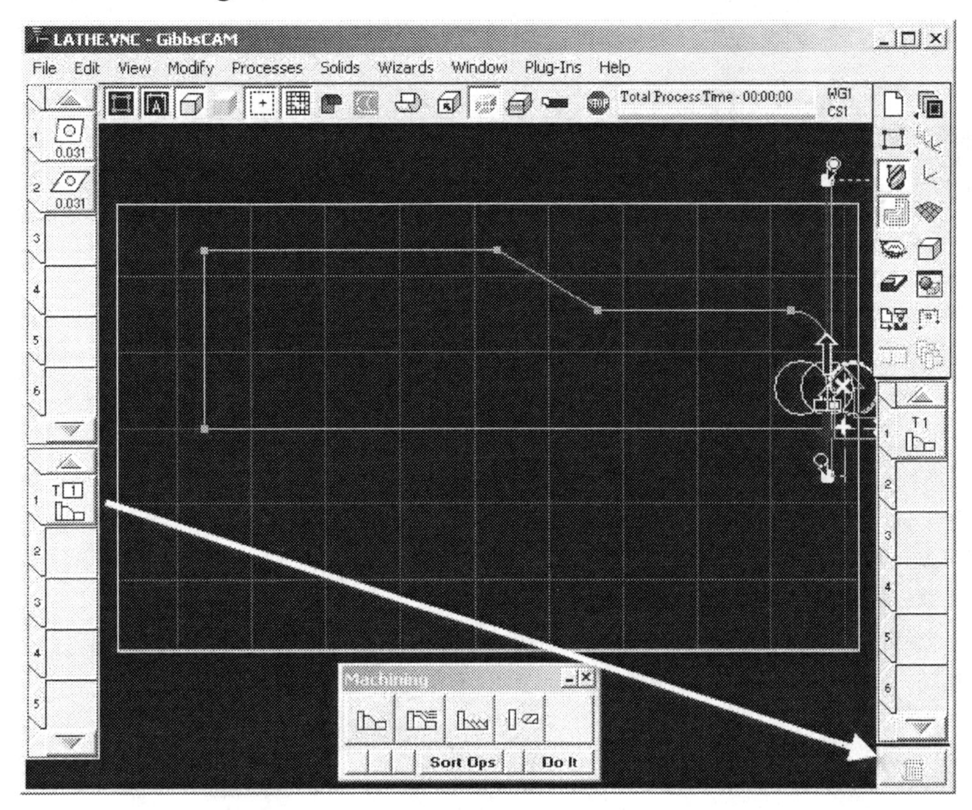


Drag Tool 1 from the Tool List to Tile 1 on the Process List.

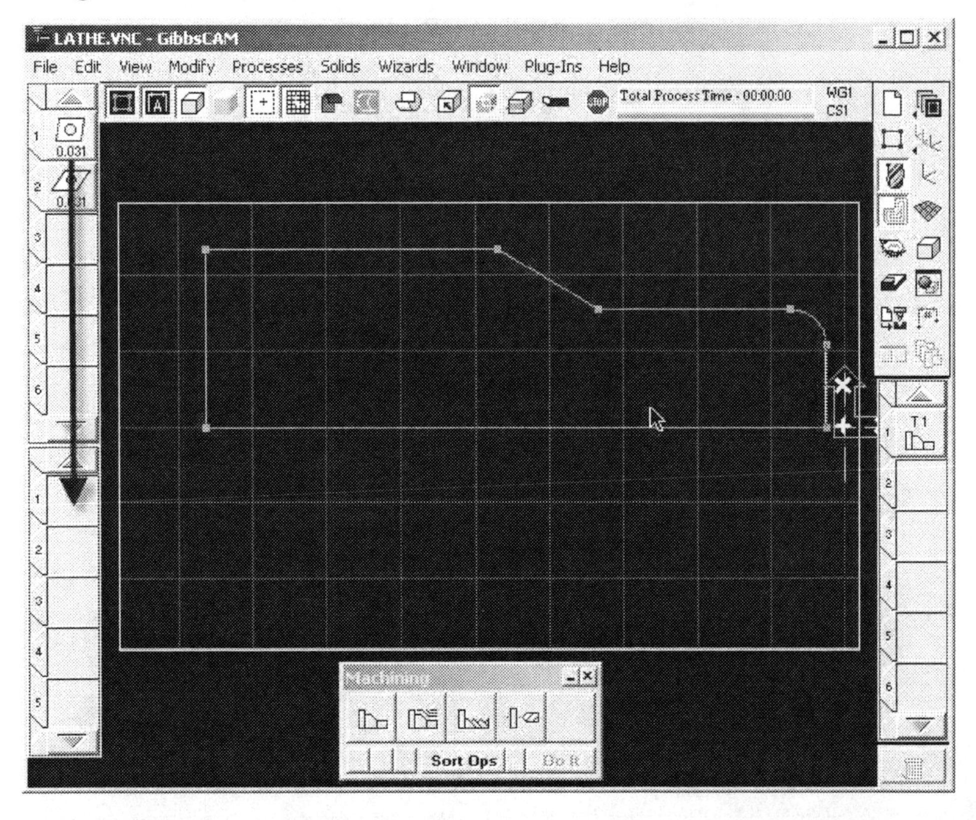

Drag the Roughing icon to Tile 1 on the Process List.

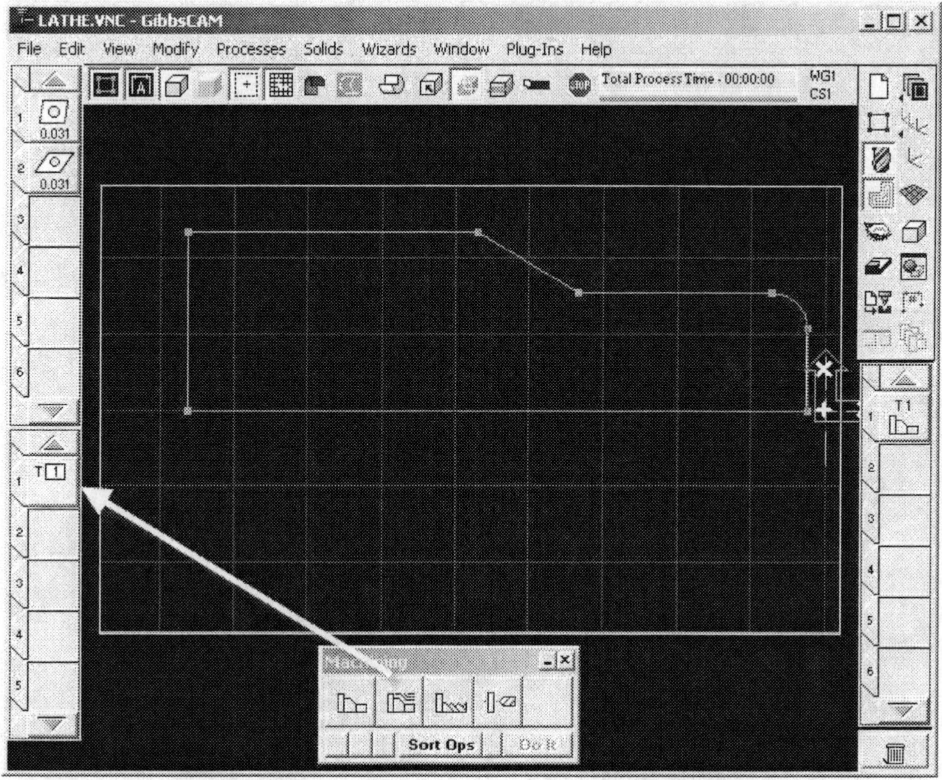


The Process dialog box will appear. This cut will occur along the outside diameter of the part. Select OD and enter the values as shown.

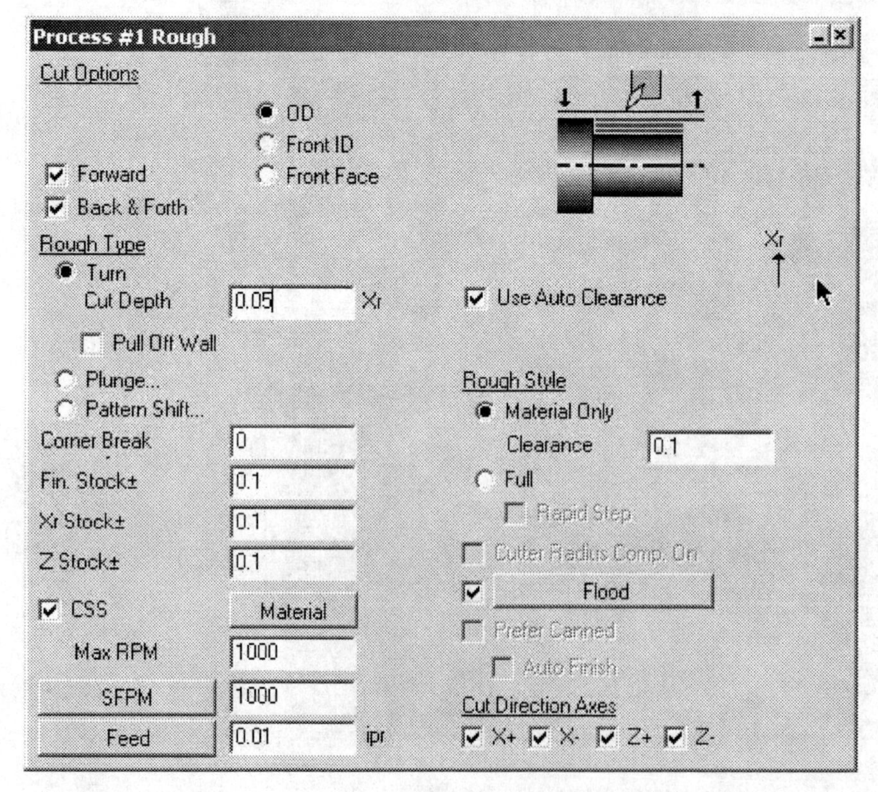

Close the Process dialog box. Click on the arc as shown in the following graphic. Select the tool position and cutting direction as shown. Drag the white dot (Start Point) to the endpoint of the arc. Drag the black square (End Feature) anywhere on the top horizontal line of the profile. Drag the black dot (End Point) to the approximate location as shown.

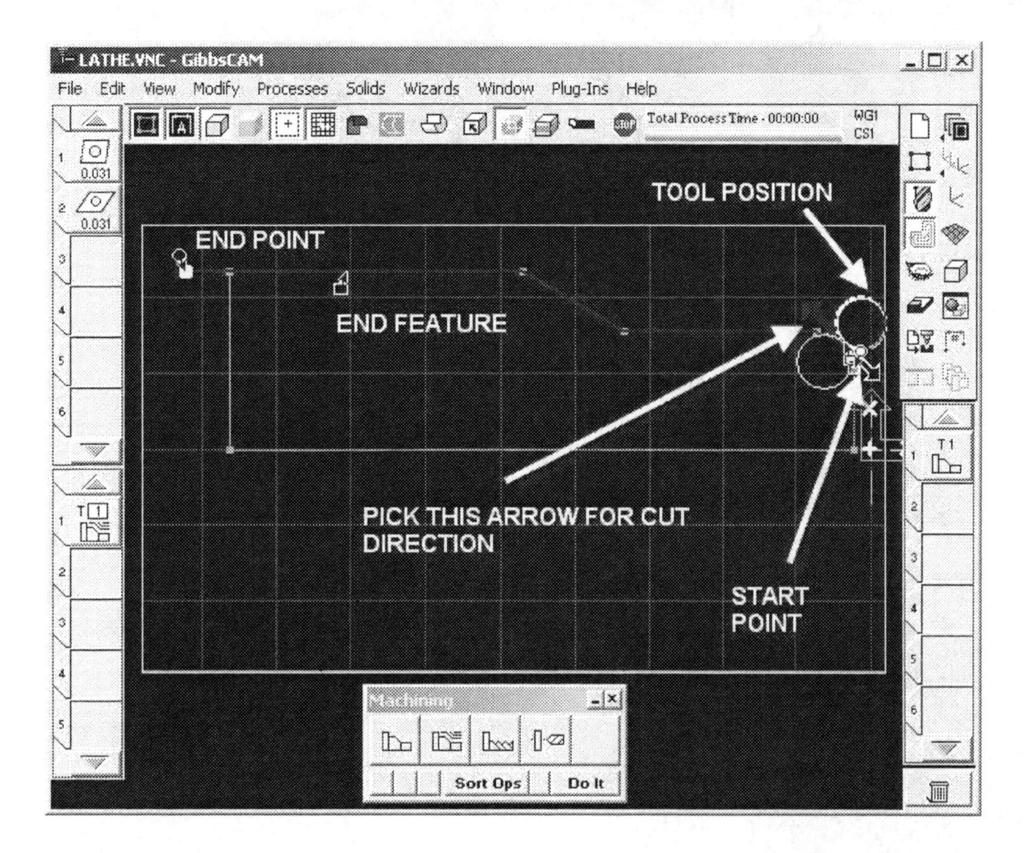


Click on **Do it**.

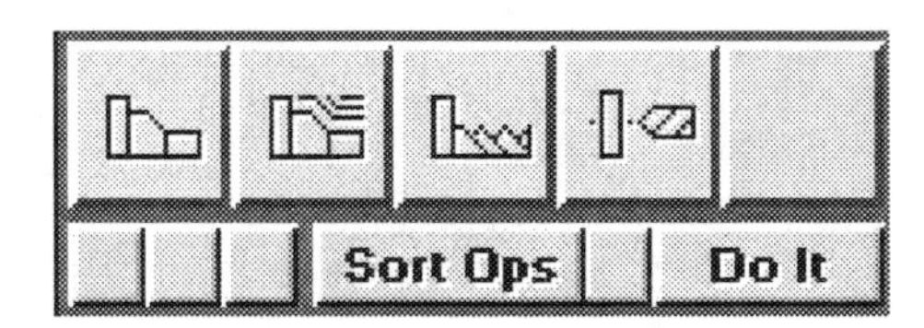


The orange lines represent the tool path as shown.

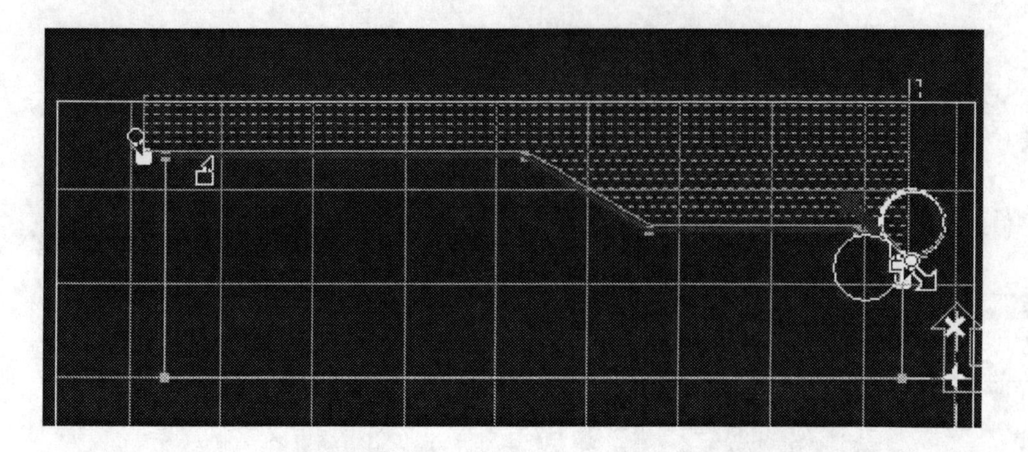

Begin the third operation by selecting Tile 3 on the Operations List.

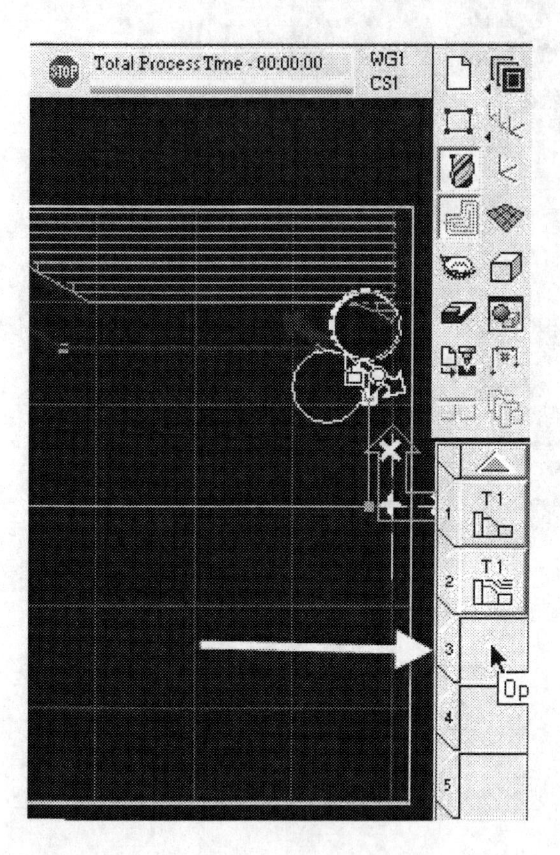

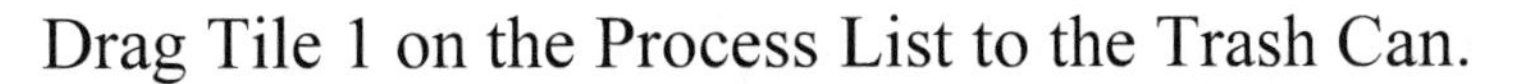

Drag Tile 1 on the Process List to the Trash Can.

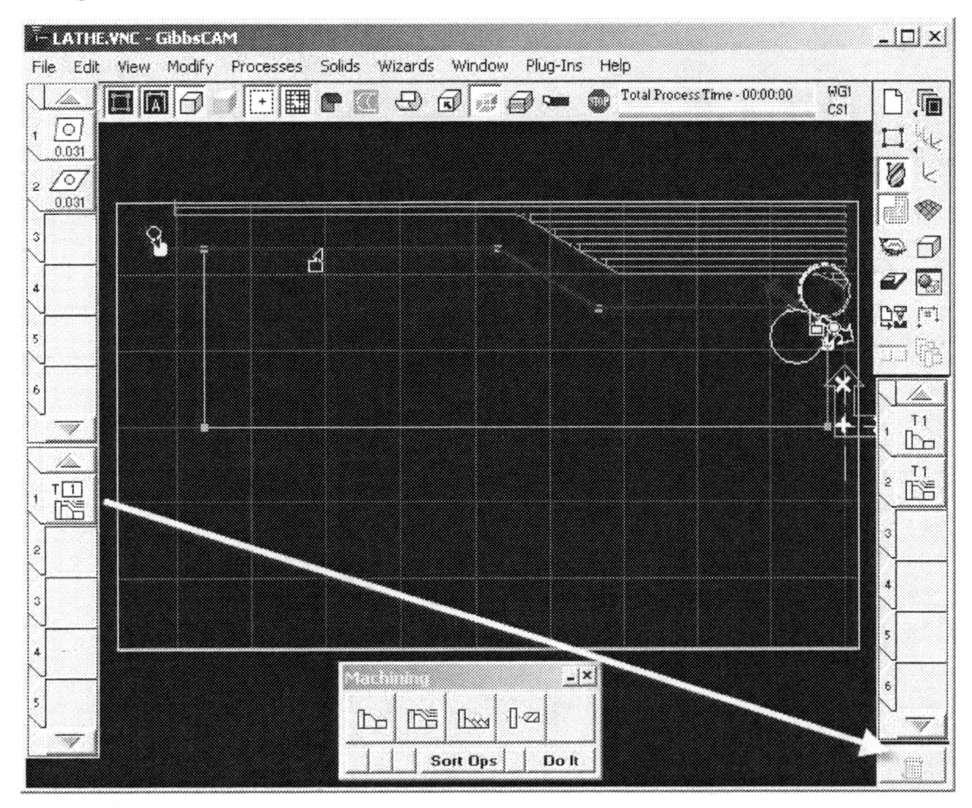


Drag Tool 2 to Tile 1 on the Process List.

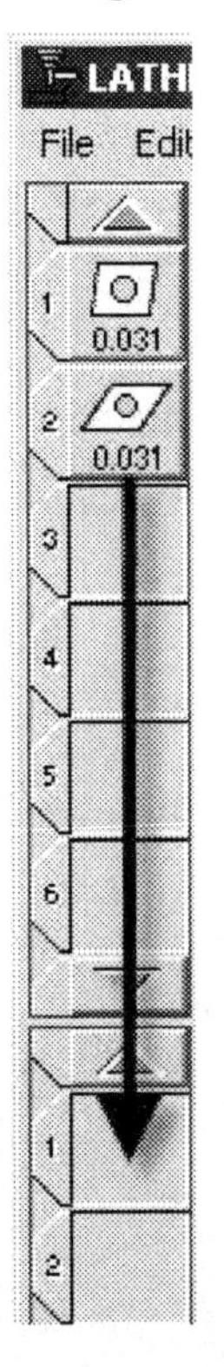

Drag the Contour icon to Tile 1 on the Process List.

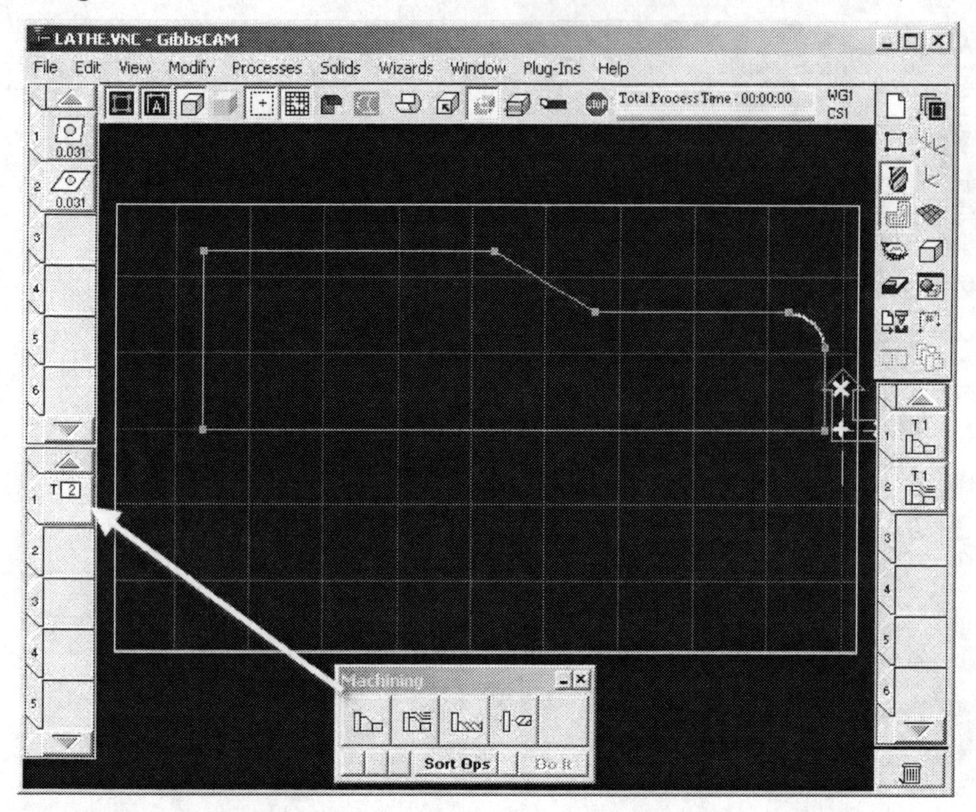


The Contour dialog box will appear.

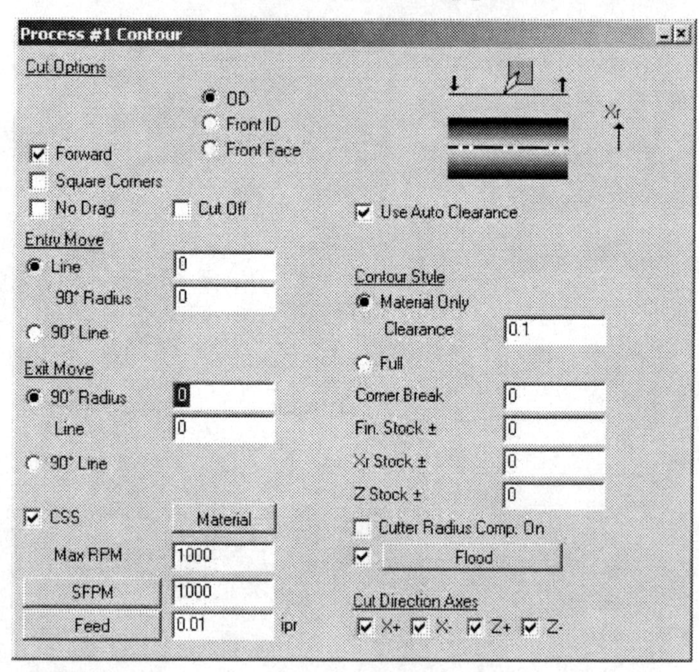

Close the Process dialog box. Select the arc and arrange machining markers as shown.

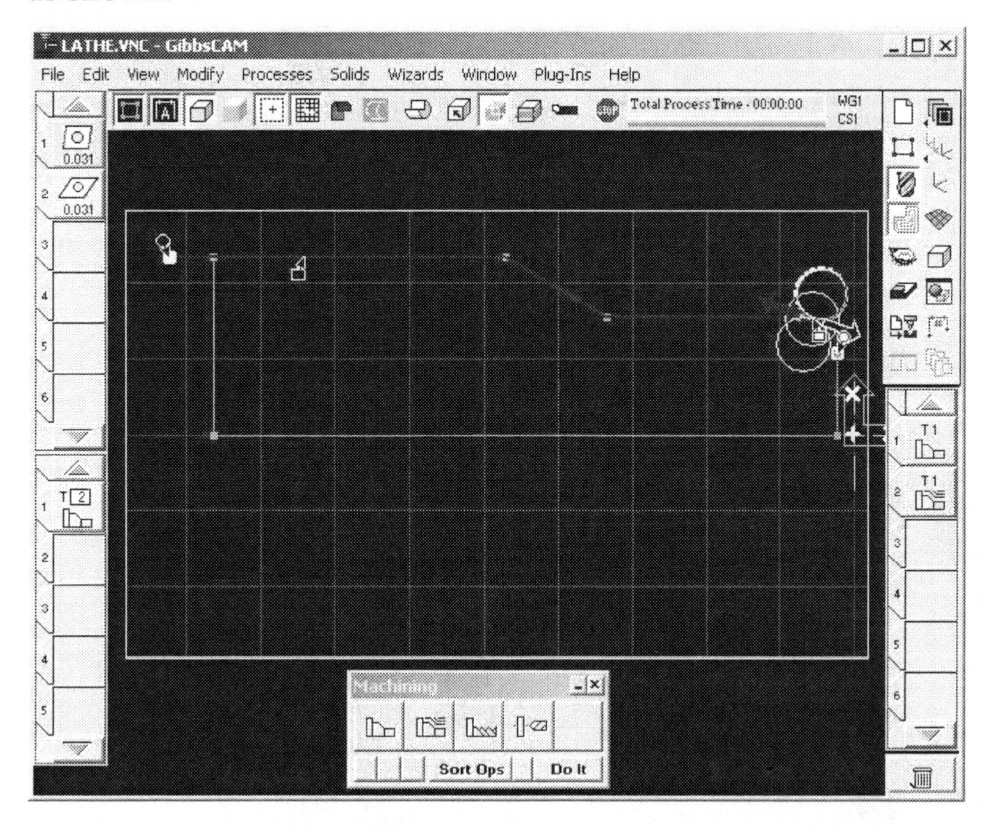


Click on **Do It**.

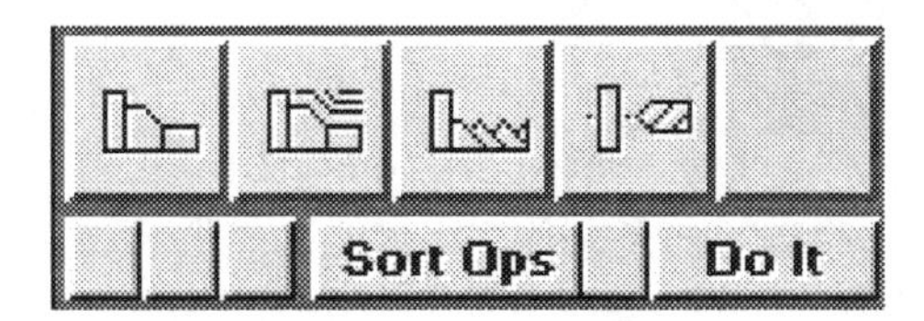


The orange lines represent the tool path.

Change the view to an isometric view.

To simulate this operation, click on the Cut Part Render icon on the Top Level Palette.

The Rendering Controls will appear. Click on the Visible Tool icon as shown. Hit the Stop button and then the Rewind button to reset the simulation. Hit the Play button to run the simulation.

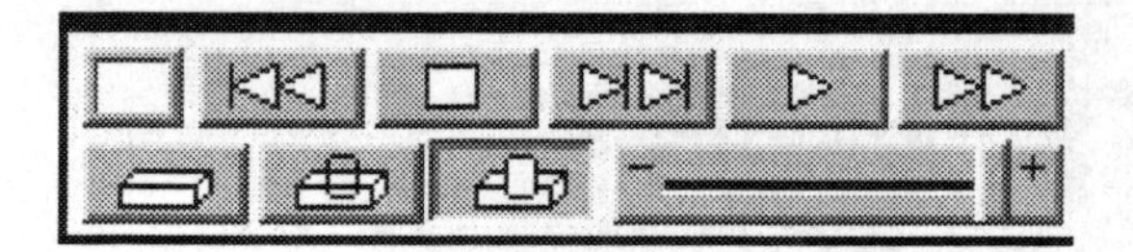

Your screen should look like the following image.

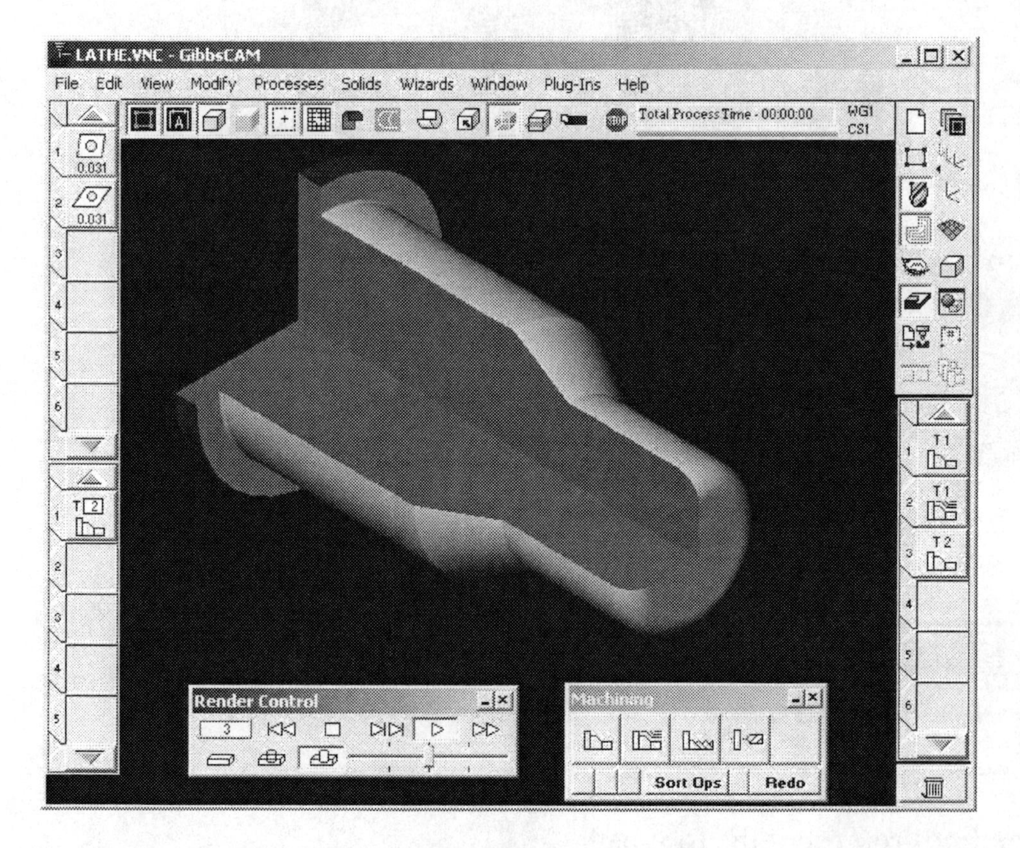


Close the Render Control dialog box.

This completes the exercise. Save the file. Choose *File-Save* from the *Main Menu*.

## Appendix A – Post Processing

Post Processing generates a CNC program from your machining operations.

After you have completed all of your machining operations and verified the accuracy of a part to your satisfaction, select the Post icon from the Top Level Palette.

Select the Post Processor Selection button and select your post processor. See your system administrator or your instructor for the location of this file.

Select the Program Name button.

The Save NC File dialog box will appear. Enter a file name and choose NC files (*.NCF) as the file type. Click on Save.

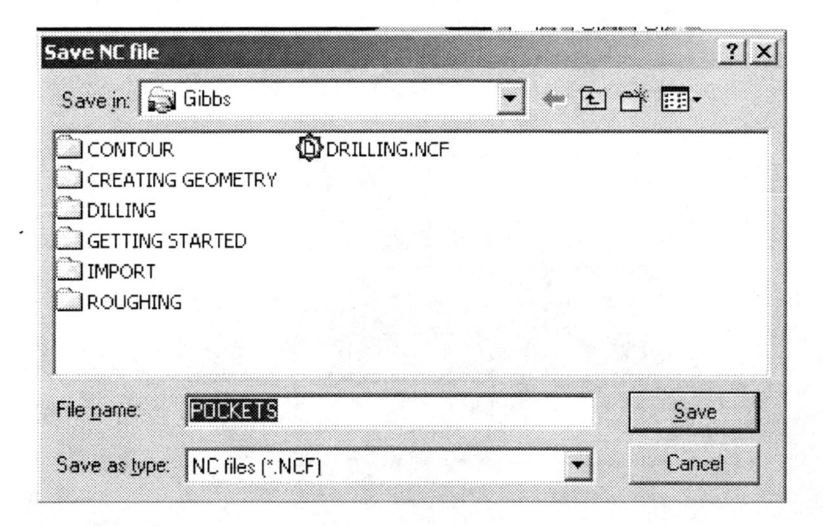

Appendix B

# Tool List Summaries

The Tool List Summary can be accessed from the *Main Menu* under the part name.

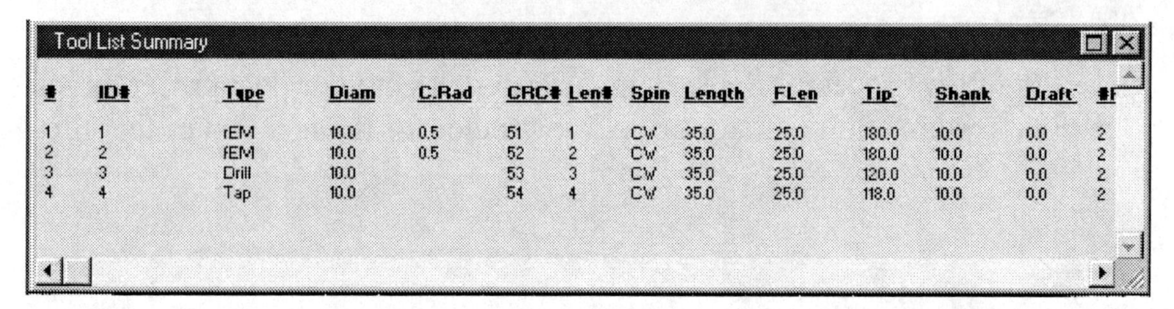
Tool List Summary

| # | ID# | Type | Diam | C.Rad | CRC# | Len# | Spin | Length | FLen | Tip° | Shank | Draft° | #F |
|---|---|---|---|---|---|---|---|---|---|---|---|---|---|
| 1 | 1 | rEM | 10.0 | 0.5 | 51 | 1 | CW | 35.0 | 25.0 | 180.0 | 10.0 | 0.0 | 2 |
| 2 | 2 | fEM | 10.0 | 0.5 | 52 | 2 | CW | 35.0 | 25.0 | 180.0 | 10.0 | 0.0 | 2 |
| 3 | 3 | Drill | 10.0 | | 53 | 3 | CW | 35.0 | 25.0 | 120.0 | 10.0 | 0.0 | 2 |
| 4 | 4 | Tap | 10.0 | | 54 | 4 | CW | 35.0 | 25.0 | 118.0 | 10.0 | 0.0 | 2 |

The Tool List Summary can be printed by selecting *File-Print-Tool List Summary* from the *Main Menu*.

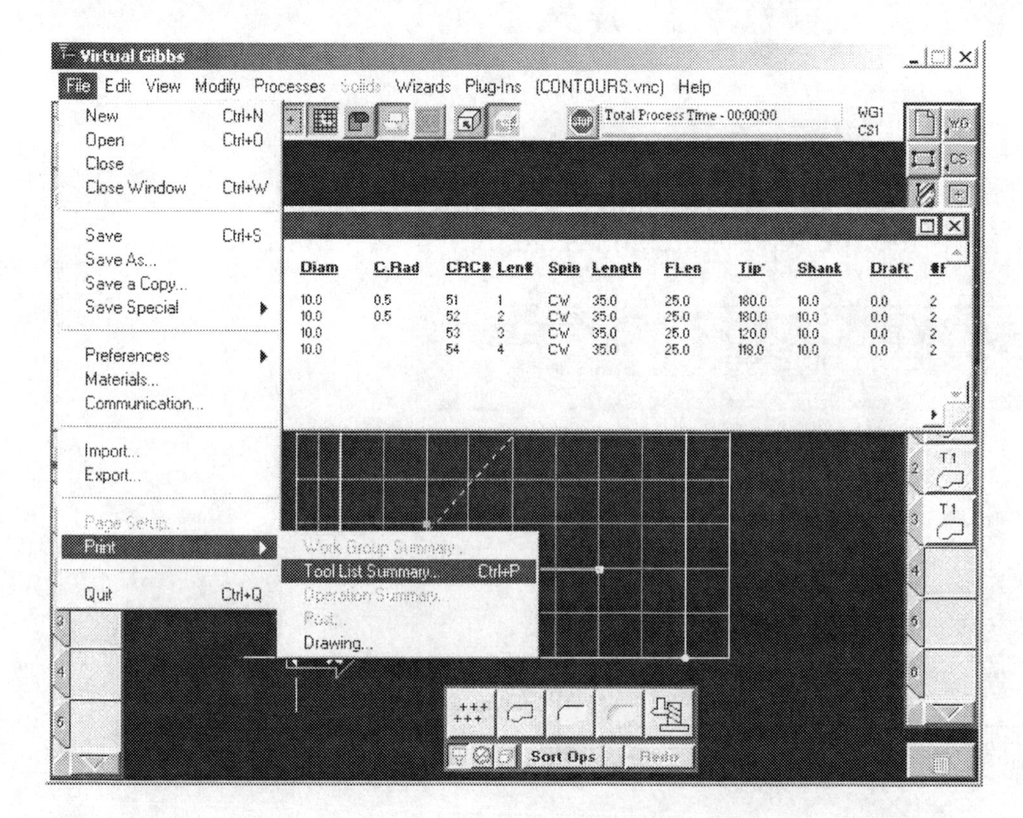

Appendix C

# Operation Summaries

The Operation Summary can be accessed from the *Main Menu* under the part name.

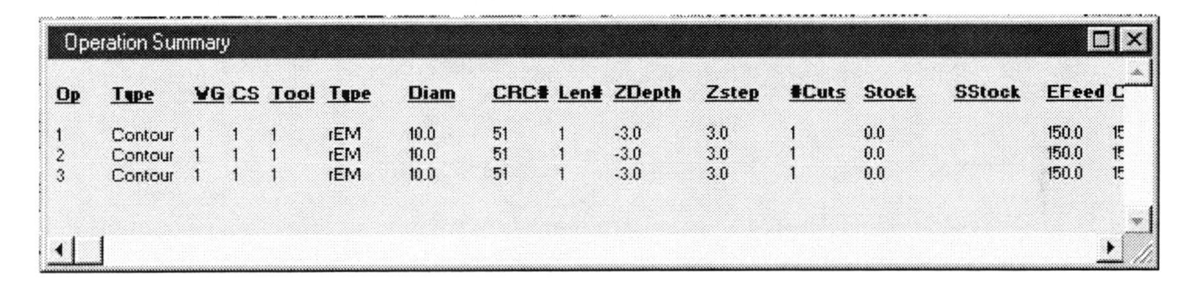

Operation Summary

| Op | Type | VG | CS | Tool | Type | Diam | CRC# | Len# | ZDepth | Zstep | #Cuts | Stock | SStock | EFeed | C |
|---|---|---|---|---|---|---|---|---|---|---|---|---|---|---|---|
| 1 | Contour | 1 | 1 | 1 | rEM | 10.0 | 51 | 1 | -3.0 | 3.0 | 1 | 0.0 | | 150.0 | 1 |
| 2 | Contour | 1 | 1 | 1 | rEM | 10.0 | 51 | 1 | -3.0 | 3.0 | 1 | 0.0 | | 150.0 | 1 |
| 3 | Contour | 1 | 1 | 1 | rEM | 10.0 | 51 | 1 | -3.0 | 3.0 | 1 | 0.0 | | 150.0 | 1 |

The Operation Summary can be printed by selecting *File-Print-Operation Summary* from the *Main Menu.*

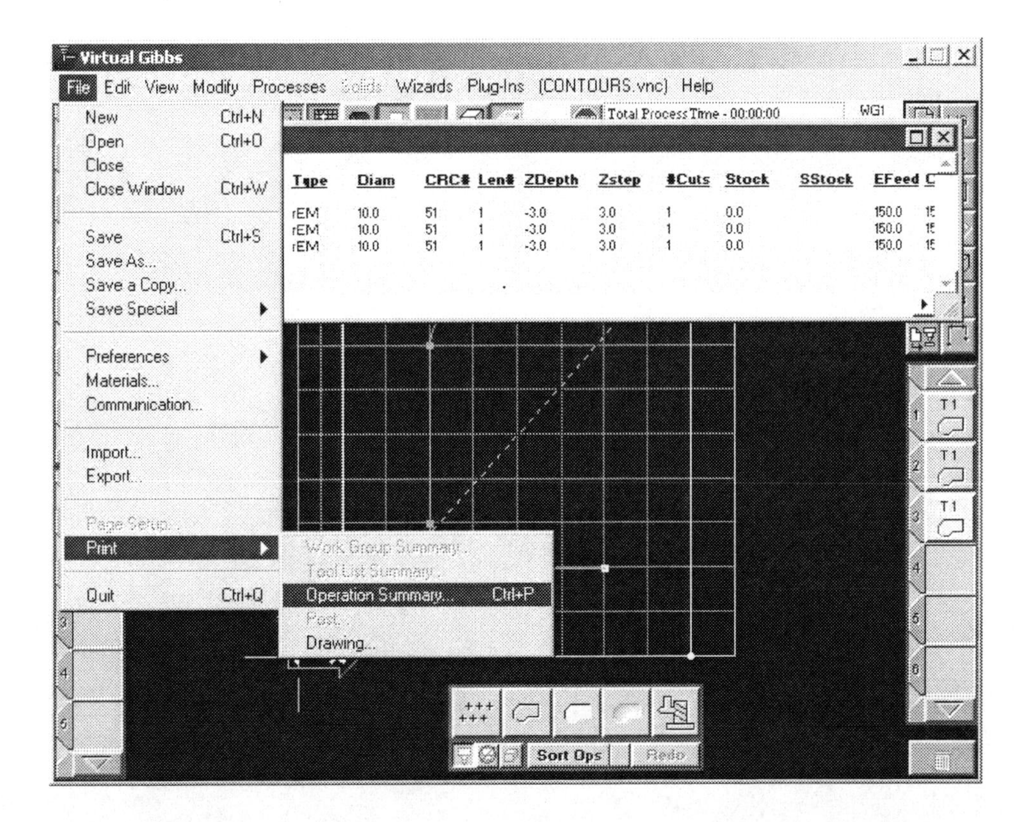